Hydrodynamics of Free Surface Flows

Hydrodynamics of Free Surface Flows

Modelling with the finite element method

Jean-Michel Hervouet

Electricité de France (EDF), France

John Wiley & Sons, Ltd

Other Wiley Editorial Offices

John Wiley & Sons Inc., 111 River Street, Hoboken, NJ 07030, USA
Jossey-Bass, 989 Market Street, San Francisco, CA 94103-1741, USA
Wiley-VCH Verlag GmbH, Boschstr. 12, D-69469 Weinheim, Germany
John Wiley & Sons Australia Ltd, 42 McDougall Street, Milton, Queensland 4064, Australia
John Wiley & Sons (Asia) Pte Ltd, 2 Clementi Loop #02-01, Jin Xing Distripark, Singapore 129809
John Wiley & Sons Canada Ltd, 6045 Freemont Blvd, Mississauga, ONT, L5R 4J3, Canada

Wiley also publishes its books in a variety of electronic formats. Some content that appears in print may not be available in electronic books.

Anniversary Logo Design: Richard J. Pacifico

British Library Cataloguing in Publication Data

A catalogue record for this book is available from the British Library

ISBN 978-0-470-03558-0 (HB)

Typeset in 10/12pt Times by Laserwords Private Limited, Chennai, India
Printed and bound in Great Britain by Antony Rowe Ltd, Chippenham, Wiltshire
This book is printed on acid-free paper responsibly manufactured from sustainable forestry in which at least two trees are planted for each one used for paper production.

Contents

List of Figures

List of Tables

List of Plates

Acknowledgements

Building a hydro-informatics system is a long-term multi-disciplinary work and the fruit of the collaboration of international teams. Before unravelling its mysteries, I would like to pay my tributes to and name all my colleagues, both French and foreign, who have accompanied me for more than twenty years on this adventure called free surface hydrodynamics, colleagues whose work and ideas have left a mark which one discovers between the lines of the program: Sylvain Aunay, Elisabeth Barros, Paul Bates, Michel Benoit, Thierry Boulet, Vincent Boyer, Céline Cheviet, Jean-Robert Courivaud, Eric David, Astrid Decoene, Thierry Denot, Arnaud Desitter, Patricia Ercoli, Antoine Garapon, Mustafa Guesmia, Claude Guilbaud, Reinhard Hinkelmann, Jean-Marc Janin, Jacek Jankowski, Rebekka Kopmann, Pierre Lang, Dominique Laurence, Catherine Le Normant, Angelo Leopardi, François Lepeintre, Marilyne Luck, Cécile Machet, Andreas Malcherek, Frédéric Marcos, Philippe Péchon, Eric Peltier, Emile Razafindrakoto, Jean-Philippe Renaud, Didier Rougé, Patrick Sauvaget, Catherine Villaret, Damien Violeau, Alain Watrin, Holger Weilbeer.

I am also indebted to the proof-readers for having meticulously revised the manuscript, in particular and sometimes once again, Céline Cheviet, Dominique Laurence, Charles Moulinec, Emile Razafindrakoto, Catherine Villaret, Damien Violeau and Chun Wang.

I express my profound gratitude to Professor Kim dan Nguyen, of the University of Caen, a long-standing friend who was the real initiator of this work, originally a dissertation for the accreditation of a thesis, and who patiently guided me through the procedures and formalities associated with it.

I duly thank "EC traduction" for the translation of the French manuscript, particularly Elisabeth Ben Slama, Hema Parthasaraty, Renuka George, Amrita Raghunandan and Shobana Varma.

I would also like to thank my wife and three children who put up with my studious silence for so many months and left me undisturbed to write this book "as huge as Harry Potter IV!" (Caroline, aged 10).

Chapter 1

Introduction

1.1 Twenty years of development at EDF

Since 1987, the year in which the Telemac project was launched, the Research and Development Department of Electricité de France (EDF) embarked upon a sustained effort to develop a hydro-informatics system dedicated to free surface flows. The subject is of considerable strategic interest. Dams, thermal and nuclear power stations on sea shores or river banks are all structures which interact with an aquatic environment that has to be protected and from which we need to protect ourselves. The quality of water, the calculation of tides and swells during storms, dam break flood waves, floods, the impact and stability of development, and sedimentology, are all crucial questions for EDF, and for all these disciplines a good command of the mechanics of free surface flows is indispensable.

Historically, physical models first ensured this command but these have progressively been replaced by numerical modelling. Today, the computer has become an unavoidable investigative tool and physical models, although still indispensable in certain key domains like spillways and dyke stability, have tended to become validation tools in other spheres.

In the last decade, numerical simulation has stood out as a vital tool for environmental hydraulics, through the concerted action of two important trends:

- Technical progress, with, on the one hand, exponential progress in the speed and memory capacity of computers and, on the other hand, the advent of new algorithms and numerical methods.

- The rise in requirements brought about, on the one hand, by a reinforcement of legislation (stocking of hydrocarbons, emergency plans, decrees on intakes and discharge, compensation water, etc.) and, on the other hand, by a series of so-called natural catastrophes. However, their anthropogenic origin (greenhouse effect, regional development) cannot be overlooked, and requires re-examination of the hypotheses on the sizing of structures.

Thus every technological advance, whether it arises from a new machine architecture or a new modelling technique, has been used immediately and advantageously to solve more and more complex problems. This situation remains unchanged and specific theoretical challenges like turbulence, or the complex physical conditions arising in sedimentology, pose problems which leave us far from our goal. From simple software solving a single type of equation, we have gradually progressed to the notion of hydro-informatics and sets of codes dealing with computerized data and capable of communicating for the processing of complex problems.

The stakes are high. The investment for a hydro-informatics system is of the order of 500000 to 1 million euros a year, but the advantages that can be obtained from it are much higher. Thus, numerical simulation of the impact of civil engineering work during a flood recently prevented an expense of 12 million euros. Another one, by helping to predict in a convincing manner the evolution of a thermal patch, helped to reduce the expense of a measurement collection campaign. However, this concerns only money matters.

To develop its hydro-informatics Telemac system, the National Laboratory of Hydraulics and Environment of EDF relied on the finite element method in which it had acquired experience in pressure flow, in particular with the N3S code. Originally, EDF had at disposal a background in finite differences on curvilinear meshing which largely predetermined the exotic architecture of the first Telemac-2D code: a theoretical framework and formalism in finite elements but with regular curvilinear meshing. The immediate enthusiasm of the first users for real non-structured meshing was quickly followed by a confirmation of their interest, and the modelling approach has therefore been maintained. However it should be noted that, in 2006, the three main numerical techniques relying on meshing, namely: finite differences, finite elements and finite volumes, are all still very widely used. Finite differences do indeed offer, thanks to their simplicity, the possibility of developing high-order numerical schemes at a low cost, while finite elements offer at the same time a strict theoretical framework and a flexibility to discretize the domain of computation, while finite volumes help to solve simply all the equations written in conservative form. Each of these three main families of techniques is subdivided into a large number of variants and, today, it is quite difficult to differentiate between the current and the outdated ones. The trend emerging currently would appear to be a gradual rejection of finite differences and a convergence on finite elements and finite volumes involving non-structured meshing. New techniques like distributive schemes are perhaps precursors of this evolution. Other techniques known as "Lagrangian", like the SPH (Smoothed Particle Hydrodynamics) method, which have already produced promising results in classically difficult domains like mixed flows, could play an important role in the future. Without predicting the form that future methods take, we shall just state that the finite element method is currently the most widely used in France by research departments and organizations in environmental hydraulics. In this work, while restricting ourselves to hydrodynamics, we intend to put forward certain technical advances achieved while building the Telemac system. We shall achieve this in a coherent framework from the establishment of the basic equations to their numerical implementation. In Chapter 2, we establish several variants of the Navier–Stokes and the Saint-Venant equations, the tracer equation and a few outlines of turbulence equations. After a short Chapter 3 on the finite element method, Chapter 4 elaborates the numerical resolution of the Saint-Venant equations. Also proposed in this chapter is a numerical scheme for the Boussinesq equations. Chapter 5 deals with the resolution of the Navier–Stokes equations, first with and then without a hydrostatic hypothesis. Chapter 6 gives a glimpse of some methods adapted to account for the hyperbolic nature of the transport equations. Chapter 7 gives in detail the computerized implementation of the finite element method utilizing modern forms of matrix storage: element-by-element storage and edge-based storage. The optimization of algorithms is dealt with mainly from the point of view of computers with vector architecture. Parallelism with distributed memory computers is dealt with later in Chapter 8. In Chapter 9 details are given of the techniques of parameter estimation in the Saint-Venant equations using an adjoint method. Finally, Chapter 10 presents some applications which include the calculation of the dam-break wave of the Malpasset dam.

1.2 Some smoother pebbles...

For a knowledgeable reader familiar with the classical techniques and the main numerical schemes used in free surface flow mechanics, and who simply wants more information about the new numerical schemes and the optimization techniques proposed in this book, here is a short résumé.

1.2.1 Saint-Venant equations

The algorithm leading to a linear system coupling water depth and velocity dates from 1988 and has since been continuously employed by hundreds of users. It has successfully replaced the Uzawa algorithm used earlier. Its strength goes beyond the theoretical limits of the equations, including its extension for solution of the Boussinesq equations. Among the essential points of this algorithm we may cite the choice of a non-conservative form, the weakly formulated impermeability conditions at solid boundaries and preconditioning called "c–u", based on a celerity–velocity form of the Saint-Venant equations, where the depth is replaced by the celerity of shallow water waves.

The special processing of exposed beds and dry areas, which does not rely on any notion of minimum water depth and assures an exact conservation of mass, is undoubtedly the advance which has contributed most to the success of the Telemac-2D software, on account of the range of application domains that it allows and the user satisfaction it achieves.

For the resolution of equations in Mercator projection, the use of local coordinates for each element is also a novel approach, which avoids changing the equations without giving rise to a topological problem.

Simultaneous use of porosity and drag force to simulate the presence of vertical obstacles like bridge piers or trees was presented for the first time in the year 2000.

1.2.2 Navier–Stokes equations

The proposed hydrostatic option, with a fractional step method and a sigma transform, is relatively classical. The solution presented for the continuity equation is new and we owe it to Jean-Marc Janin, the person in charge and trustworthy guardian for several years of the Telemac-3D software. From a staggered vertical velocity, he draws, thanks to the distributive schemes which he applies to prisms, an astonishingly conservative and monotonic model for the advection–diffusion equation of a tracer.

For non-hydrostatic Navier–Stokes equations, we attribute the pioneering research work to Jacek Jankowski who, in 1998, implemented the first version of this option in Telemac-3D. In 2001, the use of the Saint-Venant equations for the "hydrostatic" step increased the robustness tremendously, to the extent of obtaining the calculation of the solitary wave with only one layer of prisms in the three-dimensional meshing, a calculation which is undoubtedly very new.

In the same year, the algorithm for processing exposed beds, initially developed in two dimensions, was extended to three dimensions. This progress led to the calculation of an exposed beach with a non-hydrostatic option, a computation including both dry areas and torrential flow.

In 2004, the idea of the wave equation, which was already used in the Saint-Venant equations, was extended to three dimensions.

1.2.3 Finite element techniques and optimization

The discovery of the "non-assembled matrices" method in 1987 at EDF gave an impetus to the Telemac project. At the same time, the emergence of symbolic computation software enabled the rejection of Gauss quadrature. It was only a little later that the connection was made with the method known as the American EBE (Element By Element), which was older by four years. Reading the publications on this technique then served to us "on a plate" the Crout preconditioning in EBE storage. Our vectorization algorithm for assembling vectors, founded on a specific numbering of elements, was certainly original, as is also the processing of Dirichlet boundary conditions (i.e. imposed values), realized with a matrix–vector product. Interpreting the main required operations in finite elements as matrix–vector products, e.g. diagonal preconditioning, has then greatly facilitated the passage to parallelism.

Jean-Marc Janin conceived the idea of assembling matrices and sharing the terms between the elements by an exchange of elementary terms through segments. This way of storing, which already enabled a frontal matrix–vector product, was completely supplanted by a very similar edge-based storage, this idea being picked up during a congress. In the absence of precise references, all algorithms built on edge-based storage, and about which an idea is given in Section 7.3.4, have been recreated, and adapted to parallelism.

Chapter 2

Equations of free surface hydrodynamics

The Navier–Stokes equations, the cornerstone of all fluid mechanics, are, after all, only a form adapted to the fluids of the fundamental relation of dynamics. Establishing these equations has, however, been a long process since almost two centuries have passed between the publication of Newton's works in 1687 and Sir Gabriel Stokes' writing of the fundamental equations of hydrodynamics [179] which honour his name and that of the French engineer Claude-Louis Navier [147]. The theory of elasticity and the tensor calculus were, along with others, the prerequisites to a correct writing of the Navier–Stokes equations. The vast literature published since shows at the same time the importance of these equations in numerous scientific areas and their difficulty due in particular to the phenomenon of turbulence. At the beginning of the third millennium, we still do not know, except in some special cases, if these equations admit regular solutions! (In particular, see the article by Charles L. Fefferman "Existence and smoothness of the Navier–Stokes equations" on the Internet site http://www.claymath.org.) We are going to establish below the different forms of these Navier–Stokes equations which will be dealt with in the following chapters, and we shall then average them over the vertical to obtain the Saint-Venant equations.

2.1 Notations and concepts in geometry

The domain of study Ω in the referential $R = (O, x, y, z)$ (Oz along the vertical) is limited below by the bottom described by the equation $z = Z_f(x, y, t)$, or in most situations only $z = Z_f(x, y)$, and above by the free surface described by the equation $z = Z_s(x, y, t)$. It is limited laterally by vertical lines. Its boundary is Γ. The projection of Ω on the horizontal plane (O, x, y), that will be later our two-dimensional domain, is denoted here Ω2D (see Figure 2.1). The domain of calculation is thus defined by:

$$\Omega = \left\{ (x, y, z) \in R^3 \text{ such that } (x, y) \in \Omega 2\text{D and } Z_f(x, y, t) \leq z \leq Z_s(x, y, t) \right\}$$

Hydrodynamics of Free Surface Flows: Modelling with the Finite Element Method J.-M. Hervouet

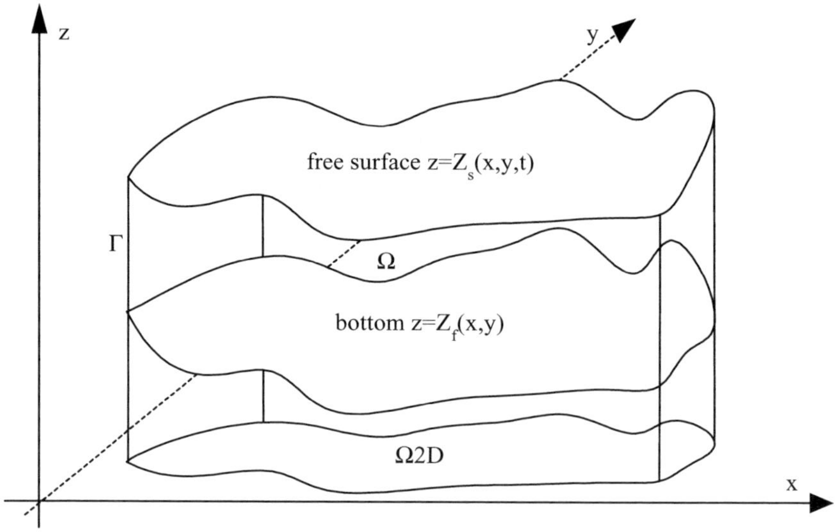

Figure 2.1: calculation domain

Gravity is directed towards decreasing z.

The height of the water column, or water depth, will be denoted as h and is equal to $Z_s - Z_f$.

The bottom elevation is given data. The water depth is generally an unknown quantity.

Time will be denoted as t.

Let us now define the normal vectors to the bottom and to the free surface.

The free surface is a univalent function of coordinates x and y, which excludes calculation of waves on the verge of breaking, and varies with time. Its equation is in the form of $z = Z_s(x, y, t)$. It can also be written as $\Phi(x, y, z, t) = 0$ with:

$$\Phi(x, y, z, t) = z - Z_s(x, y, t) \tag{2.1}$$

The normal to the free surface, directed towards increasing z (hence external in relation to the volume containing water), is then the vector (not normalized):

$$\vec{n}_s = \overrightarrow{grad}(\Phi)$$

This vector has the following components:

$$\vec{n}_s \left(-\frac{\partial Z_s}{\partial x}, -\frac{\partial Z_s}{\partial y}, 1 \right) \tag{2.2}$$

The normal to the bottom would be expressed in the same way but by replacing

Z_s by Z_f, and with a minus sign to make it external to the volume of water. That is:

$$\vec{n}_f \left(\frac{\partial Z_f}{\partial x}, \frac{\partial Z_f}{\partial y}, -1 \right) \tag{2.3}$$

The bottom elevation also is a univocal function of coordinates x and y. A vertical bottom can thus only be represented by a steep slope.

2.2 Free surface Navier–Stokes equations

2.2.1 Non-hydrostatic Navier–Stokes equations

We shall establish first the mass conservation equation, or "continuity equation", which expresses the conservation of fluid mass, and then the "momentum equation" which is in fact derived from the fundamental relation of dynamics. The conservation of energy equation, which would express the variations in temperature, particularly taking into account dissipation by friction, will be put aside and temperature will later appear as a tracer solely responsible for the effects of buoyancy.

Let ρ be the density of the fluid, and $\vec{U}$ the velocity vector, whose components are (U, V, W).

Conservation of the fluid mass contained in a domain Ω is expressed as:

$$\frac{d}{dt}\left(\int_\Omega \rho \, d\Omega \right) = 0 \tag{2.4}$$

The Leibnitz theorem breaks down this term into variations observed within the domain and the flux through the boundaries:

$$\frac{d}{dt}\left(\int_\Omega \rho \, d\Omega \right) = \int_\Omega \frac{\partial \rho}{\partial t} \, d\Omega + \int_\Gamma \rho \, \vec{U}.\vec{n} \, d\Gamma \tag{2.5}$$

The flux through the boundaries $\int_\Gamma \rho \, \vec{U}.\vec{n} \, d\Gamma$ is then reduced to a volume integral $\int_\Omega div(\rho\vec{U}) \, d\Omega$ by the Gauss theorem, and, by accepting that the conservation of mass should be ensured for any domain Ω whatsoever, we obtain a local expression of the mass conservation equation:

$$\frac{\partial \rho}{\partial t} + div(\rho\vec{U}) = 0 \tag{2.6}$$

Initially this "compressible" form will be retained. When the density is considered as constant, it will be simplified into:

$$div(\vec{U}) = 0 \tag{2.7}$$

Now we need to formulate for a fluid what is written as $\vec{f} = m\frac{d\vec{U}}{dt}$ for a rigid body. This form, which is best known as Newton's second law, should be more strictly written as $\vec{f} = \frac{d}{dt}(m\vec{U})$ because for a fluid the density is, in principle, variable. The equivalent of $\frac{d}{dt}(m\vec{U})$ will be written $\frac{d}{dt}(\int_\Omega \rho\vec{U} \, d\Omega)$, i.e. the derivative with respect to time of the momentum contained in the domain Ω. As in the case of mass, this quantity is also written as:

$$\int_{\Omega} \frac{\partial(\rho\vec{U})}{\partial t}\, d\Omega + \int_{\Gamma} \rho\vec{U}\,\vec{U}.\vec{n}\, d\Gamma \tag{2.8}$$

or as:

$$\int_{\Omega} \frac{\partial(\rho\vec{U})}{\partial t}\, d\Omega + \int_{\Omega} \nabla(\rho\vec{U} \otimes \vec{U})\, d\Omega \tag{2.9}$$

$\otimes$ designates the external product between two tensors of the order 1 and ∇ is the tensor operator "nabla".

Force $\vec{f}$ is the sum of external forces, which are applied directly in the body, and the forces known as contact forces, which are applied on its surface.

External forces will be denoted as $\vec{g} + \vec{F}$, by distinguishing the acceleration of gravity $\vec{g}$ and the other forces (Coriolis acceleration etc., which will be elaborated later).

If $d\Gamma$ is a surface element of the domain Ω whose external normal is $\vec{n}$, the contact force applied to $d\Gamma$ is $d\vec{F} = \underline{\underline{\sigma}}\,\vec{n}\, d\Gamma$, where $\underline{\underline{\sigma}}$ is the stress tensor.

Forces applied to our domain Ω are consequently:

$$\int_{\Omega} \rho\,(\vec{g} + \vec{F})\, d\Omega + \int_{\Gamma} \underline{\underline{\sigma}}.\vec{n}\, d\Gamma \tag{2.10}$$

or even:

$$\int_{\Omega} \rho(\vec{g} + \vec{F})\, d\Omega + \int_{\Omega} div(\underline{\underline{\sigma}})\, d\Omega \tag{2.11}$$

So, finally, we obtain:

$$\frac{\partial(\rho\vec{U})}{\partial t} + \nabla(\rho\vec{U} \otimes \vec{U}) = div(\underline{\underline{\sigma}}) + \rho\vec{g} + \rho\vec{F} \tag{2.12}$$

We have just established the Navier–Stokes equations in the form known as "conservative", which are classically summed up as:

Conservation of mass:

$$\frac{\partial \rho}{\partial t} + div(\rho\vec{U}) = 0 \tag{2.13}$$

Momentum equation:

$$\frac{\partial(\rho\vec{U})}{\partial t} + \nabla(\rho\vec{U} \otimes \vec{U}) = div(\underline{\underline{\sigma}}) + \rho\vec{g} + \rho\vec{F} \tag{2.14}$$

Use of the mass conservation equation also helps us to write the so-called "non-conservative" form of the momentum equation:

$$\frac{\partial(\vec{U})}{\partial t} + \vec{U}\,\nabla\vec{U} = \frac{1}{\rho}\, div(\underline{\underline{\sigma}}) + \vec{g} + \vec{F} \tag{2.15}$$

The two forms are not absolutely equivalent in the case of compressible fluids and we shall also notice this difference with the Saint-Venant equations, in the presence of discontinuities (see, in particular, Section 4.10.3). Let us now consider Newtonian fluids where, for incompressible fluids, the stress tensor is expressed as:

$$\underline{\underline{\sigma}} = -p\,\underline{\underline{\delta}} + 2\,\mu\underline{\underline{D}}$$

where p is the pressure expressed in pascals, $\underline{\underline{\delta}}$ is the identity tensor, μ the coefficient of dynamic viscosity and $\underline{\underline{D}}$ the strain rate tensor:

$$\underline{\underline{D}} = \frac{1}{2}\left(\nabla\vec{U} + {}^t\,\nabla\vec{U}\right) \tag{2.16}$$

μ is a constant which depends on the fluid, which can also be written as $\rho\nu$, where ν is the coefficient of kinematic viscosity (in m^2/s).

Variations in density are due either to the temperature, or to the presence of substances in the water, salt or sediments. In any case, we shall henceforth consider that these variations are sufficiently minor for using the mass conservation equation in an incompressible form. We finally end up writing the equations in Cartesian coordinates:

Conservation of mass:

$$\frac{\partial(U)}{\partial x} + \frac{\partial(V)}{\partial y} + \frac{\partial(W)}{\partial z} = 0 \tag{2.17}$$

Conservation of momentum:

$$\frac{\partial U}{\partial t} + U\frac{\partial U}{\partial x} + V\frac{\partial U}{\partial y} + W\frac{\partial U}{\partial z} = -\frac{1}{\rho}\frac{\partial p}{\partial x} + \nu\Delta(U) + F_x \tag{2.18}$$

$$\frac{\partial V}{\partial t} + U\frac{\partial V}{\partial x} + V\frac{\partial V}{\partial y} + W\frac{\partial V}{\partial z} = -\frac{1}{\rho}\frac{\partial p}{\partial y} + \nu\Delta(V) + F_y \tag{2.19}$$

$$\frac{\partial W}{\partial t} + U\frac{\partial W}{\partial x} + V\frac{\partial W}{\partial y} + W\frac{\partial W}{\partial z} = -\frac{1}{\rho}\frac{\partial p}{\partial z} - g + \nu\Delta(W) + F_z \tag{2.20}$$

We use the conservation of mass equation and the fact that μ is constant in space for simplifying $\frac{1}{\rho}div(2\mu\underline{\underline{D}})$ in $\nu\Delta(\vec{U})$.

Let us note that we have four unknowns, p, U, V and W, and four equations. As we have not yet formed any hypothesis on pressure, these Navier–Stokes equations will be called “non-hydrostatic”.

2.2.2 Boundary conditions

The domain so far known as Ω takes on more complex forms for free surface flows. It is limited:

- By the bottom (generally impermeable).
- By the surface of the water (which is not crossed by the fluid but which evolves with time).
- By boundaries which can be vertical impermeable structures (wall, quay) but also dykes, river banks or beaches which will let the fluid domain evolve more or less freely and will pose the problem of covering and exposing dry areas.

- In the case of numerical modelling, by the boundaries which could also be purely imaginary (boundaries in the open sea, upstream and downstream limits of a river) and not correspond to any physical property. This will add one more difficulty.

To start with, the boundary conditions will be established at the bottom and at the free surface and, subsequently, at the walls and open boundaries.

Free surface and bottom

If one follows the movement of a point on the free surface, the quantity (see Section 2.1):

$$\Phi(x, y, z, t) = z - Z_s(x, y, t)$$

remains zero, and the derivative $\frac{d\Phi}{dt}$ is thus zero (by the Lagrangian hypothesis which physically means that a particle of water on the free surface remains attached to the surface as time goes on). If the velocity of a point fixed on the surface has the components U^s, V^s and W^s, the equation $\frac{d\Phi}{dt} = 0$ is written with Euler variables:

$$W^s - \frac{\partial Z_s}{\partial t} - U^s \frac{\partial Z_s}{\partial x} - V^s \frac{\partial Z_s}{\partial y} = 0 \tag{2.21}$$

i.e.:

$$\frac{\partial Z_s}{\partial t} - \vec{U}^s \, . \, \vec{n}_s = 0 \tag{2.22}$$

At the bottom one obtains in the same way:

$$W^f - \frac{\partial Z_f}{\partial t} - U^f \frac{\partial Z_f}{\partial x} - V^f \frac{\partial Z_f}{\partial y} = 0 \tag{2.23}$$

or:

$$\frac{\partial Z_f}{\partial t} + \vec{U}^f \, . \, \vec{n}_f = 0 \tag{2.24}$$

with $\vec{n}_f$ pointing downwards.

Walls and open boundaries

Strictly speaking, the boundary condition for velocity on a solid wall is a no-slip condition: $\vec{U} = 0$. However, due to the turbulence and the existence of a boundary layer, the velocity near a wall quickly becomes non-zero, and the no-slip condition is often replaced by two weaker conditions: impermeability and the specification of a tangential stress due to the friction on the wall.

On a boundary, the impermeability condition is then written as:

$$\vec{U} . \vec{n} = 0 \tag{2.25}$$

The tangential stress is $\mu \frac{\partial \vec{U}}{\partial n}$. In theory, it can only be known if the flow is known and hence should not be data. It could, however, be provided, under certain conditions, by turbulence models. When the stress is zero, there is a slip condition on the wall.

On the open boundaries, additional information is required on the pressure, the water depth, the velocity, the discharge, etc. These boundaries, which are purely artificial, will cause difficulties in numerical modelling, as we shall see later. In particular, certain choices of boundary conditions will not necessarily lead to well-defined problems. In fact, that depends on the nature of the flow and its direction, as we shall see more precisely with the theory of characteristics for the Saint-Venant equations (Section 2.3.5).

2.2.3 Hydrostatic pressure and the Boussinesq approximation

The hydrostatic pressure hypothesis, which will also be made for the Saint-Venant equations, consists of simplifying the equation for vertical velocity W, while neglecting diffusion, the source terms and acceleration, to obtain:

$$\frac{\partial p}{\partial z} = -\rho g$$

or even, by writing ρ as $\rho = \rho_o + \Delta\rho$:

$$\frac{\partial p}{\partial z} = -\rho_o g \left(1 + \frac{\Delta\rho}{\rho_o}\right) \tag{2.26}$$

from which one obtains the expression for pressure at elevation z:

$$p = p_{atm} + \rho_o g \left(Z_s - z\right) + \rho_o g \int_z^{Z_s} \frac{\Delta\rho}{\rho_o} dz' \tag{2.27}$$

The pressure at one point only depends on the atmospheric pressure on the surface and on the weight of the column of water above it.

Linearization – Boussinesq approximation

The difference in density $\Delta\rho$ in relation to a reference value ρ_o is supposedly small. It is given by an equation of state which connects density ρ of the fluid to the concentrations of tracers, called "active tracers", transported in the mass of water and which act on the hydrodynamic equations. This equation of state has the form:

$$\frac{\Delta\rho}{\rho_o} = f(T_1, ..., T_i, ..., T_n) \tag{2.28}$$

where T_i is the ith active tracer transported. The increase in density is proportional to the differences in concentration, by the intermediary of volumetric dilatation coefficients β_i:

$$\frac{\Delta\rho}{\rho_o} = -\sum_i \beta_i \left(T_i - T_i^0\right)_i \tag{2.29}$$

where β_i can be positive (temperature) or negative (salinity, sediment in suspension). T_i^0 is the reference value of the tracer i.

The pressure gradient term in the first two momentum equations is then developed to first order:

$$-\frac{1}{\rho}\frac{\partial p}{\partial x} \approx -\frac{1}{\rho_o}\left(1-\frac{\Delta\rho}{\rho_o}\right)\frac{\partial p}{\partial x} \tag{2.30}$$

$$-\frac{1}{\rho}\frac{\partial p}{\partial y} \approx -\frac{1}{\rho_o}\left(1-\frac{\Delta\rho}{\rho_o}\right)\frac{\partial p}{\partial y} \tag{2.31}$$

which gives by using Equation 2.27:

$$-\frac{1}{\rho}\frac{\partial p}{\partial x} \approx -\frac{1}{\rho_o}\left(1-\frac{\Delta\rho}{\rho_o}\right)\frac{\partial p_{atm}}{\partial x} - g\left(1-\frac{\Delta\rho}{\rho_o}\right)\frac{\partial Z_s}{\partial x} - g\frac{\partial}{\partial x}\left(\int_z^{Z_s}\frac{\Delta\rho}{\rho_o}dz'\right) \tag{2.32}$$

$$-\frac{1}{\rho}\frac{\partial p}{\partial y} \approx -\frac{1}{\rho_o}\left(1-\frac{\Delta\rho}{\rho_o}\right)\frac{\partial p_{atm}}{\partial y} - g\left(1-\frac{\Delta\rho}{\rho_o}\right)\frac{\partial Z_s}{\partial y} - g\frac{\partial}{\partial y}\left(\int_z^{Z_s}\frac{\Delta\rho}{\rho_o}dz'\right) \tag{2.33}$$

These two gradients are broken down into a slope effect of the free surface $-g\,\nabla(Z_s)$ and into two source terms due to buoyancy, the first one called barotropic and the second baroclinic, which can be written along x:

$$g\frac{\Delta\rho}{\rho_o}\frac{\partial Z_s}{\partial x} - g\frac{\partial}{\partial x}\left(\int_z^{Z_s}\frac{\Delta\rho}{\rho_o}dz'\right) \tag{2.34}$$

and along y:

$$g\frac{\Delta\rho}{\rho_o}\frac{\partial Z_s}{\partial y} - g\frac{\partial}{\partial y}\left(\int_z^{Z_s}\frac{\Delta\rho}{\rho_o}dz'\right) \tag{2.35}$$

and to which is added the effect of atmospheric pressure. They are subsequently integrated in the source terms F_x and F_y.

In view of these hypotheses and conventions, the Navier–Stokes equations with hydrostatic hypothesis are as follows:

Mass conservation:

$$\frac{\partial U}{\partial x} + \frac{\partial V}{\partial y} + \frac{\partial W}{\partial z} = 0 \tag{2.36}$$

Momentum equation:

$$\frac{\partial U}{\partial t} + U\frac{\partial U}{\partial x} + V\frac{\partial U}{\partial y} + W\frac{\partial U}{\partial z} = -g\frac{\partial Z_s}{\partial x} + \nu\Delta(U) + F_x \tag{2.37}$$

$$\frac{\partial V}{\partial t} + U\frac{\partial V}{\partial x} + V\frac{\partial V}{\partial y} + W\frac{\partial V}{\partial z} = -g\frac{\partial Z_s}{\partial y} + \nu\Delta(V) + F_y \tag{2.38}$$

Pressure:

$$p = p_{atm} + \rho_o g\,(Z_s - z) + \rho_o g\int_z^{Z_s}\frac{\Delta\rho}{\rho_o}dz' \tag{2.39}$$

2.2.4 Source terms and body forces

Friction at the bottom

The bottom shear stress acting on the fluid is opposed to the velocity (only horizontal velocity will be considered), and is equivalent to:

$$\vec{\tau} = -\mu \frac{\partial \vec{U}}{\partial n} \tag{2.40}$$

This stress is opposed to the stress applied by the fluid to the bottom, which is in the direction of the current. As stated above, knowledge of this stress requires knowledge of the flow. The turbulence models will provide the modelling, based on knowledge of the current in the vicinity of the bottom. In one dimension it is acknowledged that:

$$\tau = -\rho U^{*2} \tag{2.41}$$

which is a definition of shear velocity U^*. Dimensional analysis also shows that $\vec{\tau}$ has the form:

$$\vec{\tau} = -\frac{1}{2}\rho C_f \sqrt{U^2 + V^2}\,\vec{U}$$

The units are kg/ m/s^2, as C_f is a dimensionless friction coefficient. This general formula for shear forces supposes that $\vec{U}$ is sufficiently far from the wall. It can also be written as:

$$\nu \frac{\partial \vec{U}}{\partial n} = -\frac{1}{2} C_f \sqrt{U^2 + V^2}\,\vec{U} \tag{2.42}$$

The specification of this stress $\nu \frac{\partial \vec{U}}{\partial n}$ will naturally appear in the variational formulation of the diffusion terms in finite elements.

The bottom shear stress condition will thus be ensured:

- Either by a turbulence model which will provide a formula for the stress $\nu \frac{\partial \vec{U}}{\partial n}$, based on the expression of the roughness at the bottom and flow in the vicinity of the bottom. Most often, the turbulence model gives the shear velocity U*, or the friction coefficient C$_f$.

- Or by knowledge of a friction coefficient C_f and its associated velocity $\vec{U}$, which, due to the laws of bottom shear stress in two dimensions, will be the average velocity on the vertical (the friction coefficients on the bottom will thus be the same in two and three dimensions). This is the case of the Chézy, Strickler and Manning formulae (and a few others elaborated in Section 2.3.11):

  Chézy formula (C is the Chézy coefficient):

  $$C_f = \frac{2g}{C^2} \tag{2.43}$$

  Strickler formula (S is the Strickler coefficient):

  $$C_f = \frac{2g}{h^{1/3} S^2} \tag{2.44}$$

Manning formula (m is the Manning coefficient, equal to the inverse of the Strickler coefficient):

$$C_f = \frac{2g\,m^2}{h^{1/3}} \tag{2.45}$$

Wind forcing

Wind forcing appears in the equations through the two-dimensional condition at the surface:

$$\nu \frac{\partial \vec{u}_H}{\partial n} = \frac{\rho_{air}}{\rho} a_{wind} \vec{w} \, \|\vec{w}\| \tag{2.46}$$

with $\rho_{air} = 1.29$ kg/m^3.

$\vec{u}_H$ is the horizontal velocity at the surface and $\vec{w}$ the wind velocity 10 m above the water. The coefficient a_{wind} (dimensionless) is given by Flather [50]:

$$\begin{aligned} a_{wind} &= 0.565\ 10^{-3} \quad \text{if} \quad \|\vec{w}\| \leq 5 \text{ m/s} \\ a_{wind} &= (-0.12 + 0.137\,\|\vec{w}\|)\ 10^{-3} \quad \text{if} \quad 5 \text{ m/s} \leq \|\vec{w}\| \leq 19.22 \text{ m/s} \\ a_{wind} &= 2.513\ 10^{-3} \quad \text{if} \quad \|\vec{w}\| \geq 19.22 \text{ m/s} \end{aligned} \tag{2.47}$$

The coefficient a_{wind} hides complex phenomena and the above values are only a guideline. In fact, the influence of the wind depends on the roughness of the free surface, which is itself dependent on the wind and the distance over which it is applied (called "fetch").

Coriolis force and centrifugal force

These two forces are due to the rotation of the Earth on its own axis. We establish its expression below.

Let a be a fixed referential with its origin at the centre of the Earth and b a referential which turns with the Earth. They will be denoted $ref.\ a$ and $ref.\ b$ in the formulae. The axes Ox and Oy are placed on the plane of the equator and the axis Oz along the rotation axis. The relation between the velocities expressed in the two referentials is written as:

$$\vec{V}_a = \vec{V}_b + \vec{\omega} \wedge \vec{r}_a$$

which signifies:

$$\left(\frac{d\vec{r}_a}{dt}\right)_{ref.\ a} = \left(\frac{d\vec{r}_b}{dt}\right)_{ref.\ b} + \vec{\omega} \wedge \vec{r}_a \tag{2.48}$$

This relation is a rule for time derivatives, valid for every vector with components in referentials a and b. In fact, for every vector $\vec{i}$ of the spinning referential, b, we have:

$$\frac{d\vec{i}}{dt} = \vec{\omega} \wedge \vec{i}$$

By deriving $\overrightarrow{V}_a$ with respect to time, we get:

$$\left(\frac{d^2\overrightarrow{r}_a}{dt^2}\right)_{ref.\ a} = \left(\frac{d^2\overrightarrow{r}_b}{dt^2}\right)_{ref.\ b} + \overrightarrow{\omega} \wedge \left(\frac{d\overrightarrow{r}_b}{dt}\right)_{ref.\ b} + \overrightarrow{\omega} \wedge \left(\frac{d\overrightarrow{r}_a}{dt}\right)_{ref.\ a} \tag{2.49}$$

or, by using the value of $\left(\frac{d\overrightarrow{r}_a}{dt}\right)_{ref.\ a}$ given by the first law of composition:

$$\left(\frac{d^2\overrightarrow{r}_a}{dt^2}\right)_{ref.\ a} = \left(\frac{d^2\overrightarrow{r}_b}{dt^2}\right)_{ref.\ b} + 2\,\overrightarrow{\omega} \wedge \left(\frac{d\overrightarrow{r}_b}{dt}\right)_{ref.\ b} + \overrightarrow{\omega} \wedge (\overrightarrow{\omega} \wedge \overrightarrow{r}_a) \tag{2.50}$$

that is, by denoting the accelerations $\overrightarrow{\gamma}$:

$$\overrightarrow{\gamma}_b = \overrightarrow{\gamma}_a - 2\,\overrightarrow{\omega} \wedge \overrightarrow{V}_b - \overrightarrow{\omega} \wedge (\overrightarrow{\omega} \wedge \overrightarrow{r}_a) \tag{2.51}$$

where the Coriolis force is:

$$-2\,\overrightarrow{\omega} \wedge \overrightarrow{V}_b$$

and the centrifugal force is:

$$-\overrightarrow{\omega} \wedge (\overrightarrow{\omega} \wedge \overrightarrow{r}_a)$$

On a point of latitude λ, let us now choose a referential with Ox directed eastwards, Oy northwards and Oz towards the vertical of the place. The rotation of the Earth vector is in this referential:

$$\overrightarrow{\omega} = \begin{pmatrix} \omega_x \\ \omega_y \\ \omega_z \end{pmatrix} = \begin{pmatrix} 0 \\ \omega\cos(\lambda) \\ \omega\sin(\lambda) \end{pmatrix} \tag{2.52}$$

In the more general case where north makes an angle N with the axis Oy (counted positively on going from Oy to the north), one has:

$$\overrightarrow{\omega} = \begin{pmatrix} \omega_x \\ \omega_y \\ \omega_z \end{pmatrix} = \begin{pmatrix} \omega\cos(\lambda)\sin(N) \\ \omega\cos(\lambda)\cos(N) \\ \omega\sin(\lambda) \end{pmatrix} \tag{2.53}$$

and the Coriolis force is:

$$-2\,\overrightarrow{\omega} \wedge \overrightarrow{V} = \begin{pmatrix} -2\omega_y W + 2\omega_z V \\ 2\omega_x W - 2\omega_z U \\ -2\omega_x V + 2\omega_y U \end{pmatrix} \tag{2.54}$$

or even:

$$-2\,\overrightarrow{\omega} \wedge \overrightarrow{V} = \begin{pmatrix} -2\omega\cos(\lambda)\cos(N)W + 2\omega\sin(\lambda)V \\ 2\omega\cos(\lambda)\sin(N)W - 2\omega\sin(\lambda)U \\ -2\omega\cos(\lambda)\sin(N)V + 2\omega\cos(\lambda)\cos(N)U \end{pmatrix} \tag{2.55}$$

ω is equal to $\frac{2\pi}{86164}$: the velocity of the Earth's rotation in radians per second (86164 is the duration in seconds of the sidereal day).

By definition, the Coriolis coefficient is equal to $2\omega \sin(\lambda)$.

The influence of the Coriolis force in the Navier–Stokes equations is governed by the Rossby number:

$$R_0 = \frac{v}{2\omega \sin(\lambda) l}$$

where v and l are the characteristic velocity and length of flow. The Coriolis force is important only if $R_0 \ll 1$; that is, for large oceanic or atmospheric areas. It is negligible in rivers and lakes (and even more so in washbasins).

The vertical component of the Coriolis force and the terms due to the vertical velocity are often ignored; however, their importance was highlighted in oceanic flows on the mesoscale [135].

For a geophysical flow, the centrifugal force is not added to the Navier–Stokes equations. In fact, one considers that the free surface of a fluid at rest coincides with an equipotential surface of apparent gravity, which already integrates the effect of centrifugal force.

Heat exchanges

The thermal power liberated into the atmosphere per surface unit, denoted as P, is supposed to be proportional to $T - T_{air}$ where T is the temperature of water on the surface and T_{air} that of air; one gets $P = A(T - T_{air})$, where A is the exchange coefficient in W/m^2/°C.

Heat flux leaving the liquid domain is written:

$$\Phi = -\rho C_P \nu_T \overrightarrow{grad}(T).\overrightarrow{n} = -\rho C_P \nu_T \frac{\partial T}{\partial z} \tag{2.56}$$

with $C_P = 4.18$ J/kg/°C. ν_T is the coefficient of heat diffusion in water.

By equating the two formulations, one deduces the boundary conditions:

$$\nu_T \frac{\partial T}{\partial z} = -\frac{A}{\rho C_p}(T - T_{air}) \tag{2.57}$$

The coefficient A should include phenomena like radiation, convection of air in contact with water and latent heat produced by evaporation of water. Sweers [181] expresses the coefficient A, in W/m^2/°C, according to the temperature of water T and the wind velocity w measured at the point under consideration, in m/s:

$$A = (4.48 + 0.049T) + 2021.5\, b\,(1 + w)\,(1.12 + 0.018\,T + 0.00158\,T^2) \tag{2.58}$$

The parameter b varies depending on location. Its average value on the shores of the Channel and of the Atlantic is 0.0025. It is higher in the Mediterranean, where it reaches approximately 0.0035.

Buoyancy terms and atmospheric pressure

As mentioned earlier, the atmospheric pressure gradient and the buoyancy terms arising from pressure gradients have been integrated with the source terms. They are, respectively along Ox and Oy:

$$-\frac{1}{\rho_o}\left(1-\frac{\Delta\rho}{\rho_o}\right)\frac{\partial p_{atm}}{\partial x}+g\left(\frac{\Delta\rho}{\rho_o}\right)\frac{\partial Z_s}{\partial x}-g\frac{\partial}{\partial x}\left(\int_z^{Z_s}\frac{\Delta\rho}{\rho_o}dz'\right) \tag{2.59}$$

$$-\frac{1}{\rho_o}\left(1-\frac{\Delta\rho}{\rho_o}\right)\frac{\partial p_{atm}}{\partial y}+g\left(\frac{\Delta\rho}{\rho_o}\right)\frac{\partial Z_s}{\partial y}-g\frac{\partial}{\partial y}\left(\int_z^{Z_s}\frac{\Delta\rho}{\rho_o}dz'\right) \tag{2.60}$$

Salinity S, expressed in kg/m^3, and temperature T are special tracers which enable calculation of variations in density for the buoyancy terms. First we shall give the reference established by UNESCO [185], then some simplified formulae that one can deduce from it.

Equations of state established by UNESCO: the formulae below give a precision of $3.5\,10^{-6}$ for temperature between 2 and 40 °C, pressure between 0 and 1000 bars, and salinity between 0 and 42 kg/m^3. Salinity is in fact defined more precisely by UNESCO based on the electric conductivity of water compared to that of a solution of potassium chloride. The transposition to kg/m^3 is thus in theory only valid for this salt.

The density of pure water is first given as a function of temperature:

$$\rho_{pure\,water}=a_0+a_1T+a_2T^2+a_3T^3+a_4T^4+a_5T^5 \tag{2.61}$$

with:

$a_0 = +999.842594$
$a_1 = +6.793952\,10^{-2}$
$a_2 = -9.095290\,10^{-3}$
$a_3 = +1.001685\,10^{-4}$
$a_4 = -1.120083\,10^{-6}$
$a_5 = +6.842594\,10^{-9}$

On the basis of this value, the density depending on the temperature and the salinity, at atmospheric pressure, is given by:

$$\rho(T,S)=\rho_{pure\,water}+(b_0+b_1T+b_2T^2+b_3T^3+b_4T^4)S$$
$$+(c_0+c_1T+c_2T^2)S^{3/2}+d_0S^2 \tag{2.62}$$

$b_0 = +8.24493\,10^{-1}$
$b_1 = -4.0899\,10^{-3}$
$b_2 = +7.6438\,10^{-5}$
$b_3 = -8.2467\,10^{-7}$
$b_4 = +5.3875\,10^{-9}$
$c_0 = -5.72466\,10^{-3}$

$c_1 = +1.0227\,10^{-4}$
$c_2 = -1.6546\ 10^{-6}$
$d_0 = +4.8314\,10^{-4}$

Density function of temperature and salinity: these variations are given by the following formula:

$$\rho = \rho_{ref}[1 - (7(T - T_{ref})^2 - 750\,S)10^{-6}] \tag{2.63}$$

for 0°C< $T <$ 40°C and 0 g/l< $S <$ 42 g/l.

T_{ref} is the reference temperature of 4 °C and ρ_{ref} the reference density at this temperature when the salinity is zero, that is, $\rho_{ref} = 999.972\,\text{kg/m}^3$.

Density function of salinity: by overlooking variations in temperature, choosing the average temperature T becomes necessary for calculations. The value of ρ_0 in the barotropic and baroclinic pressure gradients is then:

$$\rho_0 = \rho_{ref}[1 - 7\ 10^{-6}(T - T_{ref})^2] \tag{2.64}$$

The temperature T which should be given will in this case be considered as constant.

One simply has:

$$\rho = \rho_0 + 0.749979\,S \tag{2.65}$$

In finite elements ρ will then have formally the same range of approximation as S.

Linear formula: a linear formula is sometimes necessary for numerical reasons; it can then be written as:

$$\rho = \rho_0\,[1 + \alpha_S(S - S_0) + \alpha_T(T - T_0)] \tag{2.66}$$

where ρ_0, S_0 and T_0 are the reference values and α_S and α_T are called the coefficient of saline contraction and coefficient of heat expansion respectively.

2.2.5 Navier–Stokes equations with sigma transform

Sigma transform

One of the difficulties with the free surface Navier–Stokes equations lies in the fact that the domain of calculation Ω varies with time. A classical example to remedy this problem is to carry out a change of vertical coordinates such that the bottom elevation becomes 0 and the free surface elevation becomes 1. For this purpose the following change of variables is used:

$$z^* = \frac{z - Z_f}{Z_s - Z_f} = \frac{z - Z_f}{h} \tag{2.67}$$

This is an affine transformation giving a transformed domain Ω^*. The elevations are shifted and multiplied by a coefficient $1/h$. For a function f, the volume integrals are thus modified in the following manner:

$$\int_{\Omega} f\,d\Omega = \int_{\Omega^*}\left(\frac{\partial z}{\partial z^*}\right)_{x^*,y^*,t} f\,d\Omega^* = \int_{\Omega^*} h\,f\,d\Omega^* \tag{2.68}$$

For more general transformations the Jacobian $\frac{\partial z}{\partial z^*}$ may not be the depth. From now on we shall denote it $\frac{\partial z}{\partial z^*} = \Delta z$. As a matter of fact, it will appear in Section 5.2.4, when we deal with the generalized sigma transform on 3D meshes, that the Jacobian is the local height of prisms, the depth h being then the sum of these heights vertically.

The other coordinates and time remain unchanged. In other words, $t^* = t$, $x^* = x$ and $y^* = y$. Care should be taken not to foolishly replace x^* and y^* everywhere by x and y because a partial derivative with respect to x^* implies that one derives with constant y^*, z^* and t. $\frac{\partial f}{\partial x^*}$ and $\frac{\partial f}{\partial x}$ do not have the same significance, and we shall use the following basic relations, where the indices show what remains constant:

$$\left(\frac{\partial f}{\partial x}\right)_{y,z,t} = \left(\frac{\partial f}{\partial x}\right)_{y,z^*,t} + \left(\frac{\partial f}{\partial z^*}\right)_{x,y,t} \left(\frac{\partial z^*}{\partial x}\right)_{y,z,t} \tag{2.69}$$

$$\left(\frac{\partial f}{\partial y}\right)_{x,z,t} = \left(\frac{\partial f}{\partial y}\right)_{x,z^*,t} + \left(\frac{\partial f}{\partial z^*}\right)_{x,y,t} \left(\frac{\partial z^*}{\partial y}\right)_{x,z,t} \tag{2.70}$$

$$\left(\frac{\partial f}{\partial z}\right)_{x,y,t} = \left(\frac{\partial f}{\partial z^*}\right)_{x,y,t} \left(\frac{\partial z^*}{\partial z}\right)_{x,y,t} \tag{2.71}$$

$$\left(\frac{\partial f}{\partial t}\right)_{x,y,z} = \left(\frac{\partial f}{\partial t}\right)_{x,y,z^*} + \left(\frac{\partial f}{\partial z^*}\right)_{x,y,t} \left(\frac{\partial z^*}{\partial t}\right)_{x,y,z} \tag{2.72}$$

By taking $f = z$ and using our notation for the Jacobian, this immediately gives properties that will be used in the next paragraph:

$$\left(\frac{\partial z^*}{\partial x}\right)_{y,z,t} = -\frac{1}{\Delta z} \left(\frac{\partial z}{\partial x}\right)_{y,z^*,t} \tag{2.73}$$

$$\left(\frac{\partial z^*}{\partial y}\right)_{x,z,t} = -\frac{1}{\Delta z} \left(\frac{\partial z}{\partial y}\right)_{x,z^*,t} \tag{2.74}$$

$$\left(\frac{\partial z^*}{\partial t}\right)_{x,y,z} = -\frac{1}{\Delta z} \left(\frac{\partial z}{\partial t}\right)_{x,y,z^*} \tag{2.75}$$

Useful formulae with the sigma transform

This new system of coordinates is going to modify certain terms in our equations, so we introduce hereafter a few definitions and present some examples that will be useful later:

Velocity of the mesh: the velocity of points linked to the mesh is:

$$W_{mesh} = \left(\frac{\partial z}{\partial t}\right)_{x^*,y^*,z^*} \tag{2.76}$$

$\overrightarrow{W}_{mesh}$ will be the vector with components $(0,0,W_{mesh})$. We shall also encounter in the derivations a quantity which can be interpreted as the velocity of the transformed mesh:

$$W^*_{mesh} = \left(\frac{\partial z^*}{\partial t}\right)_{x,y,z} \tag{2.77}$$

Equation 2.75 is thus also:

$$W_{mesh} + \Delta z W^*_{mesh} = 0 \tag{2.78}$$

and the fact that we have:

$$\frac{\partial}{\partial t}\left(\frac{\partial z}{\partial z^*}\right) - \frac{\partial}{\partial z^*}\left(\frac{\partial z}{\partial t}\right) = 0$$

leads to:

$$\frac{\partial(\Delta z)}{\partial t} - \frac{\partial(W_{mesh})}{\partial z^*} = \frac{\partial(\Delta z)}{\partial t} - div^*(\overrightarrow{W}_{mesh}) = 0$$

or:

$$\frac{\partial(\Delta z)}{\partial t} + div^*(\Delta z \overrightarrow{W}^*_{mesh}) = 0 \tag{2.79}$$

In the same way we find:

$$\frac{\partial(\frac{1}{\Delta z})}{\partial t} + div\left(\frac{\overrightarrow{W}_{mesh}}{\Delta z}\right) = 0 \tag{2.80}$$

Advection terms: the advection terms must be carefully examined in a moving grid because the advected functions vary not only under the effect of the flow, but also because they are defined on moving points. The derivative $\frac{df}{dt}$, written as:

$$\frac{df}{dt} = \left(\frac{\partial f}{\partial t}\right)_{x,y,z} + U\frac{\partial f}{\partial x} + V\frac{\partial f}{\partial y} + W\frac{\partial f}{\partial z} \tag{2.81}$$

where $U = \frac{dx}{dt}$, $V = \frac{dy}{dt}$ and $W = \frac{dz}{dt}$ are the velocity components, now becomes, using formula 2.69:

$$\frac{df}{dt} = \left(\frac{\partial f}{\partial t}\right)_{x,y,z} + U\frac{\partial f}{\partial x^*} + U\frac{\partial z^*}{\partial x^*}\frac{\partial f}{\partial z^*} + V\frac{\partial f}{\partial y^*} + V\frac{\partial z^*}{\partial y^*}\frac{\partial f}{\partial z^*} + W\frac{\partial f}{\partial z^*}\frac{\partial z^*}{\partial z} \tag{2.82}$$

or:

$$\frac{df}{dt} = \left(\frac{\partial f}{\partial t}\right)_{x,y,z} + U\frac{\partial f}{\partial x^*} + V\frac{\partial f}{\partial y^*} + \left(U\frac{\partial z^*}{\partial x^*} + V\frac{\partial z^*}{\partial y^*} + W\frac{\partial z^*}{\partial z}\right)\frac{\partial f}{\partial z^*} \tag{2.83}$$

On the other hand the velocities in the transformed mesh are defined as:

$$U^* = \frac{dx^*}{dt} = \frac{dx}{dt} = U \tag{2.84}$$

$$V^* = \frac{dy^*}{dt} = \frac{dy}{dt} = V \tag{2.85}$$

$$W^* = \frac{dz^*}{dt} \tag{2.86}$$

This vertical velocity in the transformed mesh is also equal to:

$$W^* = \left(\frac{\partial z^*}{\partial t}\right)_{x,y,z} + U\frac{\partial z^*}{\partial x} + V\frac{\partial z^*}{\partial y} + W\frac{\partial z^*}{\partial z} \tag{2.87}$$

We thus have:

$$\frac{df}{dt} = \left(\frac{\partial f}{\partial t}\right)_{x,y,z} + U\frac{\partial f}{\partial x^*} + V\frac{\partial f}{\partial y^*} + (W^* - W^*_{mesh})\frac{\partial f}{\partial z^*} \tag{2.88}$$

where the vertical velocity is relative to the mesh. However, when the mesh is moving, f^{n+1} and f^n correspond to points having the same x, y and z^*, but not the same z. The partial derivatives in time in our equations are thus in fact $\left(\frac{\partial f}{\partial t}\right)_{x^*,y^*,z^*}$, and we have:

$$\left(\frac{\partial f}{\partial t}\right)_{x,y,z} = \left(\frac{\partial f}{\partial t}\right)_{x^*,y^*,z^*} + \frac{\partial f}{\partial z^*}\left(\frac{\partial z^*}{\partial t}\right)_{x,y,z} = \left(\frac{\partial f}{\partial t}\right)_{x^*,y^*,z^*} + \frac{\partial f}{\partial z^*}W^*_{mesh}$$

Equation 2.88 immediately yields:

$$\frac{df}{dt} = \left(\frac{\partial f}{\partial t}\right)_{x^*,y^*,z^*} + U\frac{\partial f}{\partial x^*} + V\frac{\partial f}{\partial y^*} + W^*\frac{\partial f}{\partial z^*} \tag{2.89}$$

Advection equations, in the transformed mesh and taking into account the movement of the real mesh, will then have the simple form:

$$\left(\frac{\partial f}{\partial t}\right)_{x^*,y^*,z^*} + U\frac{\partial f}{\partial x^*} + V\frac{\partial f}{\partial y^*} + W^*\frac{\partial f}{\partial z^*} = 0 \tag{2.90}$$

Comparing Equations 2.88 and 2.81 also shows that:

$$U\frac{\partial f}{\partial x} + V\frac{\partial f}{\partial y} + W\frac{\partial f}{\partial z} = U\frac{\partial f}{\partial x^*} + V\frac{\partial f}{\partial y^*} + (W^* - W^*_{mesh})\frac{\partial f}{\partial z^*} \tag{2.91}$$

while Equation 2.78 also becomes:

$$U\frac{\partial f}{\partial x} + V\frac{\partial f}{\partial y} + (W - W_{mesh})\frac{\partial f}{\partial z} = U\frac{\partial f}{\partial x^*} + V\frac{\partial f}{\partial y^*} + W^*\frac{\partial f}{\partial z^*} \tag{2.92}$$

Divergence: we now want to write $div(\overrightarrow{U})$ in the transformed mesh. Using formula 2.69, we have:

$$\begin{aligned} div(\overrightarrow{U}) = & \left(\frac{\partial U}{\partial x}\right)_{y,z*,t} + \left(\frac{\partial U}{\partial z^*}\right)_{x,y,t}\left(\frac{\partial z^*}{\partial x}\right)_{y,z,t} \\ & + \left(\frac{\partial V}{\partial y}\right)_{x,z^*,t} + \left(\frac{\partial V}{\partial z^*}\right)_{x,y,t}\left(\frac{\partial z^*}{\partial x}\right)_{x,z,t} \\ & + \left(\frac{\partial W}{\partial z^*}\right)_{x,y,t}\left(\frac{\partial z^*}{\partial z}\right)_{x,y,t} \end{aligned} \tag{2.93}$$

By multiplying the previous equation by $\frac{\partial z}{\partial z^*} = \Delta z$, which depends only on x and y, but not on z or z^*, and using the fact that:

$$\Delta z \left(\frac{\partial U}{\partial x}\right)_{y,z^*,t} = \left(\frac{\partial \Delta z U}{\partial x}\right)_{y,z^*,t} - U\frac{\partial \Delta z}{\partial x}$$

we then get:

$$\begin{aligned}\Delta z\ div(\vec{U}) = &\left(\frac{\partial \Delta z U}{\partial x}\right)_{y,z^*,t} - U\frac{\partial \Delta z}{\partial x} + \left(\frac{\partial \Delta z U}{\partial z^*}\right)_{x,y,t}\left(\frac{\partial z^*}{\partial x}\right)_{y,t,z} \\ &+ \left(\frac{\partial \Delta z V}{\partial y}\right)_{x,z^*,t} - V\frac{\partial \Delta z}{\partial y} + \left(\frac{\partial \Delta z V}{\partial z^*}\right)_{x,y,t}\left(\frac{\partial z^*}{\partial y}\right)_{x,t,z} \\ &+ \left(\frac{\partial \Delta z W}{\partial z^*}\right)_{x,y,t}\left(\frac{\partial z^*}{\partial z}\right)_{x,y,t}\end{aligned} \tag{2.94}$$

This equation is simplified thanks to a calculation of the term $\left(\frac{\partial \Delta z W^*}{\partial z^*}\right)_{x,y,t}$ and using Equation 2.87 we find:

$$\left(\frac{\partial \Delta z W^*}{\partial z^*}\right)_{x,y,t} = \frac{\partial}{\partial z^*}\left(\Delta z\frac{\partial z^*}{\partial t} + \Delta z U\frac{\partial z^*}{\partial x} + \Delta z V\frac{\partial z^*}{\partial y} + \Delta z W\frac{\partial z^*}{\partial z}\right)$$

$$= \left(\frac{\partial \Delta z U}{\partial z^*}\right)_{x,y,t}\frac{\partial z^*}{\partial x} + \left(\frac{\partial \Delta z V}{\partial z^*}\right)_{x,y,t}\frac{\partial z^*}{\partial y} + \left(\frac{\partial \Delta z W}{\partial z^*}\right)_{x,y,t}\frac{\partial z^*}{\partial z}$$

$$+\Delta z\frac{\partial}{\partial z^*}\left(\frac{\partial z^*}{\partial t}\right) + \Delta z U\frac{\partial}{\partial z^*}\left(\frac{\partial z^*}{\partial x}\right) + \Delta z V\frac{\partial}{\partial z^*}\left(\frac{\partial z^*}{\partial y}\right) + \Delta z W\frac{\partial}{\partial z^*}\left(\frac{\partial z^*}{\partial z}\right)$$

Equations 2.73 to 2.75 give:

$$\frac{\partial}{\partial z^*}\left(\frac{\partial z^*}{\partial t}\right)_{x,y,z} = -\frac{1}{\Delta z}\frac{\partial \Delta z}{\partial t} \tag{2.95}$$

$$\frac{\partial}{\partial z^*}\left(\frac{\partial z^*}{\partial x}\right)_{y,z,t} = -\frac{1}{\Delta z}\frac{\partial \Delta z}{\partial x} \tag{2.96}$$

$$\frac{\partial}{\partial z^*}\left(\frac{\partial z^*}{\partial y}\right)_{x,z,t} = -\frac{1}{\Delta z}\frac{\partial \Delta z}{\partial y} \tag{2.97}$$

We now assume that the Jacobian Δz is constant vertically (which will be our case on every finite element), and we thus have:

$$\frac{\partial}{\partial z^*}\left(\frac{\partial z^*}{\partial z}\right)_{x,y,t} = 0 \tag{2.98}$$

All this leads to:

$$\left(\frac{\partial \Delta z W^*}{\partial z^*}\right)_{x,y,t} = \left(\frac{\partial \Delta z U}{\partial z^*}\right)_{x,y,t}\frac{\partial z^*}{\partial x} + \left(\frac{\partial \Delta z V}{\partial z^*}\right)_{x,y,t}\frac{\partial z^*}{\partial y} + \left(\frac{\partial \Delta z W}{\partial z^*}\right)_{x,y,t}\frac{\partial z^*}{\partial z}$$

$$-\frac{\partial \Delta z}{\partial t} - U\frac{\partial \Delta z}{\partial x} - V\frac{\partial \Delta z}{\partial y}$$

We deduce from this that:

$$\left(\frac{\partial \Delta z W}{\partial z^*}\right)_{x,y,t} \frac{\partial z^*}{\partial z} = \left(\frac{\partial \Delta z W^*}{\partial z^*}\right)_{x,y,t} + \frac{\partial \Delta z}{\partial t} + U\frac{\partial \Delta z}{\partial x} + V\frac{\partial \Delta z}{\partial y}$$
$$-\left(\frac{\partial \Delta z U}{\partial z^*}\right)_{x,y,t} \frac{\partial z^*}{\partial x} - \left(\frac{\partial \Delta z V}{\partial z^*}\right)_{x,y,t} \frac{\partial z^*}{\partial y}$$

By carrying over this value to Equation 2.94, we get:

$$\Delta z\ div(\vec{U}) = \frac{\partial \Delta z}{\partial t} + \left(\frac{\partial \Delta z U}{\partial x}\right)_{y,z^*,t} + \left(\frac{\partial \Delta z V}{\partial y}\right)_{x,z^*,t} + \left(\frac{\partial \Delta z W^*}{\partial z^*}\right)_{x,y,t} \tag{2.99}$$

Other terms in the Navier–Stokes equations, such as diffusion terms, are modified in an even more cumbersome manner. The numerical resolution of equations with the help of this sigma transform is widely used but sometimes leads to awkward numerical artefacts (creeping diffusion etc.). We shall try to avoid this wherever possible, in particular during the diffusion stage.

2.2.6 Tracer equations in 3 dimensions

The tracer represents a temperature or a physical quantity contained in water (colouring agent, sediment). If the tracer interacts with the hydrodynamics of the flow, it is called active; if not, it is passive. Active tracers are mostly temperature and salinity, sometimes sediment. Passive tracers can serve for water quality studies.

Physically, the evolution in time of the tracer depends:

- On the current which carries it (advection).
- On diffusion (which is almost always turbulent).
- On source or sink terms.

It is unnecessary to specify here in which unit the tracer is expressed.

Transport–diffusion equation

In three dimensions, the tracer equation in a conservative form is as follows:

$$\frac{\partial}{\partial t}(\rho T) + div(\rho T\vec{U} + \vec{q}) = F_{source} \tag{2.100}$$

where F_{source} is the rate of creation and $\vec{q}$ is the flux due to the molecular diffusion (and, later in the book, turbulence) as:

$$\vec{q} = -\rho\,\nu_T\,\overrightarrow{grad}(T) \tag{2.101}$$

ν_T is the diffusion coefficient (molecular or turbulent, it is not specified here) of the tracer.

In the non-conservative form, by using the mass conservation equation, one obtains the equation known as the transport–diffusion equation:

$$\frac{\partial T}{\partial t} + \left(\vec{U}\,.\,\overrightarrow{grad}\right) T = div\left(\nu_T\ \overrightarrow{grad}(T)\right) + Q \tag{2.102}$$

where Q is a source or sink term.

This equation is written again in the developed form:

$$\frac{\partial T}{\partial t}+U\frac{\partial T}{\partial x}+V\frac{\partial T}{\partial y}+W\frac{\partial T}{\partial z} = \frac{\partial}{\partial x}\left(\nu_T\frac{\partial T}{\partial x}\right)+\frac{\partial}{\partial y}\left(\nu_T\frac{\partial T}{\partial y}\right)+\frac{\partial}{\partial z}\left(\nu_T\frac{\partial T}{\partial z}\right)+Q \tag{2.103}$$

Boundary conditions

The boundary conditions of the tracers on the walls are:
No-flux boundary condition:

$$\nu\frac{\partial T}{\partial n} = 0 \tag{2.104}$$

where the notation $\frac{\partial T}{\partial n}$ represents $\overrightarrow{grad}(T).\,\vec{n}$, where $\vec{n}$ is the perpendicular pointing outwards from the boundary.
Flux boundary condition:

$$\nu\frac{\partial T}{\partial n} = aT + b \tag{2.105}$$

2.3 Saint-Venant equations

2.3.1 Presentation and brief review

On 28 August 1997 the French Ecole Nationale des Ponts et Chaussées celebrated the bicentenary of the birth of Jean-Claude Adhémar Barré, Count of Saint-Venant. At the age of 15 he joined the Ecole Polytechnique and later the Ecole des Ponts et Chaussées which he first attended in 1825. In 1834, he presented two papers to the Académie des Sciences, on theoretical mechanics and the dynamics of fluids. Upon retirement, after a career in the Technical Department of the City of Paris, he devoted his time to scientific research and entered the Académie des Sciences in 1868. According to Karl Pearson, in 1870, this "model scholar" converted his house into a hospital of the Red Cross, founded seven years earlier. This man, who used to say "desire humble perfection and do not strive to make it noticed", died in January 1886, "the soul filled with immortal hope never tarnished by doubt" as written by his student, Boussinesq.

The equations then known as the "Barré de Saint-Venant equations", published in 1871 [169], are still today of extreme importance in maritime or fluvial hydraulics, and govern free surface flows in shallow waters, hence their English name "shallow water equations". They are obtained from the Navier–Stokes equations, with the help of simplifying hypotheses and are used in spheres as varied as protection of the environment, calculation of tides and storm waves, the impact and stability of structures, sedimentology, simulation of dam-break flood waves, study of floods, etc.

One of the restrictive hypotheses of these equations is that the wavelength should be much greater than the depth. Consequently, calculating a swell does not come under their domain of validity.

In this chapter, the Saint-Venant equations are presented with their different formulations in order to highlight the modelled physical phenomena and their implied precautions in numerical analysis.

The principal operation to get the Saint-Venant from the Navier–Stokes equations is an average over the vertical. However, the existence of non-linear terms necessitates the use of some hypotheses and approximations. These hypotheses, which condition the scope of validity of the equations, should be remembered.

2.3.2 Hypotheses, approximations and calculation rules

Hydrostatic pressure

This approximation, which is resumed here, was presented earlier with the Navier–Stokes equations. According to it, acceleration arising from pressure counterbalances gravity. In such a case, the pressure is hydrostatic. This is coupled with small movements in the vertical direction.

In fact, this hypothesis on pressure is not strictly necessary for the development of the Saint-Venant equations, it is only necessary to translate the pressure in terms of the height of water.

The pressure $p(x, y, z, t)$ being due only to the weight of the column of water above the coordinates (x, y, z), we get:

$$-\frac{1}{\rho}\frac{\partial p}{\partial z} - g = 0 \tag{2.106}$$

that is:

$$p(x, y, z) = -\rho g z + cst \tag{2.107}$$

The constant cst is chosen in such a way that $p(x, y, Z_s) = p_{atm}$, where Z_s is the elevation of the free surface and p_{atm} the atmospheric pressure. If the gradient of the atmospheric pressure is not zero, it is considered as a source term. To simplify, we shall write:

$$p(x, y, z, t) = \rho g(Z_s - z) \tag{2.108}$$

In particular, at the bottom, $p = \rho g(Z_s - Z_f) = \rho g h$.

Negligible vertical velocity

Vertical velocity will be negligible in the Saint-Venant equations and will not have an equation. This approximation goes together with the hypothesis of hydrostatics according to which vertical accelerations should be negligible.

Impermeability of the surface and the bottom

It is assumed that there is no transfer of water mass through the bottom and the surface, and that a particle of water situated on one of these two interfaces will remain there. According to the equations established in Section 2.2.2 we write:

$$\frac{\partial Z_s}{\partial t} + U^s \frac{\partial Z_s}{\partial x} + V^s \frac{\partial Z_s}{\partial y} - W^s = 0 \tag{2.109}$$

and:

$$\frac{\partial Z_f}{\partial t} + U^f \frac{\partial Z_f}{\partial x} + V^f \frac{\partial Z_f}{\partial y} - W^f = 0 \tag{2.110}$$

Note: later, we shall study the source terms due to the arrival of a discharge through the bottom or the surface. In this case there will be no more impermeability.

The Leibnitz rule

The derivative of an integral with variable boundaries gives rise to a derivative within the integral and flux terms according to the formula:

$$\frac{\partial}{\partial x}\int_{Z_f}^{Z_s} F\,dz = \int_{Z_f}^{Z_s} \frac{\partial F}{\partial x}\,dz + F(x,y,Z_s)\frac{\partial Z_s}{\partial x} - F(x,y,Z_f)\frac{\partial Z_f}{\partial x} \tag{2.111}$$

If it is stated that:

$$f = \frac{1}{h}\int_{Z_f}^{Z_s} F\,dz$$

where f represents the average of F over the vertical, by applying the above formula to the derivatives in relation to x, y and z, we get:

$$\overrightarrow{grad}(hf) = \int_{Z_f}^{Z_s} \overrightarrow{grad}(F)\,dz - F(x,y,Z_s)\,\vec{n}_s - F(x,y,Z_f)\,\vec{n}_f \tag{2.112}$$

where $\vec{n}_s$ and $\vec{n}_f$ are vectors, which are not unit vectors, directed upwards in the first case and downwards in the second.

Similarly, for a vector:

$$div(h\vec{f}) = \int_{Z_f}^{Z_s} div(\vec{F})dz - \vec{F}(x,y,Z_s)\,.\,\vec{n}_s - \vec{F}(x,y,Z_f)\,.\,\vec{n}_f \tag{2.113}$$

For a derivative in time, and bearing in mind that:

$$\frac{\partial Z_s}{\partial t} - \vec{U}^s\,.\,\vec{n}_s = 0$$

and:

$$\frac{\partial Z_f}{\partial t} + \vec{U}^f\,.\,\vec{n}_f = 0$$

we obtain:

$$\frac{\partial(hf)}{\partial t} = \int_{Z_f}^{Z_s} \frac{\partial F}{\partial t} dz + F(x,y,Z_s)\vec{U}^s.\vec{n}_s + F(x,y,Z_f)\vec{U}^f.\vec{n}_f \tag{2.114}$$

2.3.3 Depth-averaging Navier–Stokes equations

Principle

The Navier–Stokes equations with constant density and hydrostatic pressure are now going to be averaged vertically, by integration from the bottom to the surface. Apart from the already-defined height of the water, two new variables are going to appear:

$$u = \frac{1}{h}\int_{Z_f}^{Z_s} U dz \text{ and } v = \frac{1}{h}\int_{Z_f}^{Z_s} V dz \tag{2.115}$$

These averages of horizontal components of the three-dimensional velocity vector over the vertical will themselves be called velocities, or "components of the velocity vector". The vector with components u and v will be denoted $\vec{u}$.

Average of the continuity equation

The continuity equation becomes:

$$\int_{Z_f}^{Z_s} \left(\frac{\partial U}{\partial x} + \frac{\partial V}{\partial y} + \frac{\partial W}{\partial z}\right) dz = 0 \tag{2.116}$$

With the Leibnitz rule it also becomes:

$$\int_{Z_f}^{Z_s} (\frac{\partial U}{\partial x} + \frac{\partial V}{\partial y} + \frac{\partial W}{\partial z})dz = W(Z_s) - W(Z_f)$$

$$+\frac{\partial}{\partial x}\int_{Z_f}^{Z_s} U dz - U(x,y,Z_s)\frac{\partial Z_s}{\partial x} + U(x,y,Z_f)\frac{\partial Z_f}{\partial x}$$

$$+\frac{\partial}{\partial y}\int_{Z_f}^{Z_s} V dz - V(x,y,Z_s)\frac{\partial Z_s}{\partial x} + V(x,y,Z_f)\frac{\partial Z_f}{\partial y} \tag{2.117}$$

That is, taking into account the impermeability conditions:

$$\frac{\partial}{\partial x}(hu) + \frac{\partial}{\partial y}(hv) + \frac{\partial Z_s}{\partial t} - \frac{\partial Z_f}{\partial t} = 0 \tag{2.118}$$

or:

$$\frac{\partial h}{\partial t} + div(h\vec{u}) = 0 \tag{2.119}$$

Here the divergence is calculated in two dimensions.

We immediately notice the analogy with the mass conservation equation of compressible fluids:

$$\frac{\partial \rho}{\partial t} + div(\rho \vec{u}) = 0$$

In the Saint-Venant equations the water depth plays the role of density.

Average of the momentum equation

We average the momentum equation found in Section 2.2.3:

$$\frac{\partial(\vec{U})}{\partial t} + \vec{U}\,\nabla\vec{U} = -\frac{1}{\rho}\overrightarrow{grad}(p) + \frac{1}{\rho}\,div(\underline{\underline{\sigma}}) + \vec{g} + \vec{F}$$

and apply from this point the hypothesis of hydrostatic pressure (Equation 2.108). Let us look at the average of each of the terms of this equation, using the example of the component U.

Derivative in time:

$$\int_{Z_f}^{Z_s} \frac{\partial U}{\partial t} dz = \frac{\partial(hu)}{\partial t} - U(Z_s)\frac{\partial Z_s}{\partial t} + U(Z_f)\frac{\partial Z_f}{\partial t} \tag{2.120}$$

Diffusion terms: for the diffusion term $\frac{1}{\rho}div(\underline{\underline{\tau}})$, we suppose that the density does not vary along the vertical; it can then be removed from the integrals and the Leibnitz formula gives:

$$\frac{1}{\rho}\int_{Z_f}^{Z_s} div(\underline{\underline{\tau}})\,dz = \frac{1}{\rho} div\left(\int_{Z_f}^{Z_s} \underline{\underline{\tau}}\,dz\right) + \frac{1}{\rho}\,\underline{\underline{\tau}}_s \vec{n}_s + \frac{1}{\rho}\,\underline{\underline{\tau}}_f \vec{n}_f \tag{2.121}$$

(normal vectors at the bottom and at the surface are still not normalized).

The last two terms of Equation 2.121 represent the stress at the surface and the bottom, due for example to the wind and to the friction. They will be elaborated later, with all the source terms.

The two components of $\frac{1}{\rho}div(\int_{Z_f}^{Z_s} \underline{\underline{\tau}}\,dz)$ are:

$$\frac{1}{\rho} div\left(\int_{Z_f}^{Z_s} \mu\,\overrightarrow{grad}(U)\,dz\right)$$

and:

$$\frac{1}{\rho} div\left(\int_{Z_f}^{Z_s} \mu\,\overrightarrow{grad}(V)\,dz\right)$$

respectively.

Before arriving at the final form of diffusion terms, we should consider that:

$$\int_{Z_f}^{Z_s} \mu\ \overrightarrow{grad}(U)\ dz$$

is hardly different from:

$$h\,\mu\,\overrightarrow{grad}(u)$$

$\frac{1}{\rho}div(\int_{Z_f}^{Z_s}\underline{\underline{\tau}}\,dz)$ is finally taken as equal to the vector of components:

$$\frac{1}{\rho}div(h\mu\overrightarrow{grad}(u)) \text{ and } \frac{1}{\rho}div(h\mu\overrightarrow{grad}(v))$$

or even, by considering that the density is constant for calculating this term:

$$div(h\nu\overrightarrow{grad}(u)) \text{ and } div(h\nu\overrightarrow{grad}(v))$$

Advections terms: these are for the equation of U:

$$U\frac{\partial U}{\partial x}+V\frac{\partial U}{\partial y}+W\frac{\partial U}{\partial z}$$

which can also be stated as:

$$\frac{\partial(UU)}{\partial x}+\frac{\partial(UV)}{\partial y}+\frac{\partial(UW)}{\partial z}$$

thanks to the continuity equation. Let us calculate the average of each of these three terms over the vertical.

The third term is the simplest:

$$\int_{Z_f}^{Z_s}\frac{\partial(U(x,y,z)W(x,y,z))}{\partial z}\,dz=U(x,y,Z_s)W(x,y,Z_s)-U(x,y,Z_f)W(x,y,Z_f) \tag{2.122}$$

We then have:

$$\int_{Z_f}^{Z_s}\frac{\partial U^2}{\partial x}dz=\frac{\partial}{\partial x}\int_{Z_f}^{Z_s}U^2dz-U^2(Z_s)\frac{\partial Z_s}{\partial x}+U^2(Z_f)\frac{\partial Z_f}{\partial x} \tag{2.123}$$

$$\int_{Z_f}^{Z_s}\frac{\partial(UV)}{\partial y}dz=\frac{\partial}{\partial y}\int_{Z_f}^{Z_s}UV dz-U(Z_s)V(Z_s)\frac{\partial Z_s}{\partial y}+U(Z_f)V(Z_f)\frac{\partial Z_f}{\partial y} \tag{2.124}$$

These terms give rise to non-linearity. Let us develop for example the expression containing UV:

$$\begin{aligned}\frac{\partial}{\partial y}\int_{Z_f}^{Z_s}UV dz &= \frac{\partial}{\partial y}\int_{Z_f}^{Z_s}(u+U-u)(v+V-v)dz\\ &= \frac{\partial}{\partial y}(huv)+\frac{\partial}{\partial y}\int_{Z_f}^{Z_s}(U-u)(V-v)\,dz\end{aligned} \tag{2.125}$$

The term $\frac{\partial}{\partial y}\int_{Z_f}^{Z_s}(U-u)(V-v)\,dz$ is not zero when there are heterogeneities of velocity vertically. These terms, called dispersion terms, are generally interpreted as additional diffusion. Basically they take into account fluctuations around an average (here spatial) and are, on these grounds, formally similar to the Reynolds stress. In reality, these terms arise from the fact that transport of physical quantities is carried out by real current and not by the current averaged over the vertical. Depending on the depth, all particles are not moved at the same velocity. In the Saint-Venant equations, on the contrary, a depth-averaged current is chosen. Here, the form of diffusion is adopted, resulting in a tensor naturally being added to the Reynolds tensor. In fact, if $u"= U - u$ and $v"= V - v$, these dispersion terms are, respectively along x and along y:

$$\frac{1}{h}\frac{\partial}{\partial x}\int_{Z_f}^{Z_s} u"u"dz + \frac{1}{h}\frac{\partial}{\partial y}\int_{Z_f}^{Z_s} u"v"dz \tag{2.126}$$

$$\frac{1}{h}\frac{\partial}{\partial x}\int_{Z_f}^{Z_s} u"v"dz + \frac{1}{h}\frac{\partial}{\partial y}\int_{Z_f}^{Z_s} v"v"dz \tag{2.127}$$

In this form, it can be seen that the tensor $u"_i u"_j$ is naturally added to the stress tensor. One manner of taking it into consideration would be to modify the value of viscosity ν, or turbulent flow viscosity ν_t, and replace it by an effective viscosity integrating the dispersion phenomenon.

The turbulence models should thus take dispersion into account in two dimensions. This will be seen more specifically with the k–ε model.

Pressure gradient: for pressure gradients in x and y, we write:

$$\int_{Z_f}^{Z_s} -\frac{1}{\rho}\frac{\partial}{\partial x}[\rho g(Z_s - z)]dz = -hg\frac{\partial Z_s}{\partial x}$$

and:

$$\int_{Z_f}^{Z_s} -\frac{1}{\rho}\frac{\partial}{\partial y}[\rho g(Z_s - z)]dz = -hg\frac{\partial Z_s}{\partial y}$$

This is true only if the density does not vary along the horizontal. The variations of density will be taken into account with the buoyancy source terms.

Source terms and body forces: for constant source terms along the vertical, the integration along z is a simple multiplication by the water depth h, and F_x gives hF_x.

Until their more precise study in Section 2.3.11, friction forces and the terms:

$$\frac{1}{\rho}\underline{\underline{\tau}}_s\vec{n}_s + \frac{1}{\rho}\underline{\underline{\tau}}_f\vec{n}_f$$

will be included in these body forces, and will therefore be omitted hereafter.

Recapitulation: the sum of all these terms, a number of which disappear because of the impermeability of the bottom and open surface, gives the following equation:

$$\frac{\partial(hu)}{\partial t} + \frac{\partial}{\partial x}(huu) + \frac{\partial}{\partial y}(huv) = -gh\frac{\partial Z_s}{\partial x} + hF_x + div(h\nu_e\overrightarrow{grad}(u)) \qquad (2.128)$$

In the same manner, the equation for V gives:

$$\frac{\partial(hv)}{\partial t} + \frac{\partial}{\partial x}(huv) + \frac{\partial}{\partial y}(hvv) = -gh\frac{\partial Z_s}{\partial y} + hF_y + div(h\nu_e\overrightarrow{grad}(v)) \qquad (2.129)$$

ν_e is an effective diffusion which will model turbulent viscosity and dispersion.

The W equation has not been retained (it has already been used to obtain the hydrostatic pressure hypothesis).

With the mass conservation equation, these equations constitute the two-dimensional set of Barré de Saint-Venant equations, in what is called the "conservative" form.

2.3.4 Different forms of equations

Conservative form

Equations 2.119, 2.128 and 2.129 are rewritten by stating:

$$Q_x = hu \text{ and } Q_y = hv$$

$\vec{Q}$ is the vector of the components Q_x and Q_y. The "discharge" Q is in fact per unit length, in m^2/s:

Continuity:

$$\frac{\partial h}{\partial t} + div(\vec{Q}) = 0 \qquad (2.130)$$

Momentum:

$$\frac{\partial Q_x}{\partial t} + div(Q_x\vec{u}) = -gh\frac{\partial Z_s}{\partial x} + hF_x + div(h\nu_e\overrightarrow{grad}(u)) \qquad (2.131)$$

$$\frac{\partial Q_y}{\partial t} + div(Q_y\vec{u}) = -gh\frac{\partial Z_s}{\partial y} + hF_y + div(h\nu_e\overrightarrow{grad}(v)) \qquad (2.132)$$

Source terms in the continuity equation:

In the presence of water intake or release at the free surface, inside the domain or at the bottom, or if one takes into account rain or infiltrations in the ground, the right-hand side of the continuity equation is not zero but equal to a term that we shall denote as "*Sce*" for "Source" (it can also be a sink). *Sce* is expressed in m/s and is positive for an injection of water into the domain. It is for example the rainfall, expressed in metres per second. We should multiply *Sce* by a surface to obtain a real discharge.

The mass conservation equation is thus written as:

$$\frac{\partial h}{\partial t} + div(\vec{Q}) = Sce \qquad (2.133)$$

Similarly in the momentum equations we should take into account the supply from the source, a vector of components $Mt(x)$ and $Mt(y)$ (Mt for "Momentum"), in m^2/s^2. These terms are equal to $u_{sce}Sce$ and $v_{sce}Sce$, where u_{sce} and v_{sce} are the components of the velocity at source.

Non-conservative form

In theory, only a strict processing of conservative equations allows one to really take into account discontinuities, which can appear in the solutions of shallow water equations (hydraulic jump). Other reasons (stability of schemes, use of the method of characteristics) have led to the adoption of a different formulation, with unknowns h, u and v, called the "depth–velocity" formulation. This formulation is said to be "non-conservative". Let us immediately mention that this name does not prevent us from constructing numerical schemes which conserve the mass of water, as will be shown subsequently.

On the basis of the conservative form of equations, the derivatives of function products are developed in the momentum and continuity equations.

The continuity equation becomes:

$$\frac{\partial h}{\partial t} + \vec{u}\,.\,\overrightarrow{grad}(h) + h\,div(\vec{u}) = Sce \tag{2.134}$$

and the first component of the momentum equation:

$$h\frac{\partial u}{\partial t} + u\frac{\partial h}{\partial t} + u\frac{\partial (hu)}{\partial x} + hu\frac{\partial u}{\partial x} + u\frac{\partial (hv)}{\partial y} + hv\frac{\partial u}{\partial y}$$

$$= -gh\frac{\partial Z_s}{\partial x} + hF_x + div(h\nu_e\overrightarrow{grad}(u)) + u_{sce}Sce \tag{2.135}$$

that is, by using the continuity equation:

$$h\frac{\partial u}{\partial t} + hu\frac{\partial u}{\partial x} + hv\frac{\partial u}{\partial y} = -gh\frac{\partial Z_s}{\partial x} + hF_x + div(h\nu_e\overrightarrow{grad}(u)) + (u_{sce} - u)Sce \tag{2.136}$$

The two members of the equation are divided by h. By applying the same operation to v, we finally obtain:

$$\frac{\partial u}{\partial t} + u\frac{\partial u}{\partial x} + v\frac{\partial u}{\partial y} = -g\frac{\partial Z_s}{\partial x} + F_x + \frac{1}{h}div(h\nu_e\overrightarrow{grad}(u)) + (u_{sce} - u)\frac{Sce}{h} \tag{2.137}$$

$$\frac{\partial v}{\partial t} + u\frac{\partial v}{\partial x} + v\frac{\partial v}{\partial y} = -g\frac{\partial Z_s}{\partial y} + F_y + \frac{1}{h}div(h\nu_e\overrightarrow{grad}(v)) + (v_{sce} - v)\frac{Sce}{h} \tag{2.138}$$

The conservative and non-conservative forms are not equivalent in the presence of discontinuities.

Remark: if there is a source at the bottom, the terms $(u_{sce}-u)\frac{Sce}{h}$ and $(v_{sce}-v)\frac{Sce}{h}$ are added to the right-hand sides. These terms are zero if the velocity at the source is the same as that of the current. They will be included in the expressions F_x and F_y and will therefore not appear hereafter.

Celerity of waves with Saint-Venant equations

The two main phenomena modelled by the Saint-Venant equations are advection, or transport by the current, and propagation of long waves. The propagation is easily highlighted for small disturbances for which the linearized equations are valid. The advection terms also are ignored and the hypothesis of a horizontal bottom is formed, which gives, in one dimension:

$$\frac{\partial u}{\partial t} + g\frac{\partial h}{\partial x} = 0 \tag{2.139}$$

and:

$$\frac{\partial h}{\partial t} + h_0\frac{\partial u}{\partial x} = 0 \tag{2.140}$$

where h_0 is the average depth.

The combination:

$$-h_0\frac{\partial}{\partial x}\left(\frac{\partial u}{\partial t} + g\frac{\partial h}{\partial x}\right) + \frac{\partial}{\partial t}\left(\frac{\partial h}{\partial t} + h_0\frac{\partial u}{\partial x}\right) = 0 \tag{2.141}$$

gives:

$$\frac{\partial^2 h}{\partial t^2} - gh_0\frac{\partial^2 h}{\partial x^2} = 0 \tag{2.142}$$

and similarly we can get:

$$\frac{\partial^2 u}{\partial t^2} - gh_0\frac{\partial^2 u}{\partial x^2} = 0 \tag{2.143}$$

These two equations express that u and h propagate at the velocity $+c$ or $-c$ with $c^2 = gh_0$. From Euler, we know in fact that the solution of an equation of the type:

$$\frac{\partial^2 f}{\partial t^2} - c^2\frac{\partial^2 f}{\partial x^2} = 0 \tag{2.144}$$

is:

$$f(x,t) = A(x - ct) + B(x + ct) \tag{2.145}$$

the form of the A and B functions being determined by the initial conditions and the boundary conditions. In the absence of a current, the celerity of waves in the Saint-Venant equations is hence $c = \sqrt{gh}$. The celerity of long waves with low amplitude in linear theory reappears here. For a depth of 2000 m, the celerity is thus 500 km/h, a value which is effectively attained by tidal waves and tsunamis.

Celerity–velocity formulation

On the basis of the non-conservative formulation, which has just been obtained, another formulation is brought out. On the one hand it shows the analogy of the Saint-Venant equations with the Euler equations, and on the other hand it constitutes the most natural formulation for presenting the characteristic curves. Finally, it will provide at a later stage the idea of a preconditioning for the iterative methods resolving the linear system arising from the Saint-Venant equations.

This formulation is obtained by replacing the water depth h by c^2/g in the equations, to get:

Mass conservation:

$$2c\frac{\partial c}{\partial t} + 2cu\frac{\partial c}{\partial x} + 2cv\frac{\partial c}{\partial y} + c^2\left(\frac{\partial u}{\partial x} + \frac{\partial v}{\partial y}\right) = 0 \tag{2.146}$$

that is, after division by $2c$:

$$\frac{\partial c}{\partial t} + u\frac{\partial c}{\partial x} + v\frac{\partial c}{\partial y} + \frac{c}{2}div(\vec{u}) = 0 \tag{2.147}$$

Momentum:

$$\frac{\partial u}{\partial t} + u\frac{\partial u}{\partial x} + v\frac{\partial u}{\partial y} + 2c\frac{\partial c}{\partial x} = -g\frac{\partial Z_f}{\partial x} + F_x + \frac{1}{h}div(h\nu_e\overrightarrow{grad}(u)) \tag{2.148}$$

$$\frac{\partial v}{\partial t} + u\frac{\partial v}{\partial x} + v\frac{\partial v}{\partial y} + 2c\frac{\partial c}{\partial y} = -g\frac{\partial Z_f}{\partial y} + F_y + \frac{1}{h}div(h\nu_e\overrightarrow{grad}(v)) \tag{2.149}$$

This form of equation is often simpler for building exact solutions (e.g. one-dimensional dam break) and is more fundamental; in particular, gravitational acceleration no longer plays a part. Finally, all the unknowns u, v and c have the same dimension.

Analogy with Euler equations

The Euler equations govern the flow of compressible inviscid fluids with constant entropy; they are written in two dimensions:

$$\frac{\partial c}{\partial t} + u\frac{\partial c}{\partial x} + v\frac{\partial c}{\partial y} + \frac{(\gamma - 1)\,c}{2}div(\vec{u}) = 0 \tag{2.150}$$

$$\frac{\partial u}{\partial t} + u\frac{\partial u}{\partial x} + v\frac{\partial u}{\partial y} + \frac{2}{\gamma - 1}c\frac{\partial c}{\partial x} = 0 \tag{2.151}$$

$$\frac{\partial v}{\partial t} + u\frac{\partial v}{\partial x} + v\frac{\partial v}{\partial y} + \frac{2}{\gamma - 1}c\frac{\partial c}{\partial y} = 0 \tag{2.152}$$

γ is the heat capacity ratio, the ratio of the specific heat at constant pressure to the specific heat at constant volume. Here, c is the celerity of the waves in the fluid, equal to $\sqrt{\frac{\gamma p}{\rho}}$, p being the pressure and ρ the density.

There is a perfect analogy between the Euler and Saint-Venant equations (for the latter, γ would be equal to 2, and for air $\gamma = 1.405$). This analogy underlines the fact that the Saint-Venant equations are formally compressible fluid equations, the role of variable density being played by the water depth. However, the analogy ends with the right-hand sides of the equations, the source terms of the Saint-Venant equations posing specific difficulties which rapidly make the respective approaches divergent.

2.3.5 The characteristic curves

The method of characteristics arises from the geometric theory of differential quasi-linear equations. It gives a physical interpretation of certain problems in fluid mechanics and sometimes allows an analytical solution to be built from it. Moreover, it indicates the number and nature of the boundary conditions required, and can detect ill-posed problems. Concepts of the characteristics and the Riemann invariants in the Saint-Venant equations are thus fundamental. We shall choose the linearized celerity–velocity formulation in one dimension, without diffusion or source terms, and in a flat-bottomed canal:

$$\frac{\partial u}{\partial t} + u\frac{\partial u}{\partial x} + c\frac{\partial(2c)}{\partial x} = 0 \tag{2.153}$$

$$\frac{\partial(2c)}{\partial t} + u\frac{\partial(2c)}{\partial x} + c\frac{\partial u}{\partial x} = 0 \tag{2.154}$$

The sum of the two equations gives:

$$\frac{\partial(u+2c)}{\partial t} + (u+c)\frac{\partial(u+2c)}{\partial x} = 0 \tag{2.155}$$

and the difference:

$$\frac{\partial(u-2c)}{\partial t} + (u-c)\frac{\partial(u-2c)}{\partial x} = 0 \tag{2.156}$$

These equations have the form:

$$\frac{\partial A}{\partial t} + \frac{dx}{dt}\frac{\partial A}{\partial x} = 0 \tag{2.157}$$

with $A = u + 2c$ or $u - 2c$, which also signifies $\frac{dA}{dt} = 0$ along the curves represented by the equation $\frac{dx}{dt} = u + c$ or $u - c$. We thus have:

$$\frac{d(u+2c)}{dt} = 0$$

on the curves named C+, represented by the equation:

$$\frac{dx}{dt} = u + c$$

and:

$$\frac{d(u-2c)}{dt} = 0$$

on the curves named C−, represented by the equation:

$$\frac{dx}{dt} = u - c$$

On the curves C+, the quantity $u + 2c$ is invariant.
On the curves C−, the quantity $u - 2c$ is invariant.

The curves C+ and C− are called characteristic curves. The quantities $u+2c$ and $u-2c$ are called the Riemann invariants. The existence of real characteristic curves places the Saint-Venant equations in the family of hyperbolic differential equations.

Two cases then appear according to the sign of $u-c$, u being taken as positive.

If $u < c$, the flow is said to be subcritical. The waves are faster than the current. The flow at any point depends on the boundary conditions upstream and downstream.

If $u > c$, the flow is said to be supercritical. The waves are slower than the current. The flow at any point depends only on the upstream boundary conditions; it only has an influence downstream.

If $u = c$, the flow is said to be critical.

Thus, the boundary conditions change according to the nature of flow. This phenomenon is very important in numerical modelling, and leads to the failure of quite a few methods.

The subcritical or supercritical nature of a flow is given by the Froude number:

$$F = \frac{u}{c}$$

in one dimension or:

$$F = \frac{\sqrt{u^2+v^2}}{c}$$

in two dimensions.

If $F < 1$, the flow is subcritical or fluvial.

If $F = 1$, the flow is critical.

If $F > 1$, the flow is supercritical.

2.3.6 Notions on hydraulic jumps

The Saint-Venant equations, due to their compressible nature, accept discontinuous solutions, which are called hydraulic jumps. We shall distinguish between steady and unsteady hydraulic jumps.

Steady hydraulic jumps; conjugate depths relation

Let us consider in one dimension permanent flow in a canal with constant width and a flat bottom. In the absence of a source term, diffusion and friction, the Saint-Venant equations express:

- Conservation of the flow rate per unit of width $Q = hu$.
- The momentum balance, conservation of $hu^2 + g\frac{h^2}{2}$, deducted from the conservative momentum equation, and which can also be derived from the Euler theorem.

Assuming the existence of a discontinuity, with quantities h_1 and u_1 upstream, and h_2 and u_2 downstream (see Figure 2.2), it can be verified that:

$$h_1\,u_1 = h_2\,u_2 = Q \tag{2.158}$$

$$h_1u_1^2 + g\frac{h_1^2}{2} = h_2u_2^2 + g\frac{h_2^2}{2} \tag{2.159}$$

An evident solution would certainly be $h_1 = h_2$ and $u_1 = u_2$, but it is not the only one.

By replacing u_1 by Q/h_1 and u_2 by Q/h_2 in the second equation, we get:

$$h_1h_2(h_1 + h_2) = \frac{2Q^2}{g} \tag{2.160}$$

This equation is the conjugate depths relation (or hydraulic jump relation), which depends on the flow rate. It can also be written as:

$$h_1h_2(h_1 + h_2) = 2h_c^3 \tag{2.161}$$

where h_c is the critical depth, such that:

$$h_c^3 = \frac{2Q^2}{g}$$

There also exists a relation for the conjugate velocities:

$$\frac{u_1^2u_2^2}{u_1 + u_2} = \frac{gQ}{2} = \frac{u_c^3}{2} \tag{2.162}$$

where u_c is the velocity of the critical flow:

$$u_c = \sqrt{gh_c}$$

The relation for the conjugate Froude numbers is more interesting as it is independent of any other parameter. In fact, by noticing that $gh_1^3 = \frac{Q^2}{F_1^2}$ and $gh_2^3 = \frac{Q^2}{F_2^2}$, the depths relation gives:

$$F_1^{2/3}\,F_2^{2/3} = \frac{1}{2}\left(\frac{1}{F_1^{2/3}} + \frac{1}{F_2^{2/3}}\right) \tag{2.163}$$

or:

$$F_2^2 = \frac{(1 + \sqrt{1 + 8F_1^2})^3}{64F_1^4} \tag{2.164}$$

F_1 and F_2 play a symmetrical role in these formulae. When $F1 > 1$, $F2 < 1$ and vice versa. This signifies that a steady hydraulic jump is a transition between subcritical flow and supercritical flow.

A head loss always takes place across a hydraulic jump: the difference in head between the upstream and downstream ends of the jump, in units of height, is:

$$\Delta H = H_1 - H_2 = h_1 + \frac{u_1^2}{2g} - h_2 - \frac{u_2^2}{2g} \tag{2.165}$$

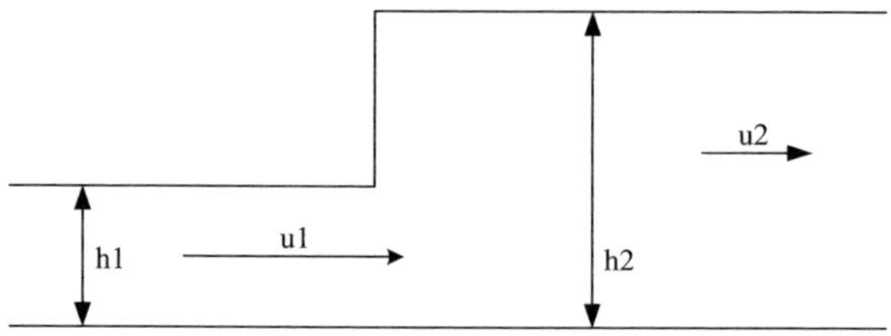

Figure 2.2: steady hydraulic jump

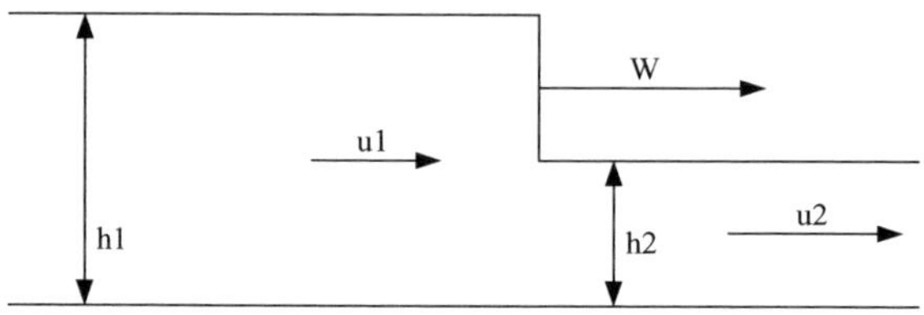

Figure 2.3: propagation of a jump on a wet bottom

which can be expressed according to the water depths on either side of the jump as:

$$\Delta H = \frac{(h_2 - h_1)^3}{4h_1 h_2} \tag{2.166}$$

This value is necessarily positive as it corresponds to a head loss. Therefore, a supercritical flow prevails upstream, while a subcritical flow prevails downstream. We shall return to the theoretical study of hydraulic jumps when we examine the conservation of quantities like energy, head and momentum at the discrete level (Sections 4.10.3 and 4.10.4).

Moving hydraulic jump

Let us assume that the jump moves with a velocity W (see Figure 2.3). We choose a referential which is attached to the hydraulic jump. In this referential the flow is steady and the Saint-Venant equations are still valid. The conservation of mass and momentum on either side of the jump are then written as:

$$h_1(u_1 - W) = h_2(u_2 - W) \tag{2.167}$$

$$h_1(u_1 - W)^2 + g\frac{h_1^2}{2} = h_2(u_2 - W)^2 + g\frac{h_2^2}{2} \tag{2.168}$$

From the first equation we derive:

$$W = \frac{h_1 u_1 - h_2 u_2}{h_1 - h_2} \tag{2.169}$$

which gives the velocity of the jump according to the upstream and downstream characteristics of the flow. It is interesting to notice that if the velocity downstream u_2 is zero, it can be shown that:

$$W^2 = \frac{g h_1 (h_1 + h_2)}{2 h_2}$$

Now if the height of the hydraulic jump tends to zero, so that we can take $h_1 = h_2 = h$, we find that $W^2 = gh$; W is thus simply the celerity of long waves. If a wave is considered as a series of infinitely small hydraulic jumps, we find its celerity here with simple considerations on the conservation of mass and momentum.

Note: W does not always have the same direction as u_1 and u_2 (the case of a hydraulic bore).

2.3.7 Saint-Venant equations in Mercator projection

Storm surges and storm-induced currents, differences between the actual levels of the sea and currents and their values if the tides alone were considered, are due to the variations of atmospheric pressure and the effect of the wind on large stretches of the ocean. The physical conditions of these phenomena ensure that vertical acceleration can be ignored and that flow is almost horizontal. Calculation of storm surges and storm-induced currents generally takes place by solving the Saint-Venant equations. To carry out numerical simulations, vast maritime areas are discretized on the basis of maps which are Mercator projections of portions of the Earth's sphere. It is then necessary to find out how this Mercator projection modifies the Saint-Venant equations.

Mercator projection

The Mercator projection belongs to the family of conformal projections which ensure the local conservation of angles, but not distances. In other words, the scale of the map is not the same at every point, but at a given point the scale is the same in all directions. The domain to be represented in plane form is on the surface of a sphere, whose points are identified by their latitude λ and their longitude φ (see Figure 2.4). Henceforth, it is assumed that the latitude and longitude are angles expressed in radians, counted positively in the northern hemisphere and to the east of the Greenwich Meridian and negatively otherwise.

On the basis of a sphere, we arrive at a plane called the Mercator plane, with the help of the projection, which to any point M of the sphere associates a point M_p of a cylinder tangent to the Earth at the equator (see Figure 2.5).

The coordinates x_p and y_p on this plane are deduced from the geographical coordinates by:

$$x_p = R\varphi$$

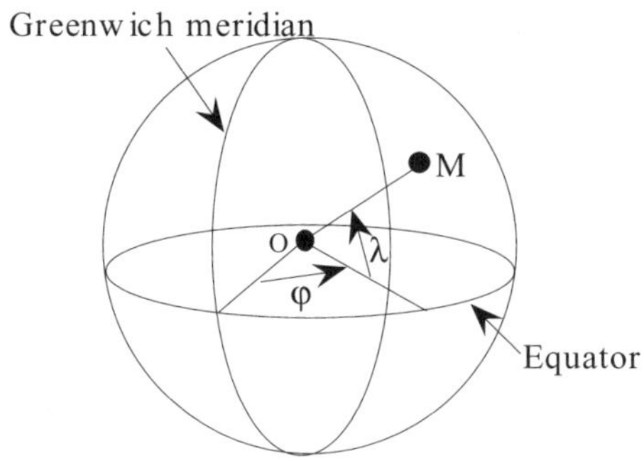

Figure 2.4: longitude and latitude

$$y_p = R\tan(\lambda)$$

However, the coordinates written in this referential do not correspond to a conformal projection because:

- for an east–west displacement, dx, on the surface of the Earth, we obtain a displacement dx_p in this referential such as:

$$dx = \cos(\lambda)\,dx_p \tag{2.170}$$

- while for a north–south displacement, dy, we obtain a displacement dy_p such as:

$$dy = \cos^2(\lambda)\,dy_p \tag{2.171}$$

This distortion, which is greater in the north–south direction than in the east–west direction, makes the projection non-conformal. We get a conformal projection by a simple change of scale in the y direction. The new ordinate Y is then deduced from y_p, then from λ, by:

$$dY = \cos(\lambda)\,dy_p = \frac{R}{\cos(\lambda)}d\lambda \tag{2.172}$$

Moreover, it is possible to carry out a change in the origin at one point (φ_0, λ_0). Along the horizontal axis we immediately obtain:

$$X = R(\varphi - \varphi_0) \tag{2.173}$$

Along the vertical axis, we obtain by integration:

$$Y = R\left(\ln\left[\tan\left(\frac{\lambda}{2} + \frac{\pi}{4}\right)\right] - \ln\left[\tan\left(\frac{\lambda_0}{2} + \frac{\pi}{4}\right)\right]\right) \tag{2.174}$$

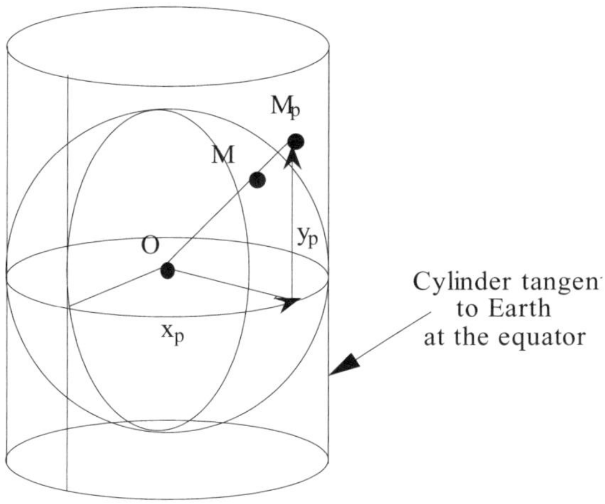

Figure 2.5: first phase of the Mercator projection

These coordinates X and Y are coordinates taken on maps and input to the software for numerical simulation. Hence, one should know how to express the Saint-Venant equations in these coordinates. X and Y being functions of spherical coordinates, it is therefore practical to express the Saint-Venant equations in these spherical coordinates at the very beginning. For this, we use the divergence and gradient operators expressed in these coordinates:

$$div(\vec{F}) = \frac{1}{R\cos(\lambda)}\frac{\partial F_\varphi}{\partial \varphi} + \frac{1}{R\cos(\lambda)}\frac{\partial (F_\lambda \cos(\lambda))}{\partial \lambda} \tag{2.175}$$

$$\overrightarrow{grad}(f) = \frac{1}{R\cos(\lambda)}\frac{\partial f}{\partial \varphi}\vec{e}_\varphi + \frac{1}{R}\frac{\partial f}{\partial \lambda}\vec{e}_\lambda \tag{2.176}$$

Given the nature of the physical quantities and of the projection, the scalar and vectorial quantities (depth, velocity) are unchanged. We have for example $F_\varphi = F_x$ and $F_\lambda = F_y$.

The derivatives of the functions with respect to φ and λ are then replaced by the derivatives along X and Y, the Mercator coordinates:

$$\frac{\partial f}{\partial \varphi} = \frac{\partial f}{\partial X}\frac{\partial X}{\partial \varphi} + \frac{\partial f}{\partial Y}\frac{\partial Y}{\partial \varphi} = R\frac{\partial f}{\partial X} \tag{2.177}$$

$$\frac{\partial f}{\partial \lambda} = \frac{\partial f}{\partial X}\frac{\partial X}{\partial \lambda} + \frac{\partial f}{\partial Y}\frac{\partial Y}{\partial \lambda} = \frac{R}{\cos(\lambda)}\frac{\partial f}{\partial Y} \tag{2.178}$$

From this, new expressions of divergence and gradient are deduced:

$$\begin{aligned} div(\vec{F}) &= \frac{1}{\cos(\lambda)}\frac{\partial F_x}{\partial X} + \frac{1}{\cos(\lambda)}\frac{\partial F_y}{\partial Y} - \frac{\tan(\lambda)}{R}F_y \\ &= \frac{1}{\cos^2(\lambda)}\frac{\partial}{\partial X}(F_x\cos(\lambda)) + \frac{1}{\cos^2(\lambda)}\frac{\partial}{\partial Y}(F_y\cos(\lambda)) \end{aligned} \quad (2.179)$$

$$\overrightarrow{grad}(f) = \frac{1}{\cos(\lambda)}\begin{pmatrix} \frac{\partial f}{\partial X} \\ \frac{\partial f}{\partial Y} \end{pmatrix} \quad (2.180)$$

With the Mercator coordinates, the mass conservation equation which has to be solved in a non-conservative form is thus:

$$\frac{\partial h}{\partial t} + \frac{\vec{u}}{\cos(\lambda)}\,.\,\overrightarrow{grad}(h) + \frac{h}{\cos(\lambda)}div(\vec{u}) - \frac{hv}{R}\tan(\lambda) = 0 \quad (2.181)$$

These equations could be solved in this form, but, in finite elements, this would pose a problem with the approximation space. As a matter of fact, if λ is a polynomial function, what becomes of $\cos(\lambda)$? If $\cos(\lambda)$ and $\sin(\lambda)$ are polynomials, what is $\tan(\lambda)$? These questions are fundamental when mass conservation has to be proved.

Now, the mass conservation equation can also be put at the continuum level as:

$$\frac{\partial(h\cos^2(\lambda))}{\partial t} + div(\cos(\lambda)\,h\,\vec{u}) = 0 \quad (2.182)$$

where *div* is now the Cartesian divergence operator.

We then observe that integration into the Mercator domain of this equation (see Section 4.5.3 on variational formulation) gives the terms:

$$\frac{\partial}{\partial t}\int_\Omega h\cos^2(\lambda)\,dXdY \quad \text{and} \quad \int_\Omega div(h\,\vec{u}\cos(\lambda)\,)\,dXdY$$

this last term being equal to:

$$\int_\Gamma h\,\vec{u}\,.\,\vec{n}\,\cos(\lambda)\,d\Gamma$$

In these terms, each length dX, dY or $d\Gamma$ is multiplied by $\cos(\lambda)$, which restores a real local scale. The volumes or flux calculated are thus real quantities. In the same way, the gradient terms would give real slopes.

As in the case of all operators, the coordinates therefore appear multiplied by $\cos(\lambda)$. The idea is to carry out this operation once and for all before the start of calculations. This can be formalized by choosing a constant latitude λ per element. The advantage is that the equations to be worked out are then strictly identical to the Cartesian equations (this is actually due to the local character of the operators and is not a general property). This choice of constant latitude per element is naturally an approximation, but it enables us to retain strict finite element formalism, the approximation being somehow transferred to the coordinates.

During a computation using spherical coordinates with a meshing in the Mercator projection, we shall build up a set of coordinates multiplied by $\cos(\lambda)$. If we use the coordinates modified in this manner, the Mercator projection is naturally taken into consideration and no other modification appears again in the Saint-Venant equations. All the mass and flow calculations are coherent and provide real values.

The error in taking a constant latitude per element can be evaluated by finding the form which the Saint-Venant equations should take with the coordinates X' and Y' such as $X' = X\cos(\lambda)$ and $Y' = Y\cos(\lambda)$. The difference at the continuum level between the equations makes the ignored terms, which are of the order of $1/R$, appear. They take into consideration the variations in latitude within the element.

Important notes:

- The multiplication of coordinates of the points by the cosine of a constant latitude per element leads to a paradoxical situation: a point which belongs to several elements has as many sets of different coordinates which should be stored element by element. In fact this does not pose a technical problem and does not lead to any contradiction. All calculations of vectors or matrices in finite elements start at the elementary level, where coherence is required only between the coordinates of the points of the element. It should even be possible, without any topological problem, to build a "two-dimensional" meshing of the Earth for calculating world tides, each element having coordinates corresponding to a local projection (poles should be excluded, however).

- The Saint-Venant equations in spherical coordinates have been obtained here by simply changing the differential operators. A more correct procedure has been used by Gray and others [61]. It consists of averaging over the vertical the Navier–Stokes equations written in spherical coordinates. Certain differences then appear, in particular in terms depending on vertical velocity (which, logically, are overlooked in the Saint-Venant equations), but also dispersion terms arising from curvature terms, which modify *a priori* the form of the operator of diffusion. These authors, however, indicate that this difference is important only for diffusion coefficients greater than $10^4 m^2/s$, a value which exceeds the magnitude currently accepted for diffusion, even if one includes the dispersion phenomenon.

2.3.8 Saint-Venant equations with porosity

A local porosity can be used to represent the density of buildings in an urban zone, or trees in a forest, or any obstacles where water cannot flow. This concerns a technique known as homogenization, where a physical phenomenon is not described completely, but is replaced by a continuous property of the area. The influence of obstacles smaller than the size of the mesh can thus be represented. Porosity can also be used to deal with partially wet elements. In such a case it is equal to the percentage of the surface covered by water. One of the first uses of porosity in the Saint-Venant equations has been published by Defina *et al.* [38]. However, the proposed formulation was not satisfactory for the conservation of mass. For a perfect conservation of mass, we are prompted to propose a mass conservation equation such as:

$$\frac{\partial(\Theta h)}{\partial t} + div(\Theta h \vec{u}) = 0 \tag{2.183}$$

where Θ is the local porosity. The total quantity of water in the domain Ω is now written as $\int_\Omega \Theta h \, d\Omega$. If the porosity is 70%, the total quantity of water in the domain is 70% of what it would be with classical equations, without obstacles. For example, this porosity takes into account buildings that occupy 30% of the space on the ground.

Then there is the question of porosity in the momentum equation. Bearing in mind the significance of porosity, in our case (a factor limiting the available space on the ground), the conservative momentum equation is written as (an example of the component along x):

$$\frac{\partial(\Theta hu)}{\partial t} + div(\Theta hu\vec{u}) = -g\Theta h\frac{\partial Z_s}{\partial x} + \Theta hF_x + div(\Theta h\nu_t \overrightarrow{grad}(u)) \tag{2.184}$$

where one ends up replacing depth h everywhere by Θh. If the continuity equation is used with porosity to get a non-conservative form, we have:

$$\frac{\partial u}{\partial t} + \vec{u}.\overrightarrow{grad}(u) = -g\frac{\partial Z_s}{\partial x} + F_x + \frac{1}{\Theta h}div(\Theta h\nu_t \overrightarrow{grad}(u)) \tag{2.185}$$

Hence only the diffusion term has changed.

Plate 1 illustrates the effect of porosity in a rectangular canal and explains its geometry. In fact it concerns a single dimension case and the curves shown are longitudinal sections in the axis of the canal. Flow from the left to the right takes place at a constant discharge of 100 m^3/s and the depth at the exit is 5 m, such that the velocity is approximately 0.5 m/s in the canal (for more realism, a Manning friction coefficient of 0.025 has been added and the turbulence is modelled with the k–ε model). The porosity is 1 everywhere, except in the zone between $x = -50$ and $x = 50$ m where it varies linearly from 1 to 0.5 (up to x=0), then from 0.5 to 1, as is seen in the bottom diagram. The main effect of porosity is that the velocity increases from 0.5 m/s to 1 m/s locally to conserve a constant discharge. Another effect is that the water depth decreases to ensure the conservation of head. The effect of constriction by obstacles such as bridge piers has been reproduced here. However, this porosity does not model the effects of head loss due to the drag force of these obstacles. For this it would be necessary to add also a drag force to the equations (see Section 2.3.11).

2.3.9 Boussinesq equations

Joseph Boussinesq (1842–1929) was a student of Saint-Venant at the Ecole Nationale des Ponts et Chaussées. In 1877, he proposed an extension of the Saint-Venant equations to take into account the acceleration terms along the vertical. It is no longer assumed that pressure is hydrostatic nor is the vertical velocity totally ignored, when dealing with more curved waves than with Saint-Venant equations. In fact the celerity of waves is not $\sqrt{gh}$, but $\sqrt{gh\frac{\tanh(kh)}{kh}}$, where k is the wave number and the Saint-Venant equations correctly process only the limit case of long waves. The different theories improving on the Saint-Venant equations, namely by Boussinesq, Serre, or Korteweg and De Vries, attempt to approach the real relation of dispersion:

$$c = \sqrt{gh\frac{\tanh(kh)}{kh}} \tag{2.186}$$

In the Boussinesq theory, we consider that the vertical velocity is a quantity depending on the first order of $\frac{A}{h}$ and on the second order of $\frac{h}{L}$, where A is the amplitude of the wave and L its length. However, this approximation of the vertical velocity limits to a great extent the application domain and subsequently many add-ons or higher approximation orders have been proposed (see in particular [134] and [150]). However, the future of the Boussinesq equations seems to be in jeopardy, due to competition from the free surface non-hydrostatic Navier–Stokes equations. Hence, we shall present here only the standard Boussinesq equations for the purpose of comparison with a complete three-dimensional resolution. In all cases, the Saint-Venant continuity equation remains unchanged. Vertical velocity is finally eliminated and this results only in a modification of the momentum equation. In the standard Boussinesq equations, this new term added to the right-hand side is:

$$-\frac{H_0^2}{6}\overrightarrow{grad}\left[div\left(\frac{\partial\vec{u}}{\partial t}\right)\right]+\frac{H_0}{2}\overrightarrow{grad}\left[div\left(H_0\frac{\partial\vec{u}}{\partial t}\right)\right] \tag{2.187}$$

where H_0 is a reference water depth. We shall see how to work out this supplementary term in Section 4.7, but at this point we can make the following remarks:

- The reference depth H_0 and the scope of the Boussinesq equations do not permit a representation of uncovered beds. If the reference depth is zero (the case of an area which is initially dry), we fall back on the Saint-Venant equations.
- The supplementary term is a derivative in time, zero in a steady flow. We must specify that it is therefore useless to solve the Boussinesq equations for permanent flows.

2.3.10 Serre equations

In 1895, Korteweg and De Vries modified the Boussinesq method and brought out a new theory of curved long waves, thanks to a periodicity hypothesis leading to a solution called the "cnoidal" wave, which includes the solitary wave. In 1948, Friedrichs proposed a disturbance method based on the ratio of the average depth to the wavelength. He found the Saint-Venant solution is of zero order. In the same year, Keller discovered the Korteweg and De Vries solution to be of first order. The general theory of Friedrichs therefore represents the known solutions of long waves more accurately. One particular case of the second order is Laitone's solution for the solitary wave which will be discussed in Section 10.3. In 1953, an engineer from Ponts et Chaussées, François Serre, gave an original solution to account for the slope of the bottom. To our knowledge, there is no software in finite elements to solve the Serre equations, though a finite difference program was used at Laboratoire National d'Hydraulique in France in the 1980s. It is nevertheless important to highlight this work. The Serre equations, like those of Boussinesq, add more terms to the Saint-Venant momentum equations. But the original hypothesis, the linearity of vertical velocity between the bottom and the free surface, seems more elegant than the limited development of the Boussinesq equations. We shall elaborate these additional terms below. This demonstration can serve as a basis for future research, which could help us to establish a relation between the Serre and the Navier–Stokes non-hydrostatic equations with two meshing planes (see the application to the solitary wave in Section 10.3).

So, we do not neglect vertical velocity but we consider it linear. That is:

$$W(z) = W_0 + \frac{z - Z_f}{h} \Delta W \tag{2.188}$$

where W_0 is $\frac{dZ_f}{dt}$, the velocity at the bottom, and $W_0 + \Delta W$ the velocity at the surface $\frac{dZ_s}{dt}$. ΔW is thus also equal to $\frac{dh}{dt}$. Let us also observe that:

$$\frac{dh}{dt} = \frac{\partial h}{\partial t} + \vec{u}.\overrightarrow{grad}(h) = -h\, div(\vec{u}) \tag{2.189}$$

It follows that:

$$\frac{\Delta W}{h} = -div(\vec{u}) \tag{2.190}$$

W being linear, the diffusion term is cancelled and the W equation gives the following pressure gradient:

$$\frac{1}{\rho}\frac{\partial p}{\partial z} = -g - \left(\frac{\partial W}{\partial t} + U\frac{\partial W}{\partial x} + V\frac{\partial W}{\partial y} + W\frac{\partial W}{\partial z} \right) \tag{2.191}$$

Integrating this equation between z and the free surface Z_s gives:

$$\frac{p(z)}{\rho} = g(Z_s - z) + \int_z^{Z_s} \left(\frac{\partial W}{\partial t} + U\frac{\partial W}{\partial x} + V\frac{\partial W}{\partial y} + \frac{1}{2}\frac{\partial W^2}{\partial z} \right) dz \tag{2.192}$$

We now use the value of W from equation 2.188. Then $\int_z^{Z_s} \frac{\partial W}{\partial t} dz$ gives:

$$(Z_s - z)\frac{\partial W_0}{\partial t} + \frac{Z_s^2 - z^2}{2}\frac{\partial}{\partial t}\left(\frac{\Delta W}{h}\right) - (Z - z)\frac{\partial}{\partial t}\left(Z_f \frac{\Delta W}{h}\right) \tag{2.193}$$

By considering that U and V are constant along the vertical and equal to u and v, the terms $\int_z^{Z_s} (U\frac{\partial W}{\partial x} + V\frac{\partial W}{\partial y}) dz$ give:

$$(Z_s - z)u\frac{\partial W_0}{\partial x} + \frac{Z_s^2 - z^2}{2}u\frac{\partial}{\partial x}\left(\frac{\Delta W}{h}\right) - u(Z_s - z)\frac{\partial}{\partial x}\left(Z_f \frac{\Delta W}{h}\right) \tag{2.194}$$

and:

$$(Z_s - z)v\frac{\partial W_0}{\partial y} + \frac{Z_s^2 - z^2}{2}v\frac{\partial}{\partial y}\left(\frac{\Delta W}{h}\right) - v(Z_s - z)\frac{\partial}{\partial y}\left(Z_f \frac{\Delta W}{h}\right) \tag{2.195}$$

The term:

$$\int_z^{Z_s} \frac{1}{2}\frac{\partial W^2}{\partial z} dz$$

gives:

$$W_0 \frac{\Delta W}{h}(Z_s - z) + \frac{Z_s^2 - z^2}{2}\left(\frac{\Delta W}{h}\right)^2 - (Z_s - z)Z_f\left(\frac{\Delta W}{h}\right)^2 \tag{2.196}$$

In the expression $p(z)/\rho$ which results from the last three equations, we can form total derivatives with respect to time by the fact that for every function f:

$$\frac{1}{h}\frac{d}{dt}(fh) = \frac{df}{dt} + \frac{f}{h}\frac{dh}{dt} = \frac{df}{dt} + f\frac{\Delta W}{h} \tag{2.197}$$

It follows that:

$$\frac{p(z)}{\rho} = g(Z_s - z) + \frac{(Z_s - z)}{h}\frac{d}{dt}(hW_0 - Z_f\Delta W) + \frac{Z_s^2 - z^2}{2h}\frac{d}{dt}(\Delta W) \tag{2.198}$$

We now look at $\frac{1}{\rho}\frac{\partial p(z)}{\partial x}$:

$$\begin{aligned}\frac{1}{\rho}\frac{\partial p(z)}{\partial x} &= g\frac{\partial Z_s}{\partial x} + \frac{1}{h}\frac{\partial Z_s}{\partial x}\frac{d}{dt}(hW_0 - Z_f\Delta W) + \frac{(Z_s - z)}{h}\frac{\partial}{\partial x}\left(\frac{d}{dt}(hW_0 - Z_f\Delta W)\right)\\ &+ \frac{Z_s^2 - z^2}{2h}\frac{\partial}{\partial x}\left(\frac{d}{dt}(\Delta W)\right) - \frac{1}{h^2}(Z_s - z)\frac{\partial h}{\partial x}\frac{d}{dt}(hW_0 - Z_f\Delta W)\\ &+ \frac{Z_s}{h}\frac{\partial Z_s}{\partial x}\left(\frac{d}{dt}(\Delta W)\right) - \frac{Z_s^2 - z^2}{2h^2}\frac{\partial h}{\partial x}\frac{d}{dt}(\Delta W)\end{aligned} \tag{2.199}$$

This can be rearranged as:

$$\begin{aligned}\frac{1}{\rho}\frac{\partial p(z)}{\partial x} &= \frac{1}{h}\frac{\partial Z_s}{\partial x}\left[gh + \frac{d}{dt}(hW_0 - Z_f\Delta W) + Z_s\frac{d}{dt}(\Delta W)\right]\\ &+ \frac{(Z_s - z)}{h}\left[\frac{\partial}{\partial x}\left(\frac{d}{dt}(hW_0 - Z_f\Delta W)\right) - \frac{1}{h}\frac{\partial h}{\partial x}\frac{d}{dt}(hW_0 - Z_f\Delta W)\right]\\ &+ \frac{Z_s^2 - z^2}{2h}\left[\frac{\partial}{\partial x}\left(\frac{d}{dt}(\Delta W)\right) - \frac{1}{h}\frac{\partial h}{\partial x}\frac{d}{dt}(\Delta W)\right]\end{aligned} \tag{2.200}$$

or:

$$\int_z^{Z_s} \frac{1}{h}\frac{\partial Z_s}{\partial z}dz = \frac{\partial h}{\partial z} + \frac{\partial Z_f}{\partial z} \tag{2.201}$$

$$\int_z^{Z_s} \frac{(Z_s - z)}{h}dz = Z_s - \left(\frac{Z_s^2 - Z_f^2}{2h^2}\right) = \frac{h}{2} \tag{2.202}$$

$$\int_z^{Z_s} \frac{Z_s^2 - z^2}{2h}dz = \frac{2h^2 + 3hZ_f}{6} \tag{2.203}$$

The term:

$$\int_{z}^{Z_s} \frac{1}{\rho} \frac{\partial p(z)}{\partial x} dz$$

is therefore equal to:

$$\frac{\partial Z_s}{\partial x} \left[gh + (h + Z_f) \frac{d}{dt}(\Delta W) + \frac{d}{dt}(hW_0 - Z_f \Delta W) \right]$$

$$+ \frac{\partial h}{\partial x} \left[-\frac{1}{2} \frac{d}{dt}(hW_0 - Z_f \Delta W) - \frac{h}{3} \frac{d}{dt}(\Delta W) - \frac{Z_f}{2} \frac{d}{dt}(\Delta W) \right]$$

$$+ \frac{h}{2} \left[\frac{\partial}{\partial x} \left(\frac{d}{dt}(hW_0 - Z_f \Delta W) \right) \right] + \frac{h^2}{3} \frac{\partial}{\partial x} \left(\frac{d}{dt}(\Delta W) \right) + \frac{hZ_f}{2} \frac{\partial}{\partial x} \left(\frac{d}{dt}(\Delta W) \right) \quad (2.204)$$

The term:

$$\frac{hZ_f}{2} \frac{\partial}{\partial x} \left(\frac{d}{dt}(\Delta W) \right)$$

is replaced by:

$$\frac{h}{2} \frac{\partial}{\partial x} \left(Z_f \frac{d}{dt}(\Delta W) \right) - \frac{h}{2} \frac{d}{dt}(\Delta W) \frac{\partial Z_f}{\partial x} \quad (2.205)$$

On the other hand, we notice that:

$$\frac{d}{dt}(hW_0 - Z_f \Delta W) + Z_f \frac{d}{dt}(\Delta W) = \frac{d}{dt}(hW_0) - \Delta W \frac{d}{dt}(Z_f) \quad (2.206)$$

We thus have:

$$\int_{z}^{Z_s} \frac{1}{\rho} \frac{\partial p(z)}{\partial x} dz = \frac{\partial h}{\partial x} \left[gh + \frac{h}{2} \frac{dW_0}{dt} + \frac{2h}{3} \frac{d}{dt}(\Delta W) \right] + \frac{h^2}{3} \frac{\partial}{\partial x} \left(\frac{d}{dt}(\Delta W) \right)$$

$$+ \frac{\partial Z_f}{\partial x} \left[gh + h \frac{d}{dt}(\Delta W) + h \frac{dW_0}{dt} \right] - \frac{h}{2} \frac{d}{dt}(\Delta W) \frac{\partial Z_f}{\partial x} + h \frac{\partial}{\partial x} \left(\frac{h}{2} \frac{dW_0}{dt} \right) \quad (2.207)$$

and to conclude:

$$\int_{z}^{Z_s} \frac{1}{\rho} \frac{\partial p(z)}{\partial x} dz = gh \frac{\partial Z_s}{\partial x} + \frac{\partial}{\partial x} \left[h^2 \left(\frac{1}{3} \frac{d}{dt}(\Delta W) + \frac{1}{2} \frac{dW_0}{dt} \right) \right]$$

$$+ h \frac{\partial Z_f}{\partial x} \left(\frac{dW_0}{dt} + \frac{1}{2} \frac{d}{dt}(\Delta W) \right) \quad (2.208)$$

By reapplying $W_0 = \frac{dZ_f}{dt}$ and $\Delta W = \frac{dh}{dt}$, the additional terms in the non-conservative Saint-Venant momentum equations are:

$$-\frac{1}{h} \frac{\partial}{\partial x} \left[h^2 \left(\frac{1}{3} \frac{d^2 h}{dt^2} + \frac{1}{2} \frac{d^2 Z_f}{dt^2} \right) \right] - \frac{\partial Z_f}{\partial x} \left(\frac{d^2 Z_f}{dt^2} + \frac{1}{2} \frac{d^2 h}{dt^2} \right) \quad (2.209)$$

Usually, we state that $\alpha = \frac{d^2h}{dt^2}$ and $\beta = \frac{d^2Z_f}{dt^2}$, and we write these terms as:

$$-h\frac{\partial}{\partial x}\left(\frac{\alpha}{3}+\frac{\beta}{2}\right)-\frac{\partial Z_s}{\partial x}\left(\frac{\alpha}{2}+\beta\right)-\frac{\partial h}{\partial x}\left(\frac{\alpha}{6}\right) \tag{2.210}$$

along x and:

$$-h\frac{\partial}{\partial y}\left(\frac{\alpha}{3}+\frac{\beta}{2}\right)-\frac{\partial Z_s}{\partial y}\left(\frac{\alpha}{2}+\beta\right)-\frac{\partial h}{\partial y}\left(\frac{\alpha}{6}\right) \tag{2.211}$$

along y.

The main difficulty with these terms for numerical resolution is the presence of the derivatives:

$$\frac{d^2h}{dt^2} \text{ and } \frac{d^2Z_f}{dt^2}$$

Naïm [146] has used the Serre equations in one dimension for studying a landslide impact. The additional terms are written in the form $-\frac{dA}{dt}+B$, and the algorithm described in Section 4.7 is used with an intermediate variable $u+A$. With an inviscid fluid and a flat bottom, the Serre theory gives a cnoidal solution. A drawback of these equations is that the Schwarz condition expressing conservation of curl along a trajectory is not verified.

2.3.11 Source terms and body forces in two dimensions

Here we shall elaborate the terms called F_x and F_y so far, constituting body forces $\overrightarrow{F}$ other than pressure and gravity. Some of these forces, like the influence of the wind, are in fact surface forces in three dimensions but appear in the Saint-Venant equations as source terms applied to the whole water body, since the equations represent an average over the vertical.

We shall successively discuss:

- Bottom friction $\overrightarrow{F}^f$
- Drag force and head loss $\overrightarrow{F}^t$
- The Coriolis force $\overrightarrow{F}^c$
- The influence of the wind $\overrightarrow{F}^v$
- Atmospheric pressure $\overrightarrow{F}^a$
- The sources of momentum $\overrightarrow{F}^q$
- Spatial variations of density $\overrightarrow{F}^d$
- Tide-generating terms $\overrightarrow{F}^m$

Bottom friction

Friction on the bottom is a major unknown in environmental studies, and the friction laws themselves are mostly empirical. This is why calibrating numerical models is so important. Calibration can be done with automatic techniques such as parameter estimation (see Chapter 9). Fairly recent publications ([139] and [30]) show the first attempts at a friction parameterization in river flood models using vegetation heights derived from airborne scanning laser altimetry.

Before considering the real friction laws, let us first cite the linear law.

Linear law: this law is rarely used as it does not represent physics: friction forces are a quadratic function of velocity. However, we can use this law for modelling other phenomena. Its form is:

$$F_x^f = -b\,u$$

$$F_y^f = -b\,v$$

where b is the friction coefficient in s^{-1}.

Real friction laws: in one dimension, friction stress exerted by a plane horizontal wall on a flow which is parallel to it is expressed according to the formula:

$$\tau_{xz} = \mu\frac{\partial u}{\partial n} = -\frac{1}{2}\rho C_f u^2 \tag{2.212}$$

τ_{xz} is expressed in $\mathrm{kg/m/s^2}$. We also write:

$$\nu\frac{\partial u}{\partial n} = -\frac{1}{2}C_f u^2 \tag{2.213}$$

with:

ρ: density of the liquid ($\mathrm{kg/m^3}$)
C_f: friction coefficient (dimensionless)
u: flow velocity (m/s)

Stress acts in the direction of the current and in two dimensions has the two components:

$$\tau_{xz} = -\frac{u}{2}\rho C_f\sqrt{u^2+v^2} \tag{2.214}$$

and:

$$\tau_{yz} = -\frac{v}{2}\rho C_f\sqrt{u^2+v^2} \tag{2.215}$$

The coefficient C_f is hardly used in free surface hydraulics and is usually replaced by others. The best known is the Chézy coefficient denoted C. C and C_f are bound by the formula $C = \sqrt{\frac{2g}{C_f}}$ (dimension: $\mathrm{m^{1/2}/s}$). For Chézy, the velocity used in the friction law is the flow velocity in a canal, which is equal to the horizontal velocity only if the vertical velocity is negligible.

We have seen that in the Navier–Stokes equations the friction at the bottom is equal to $\frac{1}{\rho}\underline{\underline{\tau}}_f\vec{n}_f$. The averaging which leads to the non-conservative form of the

Navier–Stokes equations simply gives one term, $\frac{1}{\rho h}\underline{\underline{T}}_f \vec{n}_f$. The norm of the vector normal to the bottom is:

$$\sqrt{1+\left(\frac{\partial Z_f}{\partial x}\right)^2+\left(\frac{\partial Z_f}{\partial y}\right)^2}$$

which is the cosine reciprocal of the steepest slope at one point and will be denoted as $\cos(\alpha)$. We get the following Chézy formula for the shear force at the bottom (to be added to the non-conservative form of momentum equation):

$$\vec{F}^f = -\frac{\vec{u}}{\cos(\alpha)}\frac{g}{hC^2}\sqrt{u^2+v^2} \tag{2.216}$$

where C is the Chézy coefficient.

Chézy's friction law was established for a uniform flow regime but is extended to all types of flow. Finally, it is expressed in our equations as follows:

Chézy's law:

$$F_x^f = -\frac{u}{\cos(\alpha)}\frac{g}{hC^2}\sqrt{u^2+v^2} \tag{2.217}$$

$$F_y^f = -\frac{v}{\cos(\alpha)}\frac{g}{hC^2}\sqrt{u^2+v^2} \tag{2.218}$$

However, the Chézy coefficient is not really a constant; dimensional analysis and experiments show that it should depend on the depth and on the bottom roughness. The empirical law of Manning and Strickler corrects this problem and specifies the value of the Chézy coefficient. It was specifically developed to obtain a friction coefficient independent of the depth, and hence only a function of the bottom roughness, i.e. $C = Kh^{1/6}$ where K ($\mathrm{m}^{1/3}/\mathrm{s}$) is the Strickler coefficient. In one dimension the depth h must be replaced by the hydraulic radius R_H. The approximation $R_H = h$ is, in principle, true only for very large canals. K varies between about 15 and 100. Its value is 20 to 40 for a stream and 50 to 90 for a concrete canal.

Chézy's law, with $C = Kh^{1/6}$, leads to a new law known as Strickler's law:

Strickler's law:

$$F_x^f = -\frac{u}{\cos(\alpha)}\frac{g}{h^{4/3}K^2}\sqrt{u^2+v^2} \tag{2.219}$$

$$F_y^f = -\frac{v}{\cos(\alpha)}\frac{g}{h^{4/3}K^2}\sqrt{u^2+v^2} \tag{2.220}$$

Due to the difference between hydraulic radius and depth, comparison with codes in one dimension is difficult. Calibration in one dimension will not always be valid for two dimensions.

The Strickler coefficient can be linked to the bottom roughness k_s by the Ramette formula:

$$K = 8.2\frac{\sqrt{g}}{k_s^{1/6}}$$

The Manning formula, more commonly used in Anglo-Saxon countries, is a simple variant of the Strickler formula, with a coefficient m, the inverse of K. See [63] which gives advice on choice of coefficients.

Manning's law:

$$F_x^f = -\frac{u}{\cos(\alpha)}\frac{g\,m^2}{h^{4/3}}\sqrt{u^2+v^2} \tag{2.221}$$

$$F_y^f = -\frac{v}{\cos(\alpha)}\frac{g\,m^2}{h^{4/3}}\sqrt{u^2+v^2} \tag{2.222}$$

Note: because it is used in Anglo-Saxon countries, the Manning coefficient is sometimes given in imperial units.

Nikuradse's law: this law stems from a depth-averaged value of the logarithmic turbulent velocity profile (see Equation 2.271 in Section 2.4.3). The corresponding Chézy coefficient is obtained by the formula:

$$C = 7.83\,\ln\left(12\frac{h}{k_s}\right) \tag{2.223}$$

where h is the water depth and k_s the roughness factor at the bottom. This law is favoured by Klaassen and Van der Zwaard [113] for flooded plains with vegetation.

Equivalence of friction coefficients in 2D and 3D: a software user expects the friction coefficient to be independent of numerical modelling. Hence, calibration carried out in 2D should also be valid in 3D. However, we must ensure that head loss due to friction is identical in 2D and 3D. The head losses are a function of the square of the velocity, so we should be careful to retain the same velocity; that is, as we have only this one in 2D, the depth-averaged velocity.

Expressed in dimensionless coefficient terms C_f, friction appears in the Saint-Venant equations as:

$$\frac{\partial u}{\partial t} = \text{other terms} \quad - \frac{1}{2\,h}C_f\; u\,\sqrt{u^2+v^2} \tag{2.224}$$

$$\frac{\partial v}{\partial t} = \text{other terms} \quad - \frac{1}{2\,h}C_f\; v\,\sqrt{u^2+v^2} \tag{2.225}$$

If one considers a rectangular reach having depth h, length Δx and width L, the shear force acting on the fluid is the product of mass by acceleration (negative) due to friction. The components of this force are:

$$f_x = -\rho\frac{C_f}{2}\,u\,\sqrt{u^2+v^2}\,\Delta x\,L \tag{2.226}$$

$$f_y = -\rho\frac{C_f}{2}\,v\,\sqrt{u^2+v^2}\,\Delta x\,L \tag{2.227}$$

In 3D, we reason in terms of bottom shear stress whose components are:

$$\nu\frac{\partial U}{\partial n} = -\frac{C_f}{2}\sqrt{U^2+V^2}\,U \tag{2.228}$$

and:

$$\nu\frac{\partial V}{\partial n} = -\frac{C_f}{2}\sqrt{U^2+V^2}\,V \tag{2.229}$$

For the same reach, shear force now has the components:

$$f_x = -\rho\frac{C_f}{2}U\sqrt{U^2+V^2}\Delta x\, L \tag{2.230}$$

$$f_y = -\rho\frac{C_f}{2}V\sqrt{U^2+V^2}\Delta x\, L \tag{2.231}$$

We see that to have an equivalent effect, $u = U$ and $v = V$ should be verified; that is, the velocity at the bottom is equal to average velocity over the vertical. This is precisely not so when there is friction. For equivalence of friction between 2D and 3D, in both cases, the average velocity over the vertical should be retained for calculating the friction terms. Moreover, if for numerical reasons stress is made implicit as:

$$\nu\frac{\partial U}{\partial n} = \alpha U \tag{2.232}$$

with $\alpha = -\frac{C_f}{2}\sqrt{u^2+v^2}$, then, in principle, coefficient α should be corrected by a factor u/U (U equation) and v/V (V equation). In practice the same coefficient, $\frac{\sqrt{u^2+v^2}}{\sqrt{U^2+V^2}}$, will be used for both equations. This reasoning implies that coefficient C_f is the same in 2D and 3D and that it is calculated with the velocities averaged over the vertical. Plate 2 shows the free surfaces obtained in a rectangular cross-section canal with the Saint-Venant and the Navier–Stokes hydrostatic equations. The bottom is flat at elevation 0. The depth at the downstream end is 0.5 m. The discharge is 50 m^3/s and the width of the canal is 100 m, such that the velocity at the downstream end is 1 m/s. The element size in the mesh is approximately 10 m. The cross-section at the bottom of the plate (longitudinal section along the axis $y = 50$ m) confirms that the free surfaces are the same, with an identical Strickler coefficient of 50 in the two calculations.

Accounting for drag forces and head losses

In the Saint-Venant equations, forests and urban areas are traditionally represented by a friction coefficient, generally a low Strickler coefficient (i.e. representing a high friction). However, this approach has a defect since the surface friction of vertical structures increases with water depth, as opposed to friction at the bottom. Hence, the Chézy and Strickler laws are not applied correctly. In this paragraph, an original approach is presented, to take into account in the equations the effect of vertical structures like trees or bridge piers. The aim is to avoid precisely representing these obstacles in a domain of calculations but to represent them locally as a force comparable to a shear force. This will have the same effect as the drag force of obstacles. According to the action–reaction principle, the stress of the fluid on a structure is the opposite of the structure's stress on the fluid. Let us assume that we have n vertical cylinders with a diameter D, distributed in a domain with surface area A, and let us also assume that the vertical cylinders are sufficiently tall to emerge continuously. For each cylinder, the drag force is:

$$F^t = \frac{1}{2}\rho U^2 D h C_D$$

where C_D is the drag coefficient of the cylinder (as in the tables where C_D is a function of the Reynolds number, and whose value ranges between 1 and 1.2, in most applications). The stress induced by the n structures on the surface A is:

$$\tau = \frac{nF}{A}$$

If we wish to add the effect of this stress in the non-conservative Saint-Venant equations, we should add a term $-\frac{\tau}{\rho h}$ in the right-hand members of the momentum equation. The effect of the structures corresponds to an additional source term:

$$F_x^t = -\frac{n}{A}\frac{D}{2}C_D U^2 \tag{2.233}$$

which we can compare to Chézy's law:

$$F_x^t = -\frac{g}{hC^2}U^2 \tag{2.234}$$

Resolving drag force in terms of the Chézy coefficient would give a value that is dependent on the depth, which some numerical models faced with this problem actually do. We prefer to directly add the drag force as:

$$F_x^t = -\frac{n}{A}\frac{D}{2}C_D U^2 \tag{2.235}$$

The data required for a calculation are as follows:

- The number of structures per surface unit n/A.
- The diameter of structures D.
- The drag coefficient C_D.

All these data can be easily counted or measured, as opposed to a Strickler or Chézy coefficient, which can only be estimated.

If the structures have different diameters, nD will simply be replaced by the sum of all the diameters.

In terms of singular head loss, it is:

$$\Delta H = k\frac{U^2}{2g}$$

with $k = \frac{n}{A}DC_D$.

In terms of linear head loss, we have a coefficient $\frac{k}{\Delta x}$ which, when applied to a length Δx, induces a head loss equal to $\frac{n}{A}DC_D\Delta x$.

Plate 3 shows the simulation of flow around a 2 m diameter bridge pier in a canal, without representing the pier in the finite element mesh. The drag force is distributed over an area between $x = -10$ m and $x = +10$ m. In the same area, a porosity equal to 19/20 is taken into account to create an effect of constriction. The depth is 4 m, the discharge 100 m^3/s and the velocity approximately 0.625 m/s. A drag coefficient equal to 1.34 restores a head loss identical to a pier included in the meshing. Outside the area between x =−20 m and x =+20 m (i.e. 10 diameters), it cannot be confirmed that the pier has been simulated. Only the field near the pier is different. The simulation is run with a k–ε turbulence model. The 1.34 drag coefficient integrates a correction factor due to the fact that the velocity in the axis of the canal, where the pier is located, is greater than the average velocity. If it is accepted that the ratio $U_{\max}/U_{average}$ is equal to $\frac{m+1}{m}$ with $m = 9$ for a Reynolds number of about 10^6 (see Idel'Cik [103]), then the correction to be applied to the drag force is $\left(\frac{m+1}{m}\right)^2$, which is 1.23 in this case. The real coefficient C_D would in fact be $\frac{1.34}{1.23} = 1.09$, which corresponds to a value of Reynolds number between 10^4 and 3 10^5. The values are in the correct range, the main uncertainty arising probably from numerical factors (e.g. the velocity profile at the entrance of the canal is given as constant and is not yet well established in the vicinity of the pier). For more details on this case, reference [95] may be consulted.

The theory can be extended to submerged structures, on the condition that one adds to the force a multiplier coefficient $\frac{\min(h,d)}{h}$, where d is the height of the vertical structures. Plate 4 is a real application where the effect of marine turbines on tidal currents was modelled with drag forces as described here. The coloured surfaces are the velocity intensity in m/s. The mesh being heavily refined, only a few velocity vectors are shown.

Coriolis force

For the sake of compatibility with the negligible vertical velocity hypothesis, only that part of force due to horizontal velocity is retained. The 3D Coriolis force is:

$$-2\,\vec{\omega} \wedge \vec{V} = \begin{pmatrix} -2\omega\cos(\lambda)\cos(N)W + 2\omega\sin(\lambda)V \\ 2\omega\cos(\lambda)\sin(N)W - 2\omega\sin(\lambda)U \\ -2\omega\cos(\lambda)\sin(N)V + 2\omega\cos(\lambda)\cos(N)U \end{pmatrix} \tag{2.236}$$

By neglecting W and the vertical component we get:

$$F_x^c = 2\omega\sin(\lambda)\,v = fv \tag{2.237}$$

$$F_y^c = -2\omega\sin(\lambda)\,u = -fu \tag{2.238}$$

where we set $f = 2\omega\sin(\lambda)$. Across small areas, this coefficient can be considered as a constant. Across large areas (where the Mercator projection is used), this parameter varies in space.

The angular velocity of the Earth is $\omega = 7.292\ 10^{-5}$rd/s, i.e. 2π radians per sidereal day, equal to 0.997270 of a 24 hour day, that is to say 86164 seconds. For a latitude of 48°, we get $f = 1.083\ 10^{-4}$N/ms.

Influence of wind

In a similar way as for friction at the bottom, wind forcing takes the form (free surface slope has been ignored):

$$F_x^v = \frac{1}{h}\frac{\rho_{air}}{\rho_{water}} a_{wind} U_{wind} \sqrt{U_{wind}^2 + V_{wind}^2} \tag{2.239}$$

$$F_y^v = \frac{1}{h}\frac{\rho_{air}}{\rho_{water}} a_{wind} V_{wind} \sqrt{U_{wind}^2 + V_{wind}^2} \tag{2.240}$$

where a_{wind} is the wind forcing coefficient already discussed in Section 2.2.4, U_{wind} and V_{wind} the components of wind velocity in m/s (strictly speaking this is velocity relative to the fluid) and $\frac{\rho_{air}}{\rho_{water}}$ is the ratio of the densities of air and water. Numerous empirical formulae give the value of a_{wind}. Undoubtedly, its value depends on the site being studied.

Atmospheric pressure

When atmospheric pressure P_{atm} is considered (dimension: pascal = kg /m/ s^2), the term $-\frac{1}{\rho}\overrightarrow{grad}(P_{atm})$ is simply added to the right-hand side of the momentum equation. That is:

$$F_x^a = -\frac{1}{\rho}\frac{\partial P_{atm}}{\partial x} \tag{2.241}$$

$$F_y^a = -\frac{1}{\rho}\frac{\partial P_{atm}}{\partial y} \tag{2.242}$$

The pressure term is often added to the free surface which also appears through its gradient. But this is not done as care should be taken to retain a real free surface for finding uncovered areas. The atmospheric pressure gradient should not be considered in those uncovered areas where it would be a parasitic driving term.

Sources of momentum

As seen in Section 2.3.4, sources of discharge induce sources of momentum equal to:

$$F_x^q = \frac{1}{h}(u_{sce} - u)Sce \tag{2.243}$$

$$F_y^q = \frac{1}{h}(v_{sce} - v)Sce \tag{2.244}$$

u, v, u_{sce}, v_{sce} and Sce are expressed in m/s.

Important note: when these terms are not taken into account, it is implied that the intake or output velocity of the source is equal to that of the current.

Spatial variations of density

As the Saint-Venant equations result from an average over the vertical, they cannot reproduce stratifications arising from vertical variations of density, a function of salinity or temperature. However, it is possible to take into account horizontal variations of density. If salinity is denoted as S and temperature as T, ρ is a function $\rho(S,T)$, where S and T are functions of x, y and t. These variations have two effects:

- Dilatation of water.
- Differential effects in gravity.

Dilatation of water is a secondary effect which will be ignored, as in 3D. On the other hand, the differential effects of gravity due to variations in salinity can be highlighted in estuaries, and, sometimes, they have to be considered, e.g. in the North Sea. The Boussinesq approximation is used for this purpose. Buoyancy terms in the Saint-Venant equations are just the average over the vertical of the buoyancy terms of the Navier–Stokes equations, which give:

$$F_x^d = \frac{1}{h}\int_{Z_f}^{Z_s}\left[-\frac{1}{\rho_o}\left(1-\frac{\Delta\rho}{\rho_o}\right)\frac{\partial p_{atm}}{\partial x} + g\frac{\Delta\rho}{\rho_o}\frac{\partial Z_s}{\partial x} - g\frac{\partial}{\partial x}\left(\int_z^{Zs}\frac{\Delta\rho}{\rho_o}\,dz'\right)\right] dz \quad (2.245)$$

$$F_y^d = \frac{1}{h}\int_{Z_f}^{Z_s}\left[-\frac{1}{\rho_o}\left(1-\frac{\Delta\rho}{\rho_o}\right)\frac{\partial p_{atm}}{\partial y} + g\frac{\Delta\rho}{\rho_o}\frac{\partial Z_s}{\partial y} - g\frac{\partial}{\partial y}\left(\int_z^{Zs}\frac{\Delta\rho}{\rho_o}\,dz'\right)\right] dz \quad (2.246)$$

The atmospheric pressure gradient immediately gives the terms:

$$-\frac{1}{\rho_o}\left(1-\frac{\overline{\Delta\rho}}{\rho_o}\right)\frac{\partial p_{atm}}{\partial x} \quad \text{and} \quad -\frac{1}{\rho_o}\left(1-\frac{\overline{\Delta\rho}}{\rho_o}\right)\frac{\partial p_{atm}}{\partial y} \quad (2.247)$$

in which $\overline{\Delta\rho}$ is the average of $\Delta\rho$ over the vertical.

For the other two terms $\Delta\rho$ should be changed to $\overline{\Delta\rho}$. In fact a 3D density is not accessible in 2D. The only information that can be deduced from average temperature and salinity over the vertical is $\overline{\Delta\rho}$. Buoyancy terms are then integrated in:

$$F_x^d = -\frac{gh}{2\rho_0}\frac{\partial}{\partial x}(\overline{\Delta\rho}) \quad (2.248)$$

$$F_y^d = -\frac{gh}{2\rho_0}\frac{\partial}{\partial y}(\overline{\Delta\rho}) \quad (2.249)$$

This is a baroclinic pressure gradient. The barotropic pressure gradient present in 3D disappears.

Tide-generating force

Calculations and explanations related to the tide-generating force will be considered in Appendix A. For a small area, this force can be ignored as the tide is essentially originating from the boundary conditions. For an area covering the Channel and the French Atlantic coast, it was shown [91] that the M2 wave (semi-diurnal lunar) applied within the area was responsible for a tidal range of approximately 4 cm. This is exactly the magnitude of the precision level that one can reach in a calculation of tides. Tests in the same domain, but without including tides at the boundaries, have given a tidal range of several tens of centimetres. This surprising result is explained by the non-linearity of the friction term. This term interferes with the effect of the lunar force in the momentum equation. With astral potential alone, velocities remain low and the friction term is of the same magnitude as the lunar force. On the other hand, by applying an average M2 tide at the boundaries of the domain, the velocities, and hence the friction term which varies as the square of velocity, are greater. So, the friction term almost completely masks the effect of the lunar force.

2.3.12 Boundary conditions in 2D

A distinction should be made between the physical and mathematical boundary conditions. The latter are the translation of the former in algorithmic terms. It is the physical boundary conditions that will be reviewed here.

In terms of physics a distinction should be made between two types of boundaries in the calculation domain: liquid boundaries and rigid boundaries.

Rigid boundaries

This type of boundary is impermeable: no flow should cross a rigid boundary. There is friction also. The friction coefficient can be provided by the user or determined by the turbulence model. Generally $\nu_t \frac{\partial u_t}{\partial n}$ is given in the implicit form $\nu_t \frac{\partial u_t}{\partial n} = \alpha u_t$, where u_t is the (*a priori* unknown) component of the velocity tangent to the wall. For the sake of numerical schemes, one is obliged to apply a stronger condition: $\nu_t \frac{\partial u}{\partial n} = \alpha\, u$ and $\nu_t \frac{\partial v}{\partial n} = \alpha\, v$; this means applying an artificial condition on the normal component, which is cancelled anyway by the impermeability condition.

No condition on the water depth is expected for a rigid wall.

Liquid boundaries

This type of boundary requires the most sensitive treatment as it supposes the existence of a fluid domain that does not belong to the domain of calculation but can influence it. This influence should be translated into conditions at the boundaries.

Thanks to an examination of the Saint-Venant equations, without diffusion terms and in lD, and with the help of the theory of characteristics, we can distinguish four types of liquid boundaries:

- Input with supercritical flow.
- Input with subcritical flow.
- Output with supercritical flow.
- Output with subcritical flow.

Input and output are differentiated by the sign of the dot product of the velocity vector and the normal vector pointing outwards from the wall (input: negative product, output: positive product). Supercritical and subcritical flows are differentiated with the Froude number $F = \frac{u}{\sqrt{gh}}$. As seen earlier, this number expresses the ratio between the velocity of the current and the celerity of long waves c.

If $F > 1$ the flow is supercritical; if $F < 1$ it is subcritical.

According to the theory of characteristics, the boundary conditions necessary for the boundaries of the domain are (in 1D):

- Two conditions for a supercritical input.
- Two conditions for a subcritical input.
- No condition for a supercritical output.

- One condition for a subcritical output.

This theory cannot be faithfully followed in 2D, first because the velocity is a vector. Then the separation of equations into two fractional steps during the resolution brings about intermediate boundary conditions which do not always follow physics. For example, for a liquid boundary with velocity pointing inwards, the advection step requires data for all quantities transported in the water body. An additional difficulty is that the "input" or "output" nature of a liquid boundary can vary in time (e.g. the case of a boundary in the open sea with tide).

In practice, different types of boundary conditions are created to correspond to the four types of boundaries mentioned earlier:

- Imposed velocity and imposed water depth (input with supercritical flow).
- Imposed velocity and free water depth (input with subcritical flow).
- Free velocity and free water depth (output with supercritical flow).
- Free velocity and imposed water depth (output with subcritical flow).

To these basic conditions applied to the unknowns of our equations, some others, more complex ones, need to be added as follows:

Imposed flow conditions: for example, very often the exact velocity profile on a section of a river is *a priori* not known. This information would, however, be necessary at the upstream end of a domain of calculation. The only known variable is, most often, the discharge. So, a velocity profile is assumed (constant or logarithmic). The precise numerical values of velocities are then obtained from a simple rule of proportion so as to retrieve the real discharge. This procedure is still valid if different discharges have to be entered at different input points, and also if the imposed discharge is a function of the water depth.

Incident wave condition: this condition is meant for the entry (and exit) of a progressive sinusoidal wave through open boundaries of a domain. It is written, for a wave whose frequency is ω, as:

$$c\overrightarrow{u}.\overrightarrow{n} = gZ_s - g(1 - \overrightarrow{k}.\overrightarrow{n})\widehat{S} \tag{2.250}$$

where c is the wave celerity, $\overrightarrow{n}$ the normal unit vector pointing outside of the boundary of the domain and Z_s the elevation of the free surface. $\overrightarrow{k}$ is a wave unit vector of the incident signal, whose norm is equal to ω/c.

$\widehat{S} = A(M)\cos(\omega t - \varphi(M))$ is the sinusoidal incident signal, homogeneous to a length. The origin of formula 2.250 is the theory of characteristic curves and the Riemann invariants. In subcritical flow, we write that the Riemann invariant coming from outside the domain corresponds to the incident wave. In the absence of an incident wave, and in 1D, this condition means that the invariant $u + 2c$ is equal to $2c_0$, where c_0 is the existing celerity beyond the domain boundary.

This boundary condition minimizes reflection for waves leaving the domain in a direction normal to the boundary. However, the result depends on the orientation of the boundary.

On the other hand, when it concerns the entry of a sinusoidal wave, the use of this boundary condition requires perfect knowledge of the input signal (amplitude, phase, frequency, wave vector), which is not the signal measured due to the reflection of waves.

In case only total signal information is available, imposed values are preferred.

2.3.13 Tracer equation in two dimensions

In 2D the numerical value of the tracer only represents an average over the vertical. This excludes the possibility of representing temperature or density stratifications. Accounting for these phenomena requires the 3D Navier–Stokes equations, preferably the non-hydrostatic ones.

Conservative form

Averaging over the vertical, done earlier to obtain the Saint-Venant equations, gives the conservative form of the tracer equation, in 2D:

$$\frac{\partial}{\partial t}(hT) + div(hT\overrightarrow{u}) - div(h\nu_T\overrightarrow{grad}(T)) = T_{sce}Sce \tag{2.251}$$

T_{sce} is the source value of the tracer and Sce the flow source in m/s, connected to the discharge at the source. Q_{sce} is expressed in m^3/s.

Here ν_T also takes dispersion into account.

This equation with unknown hT could be dealt with directly in this form, which would be more efficient to ensure conservation of the tracer. It has the following drawbacks, however:

- The variable T can only be obtained with a division by the depth. This poses a problem with the monotonicity of T, not to mention uncovered banks where water depth $h = 0$.
- When T is constant over the whole domain, the situation is not stable, as it does not correspond to an obvious solution of the equation in hT. As a matter of fact the variable hT is not constant if the depth varies.

Non-conservative form

For all the reasons given in the preceding paragraph, the non-conservative form of the tracer equation is preferred, despite its drawbacks. It is obtained in the manner elaborated below. It has its importance, as we should be capable of doing it backwards at the discrete level for proving the conservation of the tracer in finite elements.

First, the derivatives of the products are developed:

$$h\frac{\partial T}{\partial t} + T\frac{\partial h}{\partial t} + h\overrightarrow{u}.\overrightarrow{grad}(T) + T\,div(h\overrightarrow{u}) - div(h\nu_T\overrightarrow{grad}(T)) = T_{sce}Sce \tag{2.252}$$

The term $T(\frac{\partial h}{\partial t} + div(h\overrightarrow{u}))$ should be noted. It is replaced with the help of the continuity equation by $T\ Sce$, where Sce is the source term.

The remaining equation is then divided by h to finally obtain:

$$\frac{\partial T}{\partial t} + \vec{u}.\overrightarrow{grad}(T) - \frac{1}{h} div(h\nu_T \overrightarrow{grad}(T)) = \frac{(T_{sce} - T)Sce}{h} \tag{2.253}$$

The right-hand side of this equation is zero if the value of the source tracer is equal to that of the domain where the source arrives. This is generally not the case for an injection into the domain. For an outlet from the domain, we have $T_{sce} = T$.

Boundary conditions

The boundary conditions for a tracer are:

- Imposed value.
- Free value (at an outlet, for example).
- Flux condition.

The flux condition is as in 3D:

$$\nu \frac{\partial T}{\partial n} = aT + b$$

where a and b are provided by the user.

In the case of free outlet condition and if the method of characteristics is used, the result will depend on the advection step only. The effect of diffusion and source terms is thereby ignored. In fact, the diffusion step would require a flux condition which would ultimately be an artificial boundary condition. A no-flux boundary condition is not equivalent to a free outlet condition.

2.4 Modelling of turbulence and dispersion

2.4.1 Reynolds stress

So far only laminar flow has been considered. When the Reynolds number, equal to $\frac{UL}{\nu}$, where U and L are a velocity and a length characteristic of flow, becomes greater than around 2000, physical quantities like velocity and pressure show rapid variations in time and space. The value 2000 is in fact an order of magnitude, valid for canals, and varies greatly according to flows. It also depends on the choice of characteristic quantities U and L. For large Reynolds numbers, i.e. for a majority of applications, regular or laminar solutions, are no longer stable and are replaced by chaotic solutions whose variations are unpredictable. These fluctuations are not described, only average values are of interest. These averages can be averages in time (the case of a steady flow), averages of the same experiment conducted a certain number of times (the case of unsteady flow) or averages over space (the case of homogeneous flow). Unfortunately, due to the non-linearity of advection terms, the averaged flow is no longer the solution for the Navier–Stokes equations and the equations of this averaged flow are called "Reynolds equations". If the components of velocity U_i are broken down as $\overline{U_i} + u_i'$ (average + fluctuation), the Reynolds momentum equation is written as (in the conservative form):

$$\frac{\partial(\rho\overline{U_i})}{\partial t} + \frac{\partial(\rho\overline{U_i U_j})}{\partial x_j} = -\frac{\partial p}{\partial x_i} + \frac{\partial}{\partial x_j}(\tau_{ij} + R_{ij}) + \rho F_i + \rho g_i \tag{2.254}$$

The continuity equation is unchanged, averaged values and fluctuations all being zero.

The indices i and j vary from 1 to 3. The "Einstein" tensor notation has been used and the repeated indices correspond implicitly to a summation. F_i represents the body forces other than pressure and gravity. g_i represents gravity ($g_1 = g_2 = 0$, $g_3 = -g$).

The term:

$$\tau_{ij} = \mu\left(\frac{\partial U_i}{\partial x_j} + \frac{\partial U_j}{\partial x_i}\right)$$

is the viscous part of the stress tensor.

The components of tensor R_{ij} are the correlations of velocity fluctuations:

$$R_{ij} = -\overline{\rho u_i' u_j'}$$

It is this tensor, unfortunately unknown, which makes the difference with the Navier–Stokes equations.

It is possible to form equations for the Reynolds stresses R_{ij}, but they would give rise to new unknowns, triple correlations, in greater numbers. This never-ending hierarchy of equations constitutes an open system.

To solve the Reynolds equations the problem should be closed. That is, the unknown tensor R_{ij} should be expressed on the basis of averaged physical quantities with the help of a turbulence model which can be: an algebraic relation (zero-equation model), a combination of an algebraic relation and an equation (single equation model) or two transport–diffusion equations (two-equation model), or even more (Reynolds stress model). In all cases, this will be only a model, based on dimensional analysis, empiricism, or on interpretation of physical phenomena and their respective importance.

Turbulence models with zero, one, or two equations are all based on a hypothesis of isotropic turbulence. However, in certain applications a tensor of isotropic turbulent viscosity is not adapted. For example, when the mesh is hundreds of metres in the horizontal direction, but of the order of 1 meter in the vertical direction, or when dispersion along the vertical is taken into account.

In most turbulence models, the terms:

$$\frac{1}{\rho}\frac{\partial}{\partial x_j}(\tau_{ij} + R_{ij})$$

are modelled according to the Boussinesq model [17], which states that:

$$\frac{1}{\rho}\frac{\partial}{\partial x_j}(\tau_{ij} + R_{ij}) = \frac{\partial}{\partial x_j}\left[(\nu + \nu_t)\left(\frac{\partial\overline{U_i}}{\partial x_j} + \frac{\partial\overline{U_j}}{\partial x_i}\right) - \frac{2}{3}k\,\delta_{ij}\right] \tag{2.255}$$

where ν_t is the turbulent viscosity. ν_t is variable in time and space and is provided by the turbulence model. δ_{ij} is the Kronecker tensor and k the turbulent kinetic energy equal to $\frac{1}{2}\overline{u_i'u_i'}$, which is also present in the k–ε model.

For a tracer, this Boussinesq hypothesis is expressed by the formula:

$$-\overline{u_i'T'} = \frac{\nu_t}{Prt}\frac{\partial T}{\partial x_i} \tag{2.256}$$

where Prt is the Prandtl number (if T is the temperature; with salinity, it would be the Schmidt number, σ_T). The Prandtl and Schmidt numbers range generally between 0.8 and 1.2 depending on the flow, but other values are sometimes favoured, particularly in the presence of sediments.

Henceforth, the Boussinesq hypothesis will be accepted. The different models presented later will differ by the value ν_t that they provide. In some cases, in order to take into account the effect of dispersion, turbulent viscosity will be considered as an anisotropic quantity, with different values along each direction. The coefficient ν_t is dimensionally the product of a velocity and a length. It is considered roughly to represent the product of velocity of an eddy with the distance along which it mixes momentum. This interpretation demonstrates the anisotropy of the process. On the other hand, in the Boussinesq model, the Reynolds tensor is always analogous to the deformation tensor, with an isotropic behaviour as per elastic material.

We shall now examine, in increasing complexity, the different turbulent models used in free surface hydrodynamics.

After presenting "zero-equation" models, we shall be in a position to specify the values of turbulent stress near walls due to turbulence, and then, in detail, the most commonly used two-equation model: the k–ε model.

2.4.2 Zero-equation models

These very simple models assume that turbulent viscosity is constant, or directly depend on known or easily calculable parameters.

Constant viscosity

Constant viscosity is sufficient when flow is governed by the pressure gradient and by advection, in the tide flow regime for example, and especially for modelling oceanic circulation on a large scale. In 2D, this constant viscosity should include dispersion. The values of ν_t provided in the literature (see [49]) extend from 0.12 m^2/s (transversal diffusion in the Mackenzie River) to 1500 m^2/s (longitudinal diffusion in the Missouri River).

Elder model

The Elder model for the Saint-Venant equations takes into account the role of dispersion and recommends two different coefficients, one, K_l, along the flow (longitudinal diffusion) and the other, K_t, transverse to the flow (transversal diffusion).

Elder's observations and calculations, based on the velocity profile in the logarithmic layer (see Section 2.4.3 below), led him to the values: $K_l = 6\,u^*h$ and $K_t = 0.6\,u^*h$, where u^* is the shear velocity and h the water depth.

If α is the angle $(\overrightarrow{Ox}, \overrightarrow{u})$ of the velocity vector with respect to the x axis, the Elder model requires the adoption of a viscosity tensor:

$$\overline{\overline{K}} = \begin{pmatrix} K_{xx} & K_{xy} \\ K_{xy} & K_{yy} \end{pmatrix} \tag{2.257}$$

such that $\overline{\overline{K}}\overrightarrow{u} = K_l \overrightarrow{u}$ and $\overline{\overline{K}}\overrightarrow{t} = K_t \overrightarrow{t}$, where $\overrightarrow{t}$ is a vector transversal to $\overrightarrow{u}$. By identification:

$$K_{xx} = K_l \cos^2(\alpha) + K_t \sin^2(\alpha) \tag{2.258}$$

$$K_{xy} = (K_l - K_t)\sin(\alpha)\cos(\alpha) \tag{2.259}$$

$$K_{yy} = K_l \sin^2(\alpha) + K_t \cos^2(\alpha) \tag{2.260}$$

Nezu and Nakagawa model

This model, published in 1993, gets its inspiration from the expression $\nu_t = \kappa u^* z$ in the boundary layer above the bottom. But it also takes into account that turbulent viscosity decreases to zero towards the free surface (assuming the absence of wind). Nezu and Nakagawa [148] then propose:

$$\nu_t = \kappa u^* z \left(1 - \frac{z}{h}\right) \tag{2.261}$$

This model is very useful for representing diffusion along the vertical.

Mixing length models

Standard mixing length models: this model, proposed by Prandtl [157] in 1925 gives the value of the viscosity coefficient as:

$$\nu_t = L_m^2 \sqrt{2 D_{ij} D_{ij}} \tag{2.262}$$

where $\underline{\underline{D}}$ is the strain rate tensor of average motion, with:

$$D_{ij} = \frac{1}{2}\left(\frac{\partial \overline{U_i}}{\partial x_j} + \frac{\partial \overline{U_j}}{\partial x_i}\right) \tag{2.263}$$

L_m is the “mixing length” parameter equal to κy at a distance y from the wall, with $\kappa = 0.41$ (von Karman constant), till the size of eddies is no longer influenced by the wall and remains constant (in the case of flow between two walls, the limit is situated at 1/10th of the space between the walls). This model is used to describe the velocity profile in the vicinity of a wall. It is obtained by a hypothesis of equilibrium turbulence (for details see Section 2.4.4).

Mixing length models along the vertical: still according to the Prandtl model, applied this time along the vertical, the model provides vertical viscosities to be used for velocities or tracers:

$$\nu_z = f(Ri)\, L_m^2 \sqrt{\left(\frac{\partial U}{\partial z}\right)^2 + \left(\frac{\partial V}{\partial z}\right)^2} \tag{2.264}$$

and:

$$\nu_{zT} = f_T(Ri)\, L_m^2 \sqrt{\left(\frac{\partial U}{\partial z}\right)^2 + \left(\frac{\partial V}{\partial z}\right)^2} \tag{2.265}$$

where:

$$Ri = -\frac{g}{\rho_0} \frac{\frac{\partial \rho}{\partial z}}{\left(\frac{\partial U}{\partial z}\right)^2 + \left(\frac{\partial V}{\partial z}\right)^2} \tag{2.266}$$

is the Richardson number. f and f_T are the damping functions of the velocity and tracer which help to refine the model and adapt it to special cases, particularly for stratifications. If stratification is stable, which corresponds to a positive Richardson number, f is near zero. If stratification is unstable, which corresponds to a negative Richardson number, f takes a value between 1 and ∞, depending on the authors.

There are many expressions for damping functions, which always show a tendency to decrease with Ri. One way of obtaining an expression is to start with the Reynolds tensor equations and assume a local equilibrium of turbulence, so certain terms can be ignored. The remaining terms are then modelled. We get an algebraic system dependent on the Richardson number. By choosing the models described in reference [176], we arrive at the graph in Figure 2.6 for damping functions of velocity and temperature. The function defined for temperature does not exceed 1 for $R_i = 0$ but it does for velocity. So there is a turbulent Prandtl number different from 1 in a neutral situation ($Ri = 0$). In the case of a tracer, different from a temperature, other damping functions can be used. For example:

$$f(Ri) = (1 + a\, Ri)^b$$

a and b being two constants depending on the nature of the tracer. This is a totally empirical law, but it is sometimes used in the presence of sediment.

The mixing length L_m can be modelled in different ways:

Classical Prandtl model (see [165]):

$L_m = \kappa z$ if $z \leq 0.2h$ (here, z is the distance to the bottom and h the water depth).

$L_m = 0.2\kappa h$ if $z \geq 0.2h$.

Nezu and Nakagawa model (see [148]):

$$L_m = \kappa z \sqrt{1 - \frac{z}{h}}$$

In this case the mixing length vanishes at the free surface. This model can give a logarithmic velocity profile from bottom to free surface.

Quetin model (1977, see [159]):

$$L_m = \frac{1}{\frac{1}{\kappa z} + \frac{1}{0.65d}}$$

where d is the distance to the free surface.

Tsanis model (see [183]):

$L_m = \kappa z$ if $z \leq 0.2h$ (here, z is the distance to the bottom and h the water depth).

$L_m = 0.2\kappa h$ if $z \geq 0.2h$ and $z \leq 0.8h$.

$L_m = \kappa d$ if $z \geq 0.8h$ (d is the distance to the free surface).

In the production of turbulence expression, the horizontal gradients of U and V as well as the gradients of W are assumed negligible compared to the vertical gradients. The term:

$$\sqrt{\left(\frac{\partial U}{\partial z}\right)^2 + \left(\frac{\partial V}{\partial z}\right)^2}$$

is thus only an approximation of $\sqrt{2D_{ij}D_{ij}}$. For a calculation in the vicinity of a structure (quay, spikes, etc.), the gradients overlooked here can become important. However, their presence in the vertical viscosity expression would not be of much help to perform an efficient calculation around a structure. A mixing length model for horizontal viscosity would also become necessary. Hence a precise computation of flow around structures is not possible with this turbulence modelling.

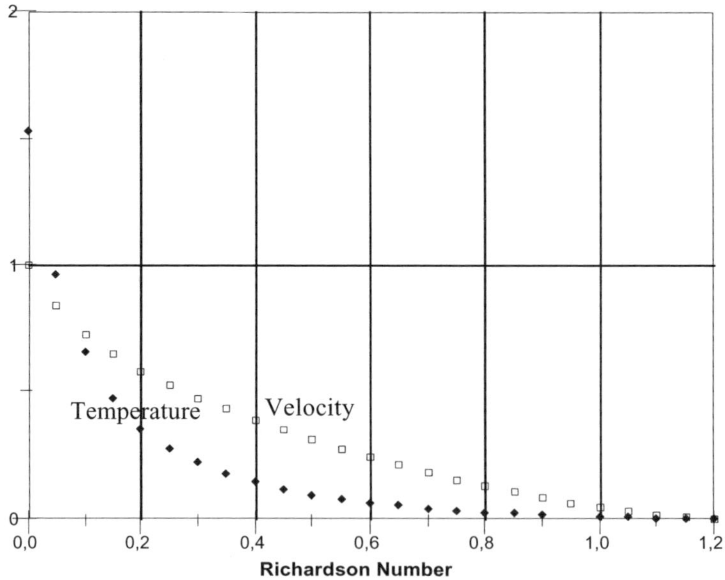

Figure 2.6: damping functions

Mixing length models for buoyant jets: the use of mixing length models mentioned earlier leads to abnormal results in the case of an outflow. In fact dissipation

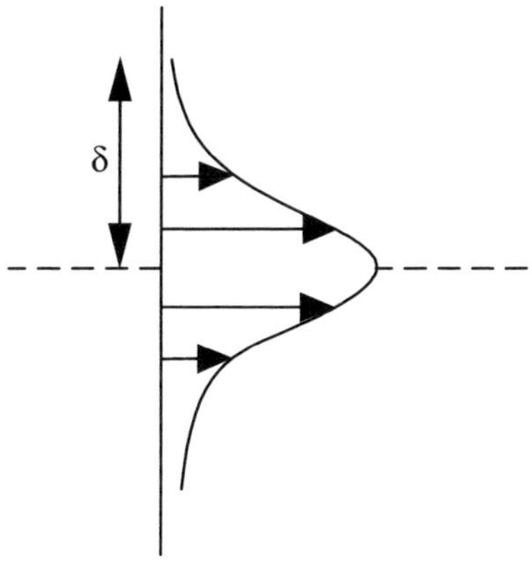

Figure 2.7: half-height of a jet

is overestimated in the vicinity of an outflow since the mixing length L_m is fixed at $0.2\,\kappa\,h$, which is approximately 8% of h. For highly shear flows like free jets in a still environment, Rodi [165] proposes a mixing length L_m of about 8%, but 8% of the half-height d for the jet:

$L_m = 0.09\,d$ for a plane jet.

$L_m = 0.075\,d$ for a circular jet.

The half-height is defined as the distance between the elevation of the jet axis and the one where velocity does not exceed 1% of its value on the axis (see Figure 2.7).

This half-height of the jet is necessary data in this case. If the jet is not symmetric, a distinction should be made between the upper half-height and lower half-height.

In practice, for each vertical, the elevation of the maximum velocity vector is determined first (elevation of the jet axis), then the upper and lower half-heights δ_{inf} and δ_{sup}. The mixing length in one point will be proportional to the local half-height. The coefficient of proportionality is always taken as $0.2\,\kappa$, i.e. 0.082, for two reasons:

- It is taken between the constants for plane and circular jets given by Rodi.
- It allows a continuity of the mixing length value when the jet has reached the surface: $L_m = 0.082h$.

At the bottom, the linear decrease of the mixing length remains unchanged. Figure 2.8 is a vertical profile showing the different stages of development of a buoyant jet. Far from the source δ_{inf} is equal to the depth.

In the case of a cross-current, determining the axis of the jet is less simple. First, the vertical plane with maximum velocity has to be determined. Then, the other velocities are projected onto this one. The elevation of the jet axis d_{inf} and d_{sup} is calculated in this plane.

It should be noted that this model was exclusively developed for buoyant jets. It is simple and well adapted to this type of application. It helps to model a plume in

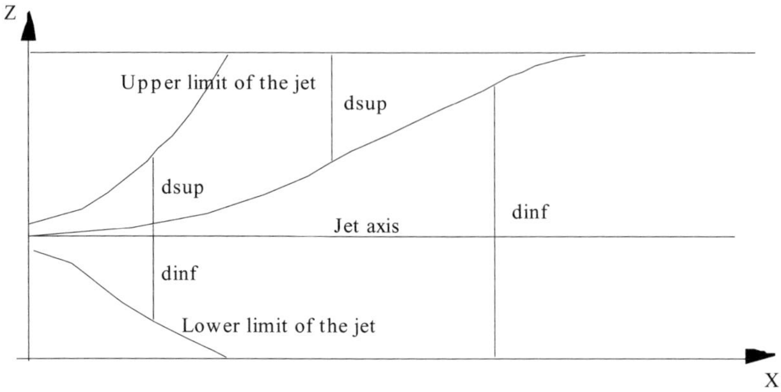

Figure 2.8: different stages of expansion of a jet

an almost satisfactory manner.

Smagorinski model

The Smagorinski model (see [175]) has a meaning only in numerical modelling and is based on the mixing length model. It belongs to the category of sub-grid turbulence models. The principle is as follows: turbulence is the solution of the Navier–Stokes equations. It would naturally appear in the numerical solutions if the size of finite elements allowed the reproduction of all mechanisms including the viscous dissipation of very small vortices. Only in the formation of smaller vortices, where turbulence is inhibited by the mesh, does modelling take place in a numerical simulation. Smagorinski's idea is to add to the molecular viscosity a turbulent viscosity deduced from a mixing length model. This mixing length corresponds to the size of the vortices smaller than that of the mesh size. We simply arrive at the following formulation:

$$\nu_t = C_s^2 \Delta^2 \sqrt{2 D_{ij} D_{ij}} \tag{2.267}$$

where C_s is a dimensionless coefficient to be calibrated and Δ the mesh size derived in 2D or 3D from the surface or from the volume of the elements. The values of C_s range from 0.1 (flow in a canal) to 0.2 (isotropic turbulence). This model is best understood in 3D. In 2D it does not take dispersion into account.

2.4.3 Turbulence stress on the walls

After describing mixing length models we are in a position to describe turbulence stress on walls. It depends on the velocity profile above the bottom or in the vicinity of walls, in what is called the boundary layer, a layer which is influenced by the wall. The important parameters are shear velocity, defined by $\tau = -\rho\,(U^*)^2$, as seen earlier, and the dimensionless distance to the wall $y^+ = \frac{yU^*}{\nu}$, where y is the distance to the wall. Traditionally, there is a distinction between:

- A viscous sub-layer, where turbulence is negligible, and where the velocity profile is linear, given as:

$$\frac{U}{U^*} = \frac{yU^*}{\nu} \tag{2.268}$$

  The thickness of this viscous sub-layer is limited by $y^+ = 10$ approximately. If the characteristic roughness size of the wall is greater than the thickness of the viscous sub-layer, this viscous sub-layer cannot develop and the flow is known to be hydraulically rough. If there is a viscous sub-layer the flow is known to be hydraulically smooth.

- An intermediate layer, between $y^+ = 10$ and $y^+ = 30$. There is no simple expression of velocity profile in this zone, but only formulae connecting the viscous sub-layer to the logarithmic layer.

- A logarithmic layer, starting at $y^+ = 30$, and until the size of eddies is no longer influenced by the wall (assuming that $L_m = \kappa y$). The turbulent viscosity is then written as $\nu_t = \kappa U^* y$. In this layer, the velocity profile takes the following form:

For hydraulically smooth flow:

$$\frac{U}{U^*} = \frac{1}{\kappa}\ln\left(\frac{yU^*}{\nu}\right) + 5.2 \tag{2.269}$$

For hydraulically rough flow:

$$\frac{U}{U^*} = \frac{1}{\kappa}\ln\left(\frac{33\,y}{k_s}\right) = \frac{1}{\kappa}\ln\left(\frac{y}{k_s}\right) + 8.5 \tag{2.270}$$

where k_s is the roughness size.

An equivalent way of writing the velocity profile for the hydraulically rough flow is:

$$\frac{U}{U*} = \frac{1}{\kappa}\ln\left(\frac{y}{y_o}\right) \tag{2.271}$$

where $y_o \approx \frac{k_s}{33}$.

For hydraulically smooth flow the linear and logarithmic laws are unified by the Reichard law:

$$\frac{U}{U_*} = \frac{1}{\kappa}\ln(1+\kappa y^+) + 7.8(1 - e^{-\frac{y^+}{11}}) - \frac{y^+}{11}e^{-0.33y^+} \tag{2.272}$$

Reference [188] contains theoretical justifications of these formulae.

At this point, it is possible to specify the laws to be adopted in numerical simulations.

Hydraulically smooth flow

The problem is finding the shear velocity U^*. It is necessary to have data on the velocity profile in the vicinity of the wall. These data are obtained from the following reasoning: assume that the velocity at the wall on a mesh is not really taken on the wall (otherwise, it would be zero), but is taken in the boundary layer at a distance y. y is chosen arbitrarily or depending on the local mesh size (e.g. the edge point is at a distance from the real wall equal to 1/10th of the size of the mesh). By knowing U and y, the Reichard law (Equation 2.272) gives U^* from the formula:

$$\frac{U^*}{U} = \frac{1}{\frac{1}{\kappa}\ln(1+\kappa y^+) + 7.8(1 - e^{-\frac{y^+}{11}}) - \frac{y^+}{11}e^{-0.33y^+}} \tag{2.273}$$

This law is implicit as y^+ depends on U^*, and U^* is found by successive iterations. We then write:

$$\nu\frac{\partial U}{\partial n} = -U^{*2} \tag{2.274}$$

Hydraulically rough flow

For hydraulically rough flow, various formulae give the friction coefficient C_f depending on the roughness at the bottom. In Section 2.3.11 we have examined the Chézy, Strickler, Manning and Nikuradse formulae. For Nikuradse formula, we can now understand how it stems from the Chézy law and an integration vertically of the logarithmic profile given in Equation 2.271.

When the roughness size k_s is known, the following laws can also be used:

Haaland formula:

$$C_f = \frac{1}{4\left[0.6\ln\left(\left(\frac{6.9\nu}{4h\sqrt{U^2+V^2}}\right)^3 + \frac{k_s}{14.8h}\right)\right]^2} \tag{2.275}$$

Ramette formula: according to Ramette [160], the roughness size k_s can also be connected to the Chézy coefficient C by the expression:

$$C = \frac{26.4}{k_s^{1/6}}\left(\frac{k_s}{h}\right)^{1/24} h^{1/6} \tag{2.276}$$

This last empirical formula is not dimensionless and requires a strict respect of SI units.

2.4.4 Equations of the k–ε model

k–ε model in 3D

This model is based on the calculation of physical quantities representing turbulence in the flow. k and ε respectively designate turbulent kinetic energy and its dissipation:

$$k = \frac{1}{2}\overline{u_i^{'}u_i^{'}} \tag{2.277}$$

$$\varepsilon = \nu \overline{\frac{\partial u_i^{'}}{\partial x_j} \frac{\partial u_i^{'}}{\partial x_j}} \tag{2.278}$$

where $u_i^{'}$ is the turbulent fluctuation of velocity and where the bar represents averaging.

k and ε equations: in 3D , the quantities k and ε are obtained by solving the following equations (see [121] and [165]):

$$\frac{\partial k}{\partial t} + \overline{U_i} \frac{\partial k}{\partial x_i} = \frac{\partial}{\partial x_i} \left(\frac{\nu_t}{\sigma_k} \frac{\partial k}{\partial x_i} \right) + P - G - \varepsilon \tag{2.279}$$

$$\frac{\partial \varepsilon}{\partial t} + U_i \frac{\partial \varepsilon}{\partial x_i} = \frac{\partial}{\partial x_i} \left(\frac{\nu_t}{\sigma_\varepsilon} \frac{\partial \varepsilon}{\partial x_i} \right) + C_{1\varepsilon} \frac{\varepsilon}{k} \left[P + (1 - C_{3\varepsilon}) G \right] - C_{2\varepsilon} \frac{\varepsilon^2}{k} \tag{2.280}$$

where:

P is the term for the production of turbulent energy:

$$P = \nu_t \left(\frac{\partial \overline{U_i}}{\partial x_j} + \frac{\partial \overline{U_j}}{\partial x_i} \right) \frac{\partial \overline{U_i}}{\partial x_j} = 2\nu_t D_{ij} D_{ij} \tag{2.281}$$

G is a source term due to the forces of gravity in the case of temperature gradients:

$$G = \beta \frac{\nu_t}{Prt} g \frac{\partial T}{\partial z}$$

with $\beta = -\frac{1}{\rho} \frac{\partial \rho}{\partial T}$, Prt being the Prandtl number.

Knowing k and ε, ν_t is deduced from the formula:

$$\nu_t = C_\mu \frac{k^2}{\varepsilon}$$

which essentially comes from dimensional analysis. We also get:

$$\nu_{tT} = C_{\mu T} \frac{k^2}{\varepsilon}$$

It is the equilibrium hypothesis between the creation of turbulence and dissipation which leads to the mixing length model. Actually, by assuming $P = \varepsilon$, and as $\nu_t \approx \frac{k^2}{\varepsilon}$, we immediately find $\nu_t \approx \frac{k^2}{P}$. Also $\nu_t \approx \sqrt{k} L_t$, where L_t is the characteristic size of vortices. We get:

$$\nu_t^2 \approx \frac{k^2}{2D_{ij} D_{ij}}$$

and $\nu_t^4 \approx k^2 L_t^4$. By dividing the second equation by the first one, member by member, we finally get:

$$\nu_t \approx L_t^2 \sqrt{2D_{ij} D_{ij}}$$

The constants of the k–ε model have been determined by comparison with simple situations. The constants C_μ and $C_{1\varepsilon}$ are obtained from data for turbulent flow

Table 2.1: constants of the k–ε model

$C\mu$	$C_{\mu T}$	Prt	$C_{1\varepsilon}$	$C_{2\varepsilon}$	$C_{3\varepsilon}$	σ_k	σ_ε
0.09	0.09	1.0	1.44	1.92	0 if $G > 0$ and 1 if $G \leq 0$	1.0	1.3

near a rigid wall. The free decay of grid turbulence helps us to find a value for $C_{2\varepsilon}$ on the basis of well-documented experimental data. The constants σ_k and σ_ε have been "optimized", based on the performance of the model in both the test cases just mentioned. Obtaining $C_{3\varepsilon}$ is elaborated in references [121] and [187]. We retain that $C_{3\varepsilon}$ is equal to 1 for a stable situation, i.e. if G is negative, and equal to 0 for unstable stratifications. σ_k is sometimes called "Prandtl's turbulent number for the kinetic energy of turbulence".

All these constants are summarized in Table 2.1.

In internal flow modelling, the term $-\frac{2}{3}k\,\delta_{ij}$ is often omitted because it plays the same role as pressure. If we want to have the true pressure, it is not possible to integrate this term which should thus appear in numerical resolution.

Boundary conditions of the k–ε model

Rigid boundaries: to define rigid boundary conditions, it is considered that turbulence is in local equilibrium at the wall, such that the production of turbulence (due to shear at the wall and at the bottom) is equal to dissipation and the velocity profile is locally logarithmic. This idea was exploited by Mellor and Yamada to establish a zero-equation turbulence model (see [141]).

With the equilibrium hypothesis, at a distance δ from the wall, the k–ε model equations are:

$$k = \frac{u*^2}{\sqrt{c_\mu}} \tag{2.282}$$

$$\varepsilon = \frac{u*^3}{k\,\delta} \tag{2.283}$$

The distance δ is chosen equal to 1/10th of the local mesh size, measured along the normal direction to the wall. In the presence of turbulence, the boundary condition for velocity is now written as:

$$\nu_t \frac{\partial u}{\partial n} = -u*^2$$

ν being replaced by ν_t.

Liquid boundaries: without additional information on the value of k and ε at a liquid boundary, a "weak" condition is applied, i.e. $\nu_t \frac{\partial k}{\partial n} = 0$ and $\nu_t \frac{\partial \varepsilon}{\partial n} = 0$. Such a condition can, however, give rise to numerical problems at the entrance of a domain.

k–ε model averaged over the vertical

The integration of the Reynolds equation over the vertical brings about dispersion terms due to the non-uniformity of vertical velocity profiles. On account of these dispersion terms, Rastogi and Rodi [161] proposed an extension of the classical k–ε model, adapted to the Saint-Venant equations.

k and ε are averaged values over the vertical, defined by:

$$k = \frac{1}{h}\int_{Z_f}^{Z_s} \frac{1}{2}\overline{u_i^{'}u_j^{'}}\,dz \tag{2.284}$$

$$\varepsilon = \frac{1}{h}\int_{Z_f}^{Z_s} \nu\overline{\frac{\partial u_i^{'}}{\partial x_j}\frac{\partial u_i^{'}}{\partial x_j}}\,dz \tag{2.285}$$

The k and ε equations take the form:

$$\frac{\partial k}{\partial t} + u_i\frac{\partial k}{\partial x_i} = \frac{1}{h}div\left(h\frac{\nu_t}{\sigma_k}\overrightarrow{grad}(k)\right) + P - \varepsilon + P_{kv} \tag{2.286}$$

$$\frac{\partial \varepsilon}{\partial t} + u_i\frac{\partial \varepsilon}{\partial x_i} = \frac{1}{h}div\left(h\frac{\nu_t}{\sigma_\varepsilon}\overrightarrow{grad}(\varepsilon)\right) + \frac{\varepsilon}{k}\left[C_{1\varepsilon}P - C_{2\varepsilon}\varepsilon\right] + P_{\varepsilon v} \tag{2.287}$$

Production terms (always positive) are now calculated with horizontal velocity gradients:

$$P = \nu_t\left(\frac{\partial u_i}{\partial x_j} + \frac{\partial u_j}{\partial x_i}\right)\frac{\partial u_i}{\partial x_j}$$

P_{kv} and $P_{\varepsilon v}$ are due to the shear force of flow along the vertical:

$$P_{kv} = C_k\frac{u_*^3}{h} \text{ and } P_{\varepsilon v} = C\varepsilon\frac{u_*^4}{h^2}$$

In these formulae:

$$C_k = \frac{1}{\sqrt{C_f}} \text{ and } C_\varepsilon = 3.6\frac{C_{2\varepsilon}\sqrt{C_\mu}}{C_f^{3/4}}$$

C_f is the dimensionless friction coefficient and u^* is the shear velocity (or friction velocity) on the bottom. By definition it is equal to $\sqrt{\frac{\tau}{\rho}}$, where τ is the friction stress on the bottom. With:

$$\overrightarrow{\tau} = -\rho\frac{C_f}{2}\sqrt{u^2+v^2}\,\overrightarrow{u} = -\rho(u^*)^2\frac{\overrightarrow{u}}{\sqrt{u^2+v^2}}$$

the shear velocity is also equal to:

$$u^* = \sqrt{\frac{C_f}{2}\left(u^2+v^2\right)}$$

The new parameters C_k and C_ε have been determined on the basis of normal flow in the centre of a rectilinear fluvial canal.

Finally, the friction coefficient C_f is deduced from the chosen friction law, as already seen in Section 2.3.11.

Boundary conditions on rigid boundaries: we consider the flow at a distance δ from the wall and continue to assume a local equilibrium of production and dissipation of turbulence. However, in 2D, production of turbulence due to the bottom should also be considered. The value ε at a distance δ from the wall for a water depth h is now given by:

$$\varepsilon_\delta = \frac{u_*^3}{\kappa\delta} + \frac{1}{\sqrt{C_f}}\frac{\widetilde{u}_*^3}{h} \tag{2.288}$$

where $\widetilde{u}_*$ is the shear velocity on the bottom. As for the production of turbulent kinetic energy, it is assumed that it comes from two contributions: the wall and the bottom. The following expression is deduced:

$$k_\delta = \frac{u_*^2}{\sqrt{C_\mu}} + \frac{C_{2\varepsilon}}{C_\varepsilon C_f}\widetilde{u}_*^2 \tag{2.289}$$

The shear velocity at the wall is calculated according to the chosen friction law and the nature of flow (hydraulically smooth or rough), as seen in Section 2.4.3.

For the sake of numerical stability, the stress on the wall is given the form:

$$\nu_t \frac{\partial u_t}{\partial n} = -a\, u_t \tag{2.290}$$

According to the same equilibrium hypothesis between production and dissipation, the logarithmic relation between shear velocity and tangential velocity at the wall u_t, gives the following relations:

$$\frac{\partial u_t}{\partial n} = \frac{u_*}{\kappa y}$$

where y is the distance to the wall, and:

$$\nu_t = \kappa y u_*$$

which finally gives:

$$a = -\frac{u_*^2}{u_t}$$

Boundary conditions on liquid boundaries: homogeneous Neumann conditions are chosen for an outlet and slip conditions for a wall:

$$\frac{\partial k}{\partial n} = 0 \text{ and } \frac{\partial \varepsilon}{\partial n} = 0 \tag{2.291}$$

As for entrance conditions, there is an absence of information that can hardly be asked of a user. Equilibrium is assumed between kinetic energy produced by friction on the bottom and its dissipation rate. The k and ε set of equations then give the following conditions for an open boundary of the domain:

$$\varepsilon = P_{kv} \text{ and } k = C_{2\varepsilon}\frac{P_{kv}^2}{P_{\varepsilon v}}$$

These values of k and ε can also be used within the domain to initialize calculations.

2.4.5 Other models

Though sufficiently difficult to implement numerically, the k–ε model has obvious limitations, particularly for eddies behind obstacles and in the presence of a high rate of deformation (tensor D_{ij}) near a stagnation point, for example. The Reynolds stress model is a higher order model. The six Reynolds stress equations are then directly solved, and the ε equation, together with the equations of scalar turbulent flux of temperature and salinity. Only this type of model is capable of really taking into account turbulence anisotropy by renouncing the Boussinesq hypothesis. However, the triple correlations of the equations are difficult to model and the boundary conditions difficult to justify. To our knowledge, this is the reason why there is still no industry-oriented application of this model in free surface hydrodynamics.

Chapter 3

Principles of the finite element method

Equations do not explode. Bertrand Russell (1872–1970)

3.1 Introduction

A simple description of the method is sketched out below. It might require further development, but a list of reference works is given if more details are needed.

The finite element method is based on two principal ideas: an interpolation method and a variational method known as "weighted residuals". Each of these methods has a similar aim: to pass from a continuous problem to a discrete problem. If a function u is defined by the equation $L(u) = 0$ on a continuous domain Ω, the interpolation method will provide an approximate function of u constructed from a finite number of real numbers. The variational method will replace the continuous equation $L(u) = 0$, valid at every point of Ω, by a finite number of equations. In short, interpolation discretizes the unknown and the variational principle discretizes the equation. This should be specified in a strict mathematical framework, which (sometimes) helps to show the existence and uniqueness of the solution. Unfortunately, in free surface fluid mechanics, these demonstrations only exist in simple cases.

The domain of computation Ω is an open and bounded domain in $\mathbb{R}^N$, $\mathbb{R}$ being the set of real numbers. N is the space dimension, equal to 1, 2 or 3. Γ is the boundary, regular (a normal vector can be defined on it), of Ω. At the continuum level, physical quantities (pressure, temperature, components of the velocity vector, etc.) are functions of Ω in $\mathbb{R}$. Moreover, we require that these functions are square integrable and $L^2(\Omega)$ denotes the set of square integrable functions. A dot product can be defined on this set as:

$$(u, v)_0 = \int_\Omega uv \, d\Omega \tag{3.1}$$

A subset of $L^2(\Omega)$ is composed of functions whose derivatives along the N directions of space are also square integrable. This guarantees that a function gradient

Hydrodynamics of Free Surface Flows: Modelling with the Finite Element Method J.-M. Hervouet

would also be square integrable. This subset is denoted $H^1(\Omega)$. A new dot product is defined in $H^1(\Omega)$ as:

$$(u,v)_1 = \int_\Omega uv + \overrightarrow{grad}(u).\overrightarrow{grad}(v)d\Omega \tag{3.2}$$

The spaces $L^2(\Omega)$ and $H^1(\Omega)$ with their dot product are called Hilbert spaces. Hilbert spaces are very similar to the Euclidean vector spaces. On both of them, and with the help of the dot product, we can define:

a norm:

$$\|u\| = \sqrt{(u,u)} \tag{3.3}$$

a distance:

$$d(u,v) = \|u - v\| \tag{3.4}$$

and an angle:

$$\cos(\alpha) = \frac{(u,v)}{\|u\|\,\|v\|} \tag{3.5}$$

but their dimension is infinite. $H^1(\Omega)$ has been described with derivatives of the first order. $H^k(\Omega)$ can be described in the same way, with derivatives of the kth order, like the set of functions whose derivatives up to kth order are all square integrable. The $H^k(\Omega)$ spaces are known as Sobolev spaces.

As for the equations being worked out, only linear equations with partial derivatives are considered (fluid mechanics equations are generally non-linear, but the resolution methods used for solving them will generally be written in a linear form). Here are some of the linear equations that we shall come across later:

Helmholtz equation:

$$\frac{\partial^2 u}{\partial x^2} + \frac{\partial^2 u}{\partial y^2} + \lambda u = 0 \tag{3.6}$$

Poisson equation:

$$\frac{\partial^2 u}{\partial x^2} + \frac{\partial^2 u}{\partial y^2} = f \tag{3.7}$$

In this chapter, we consider that our equation is in the form:

$$L(u) = f \tag{3.8}$$

Interpolation and the variational method will now be described more precisely.

3.2 Interpolation in finite elements

The general idea is to replace an unknown function u, which belongs to an infinite dimensional space, by an approximation u_h defined on a finite dimensional space N. This is an approximation and we cannot always ensure that $L(u_h) = f$. We simply try to minimize $L(u_h) - f$ and this will be the goal of the variational method.

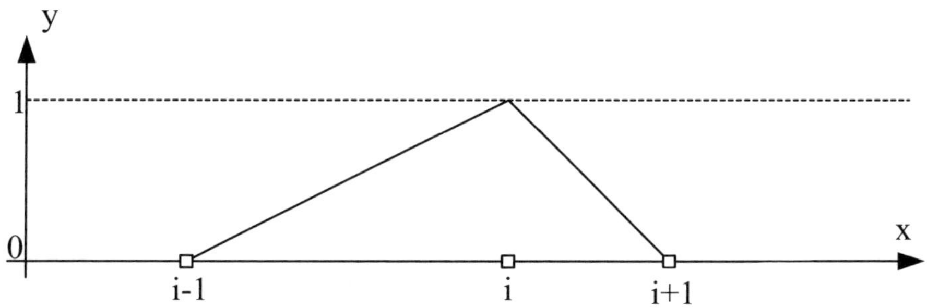

Figure 3.1: linear basis in one dimension

In finite elements, a function u is represented by n real numbers which are the exact values of u at special points of the domain known as degrees of freedom. For the other points, an interpolation is sufficient. To define a unique interpolation function, e.g. a polynomial one, with exact values of u for the n degrees of freedom would be very complex, or would lead at least to a high-order function. So, finite elements give up the univocal nature of the interpolation function and choose a locally simple interpolation function (e.g. constant, linear, quadratic). A tessellation of space into segments, triangles, quadrilaterals, tetrahedra, prisms, etc., is executed. The degrees of freedom are assigned to these elements, on the vertices, at the centre of gravity, in the middle of the sides, etc., for example. Thus, each element is described both by the coordinates of its geometric nodes and by those of its interpolation nodes. This is followed by a simple definition of interpolation within each element. For example, in one dimension, taking the vertices of segments as degrees of freedom, a linear interpolation will be chosen. The function u_h approximating u can then be written as:

$$u_h = \sum_{i=1}^{n} u_i \Psi_i \tag{3.9}$$

where Ψ_i is a basis function defined differently, but in a linear manner, on each segment ending at the point i. Outside these segments, Ψ_i is zero. Figure 3.1 shows this basis function in one dimension, on a non-regular mesh. The value of Ψ_i is 1 at point i, and 0 on all the other degrees of freedom.

The property $\sum_{i=1}^{n} \Psi_i = 1$ is very important and will be the key to several demonstrations. In two dimensions, with triangles and a linear interpolation, each basis $\Psi_i(x, y)$ will be written as $ax + by + c$, a, b and c depending on the triangle, so as to obtain the value 1 at point i and 0 at the other ones. Figure 3.2 shows the extent and the nodal values of a linear basis on a finite element mesh.

These basis functions are sometimes called "shape functions". With a linear interpolation, a linear function u will be represented exactly. It is said to be in the approximation space. The interpolation can also be quadratic, or of any other order which enables representation of the function on the degrees of freedom but also some

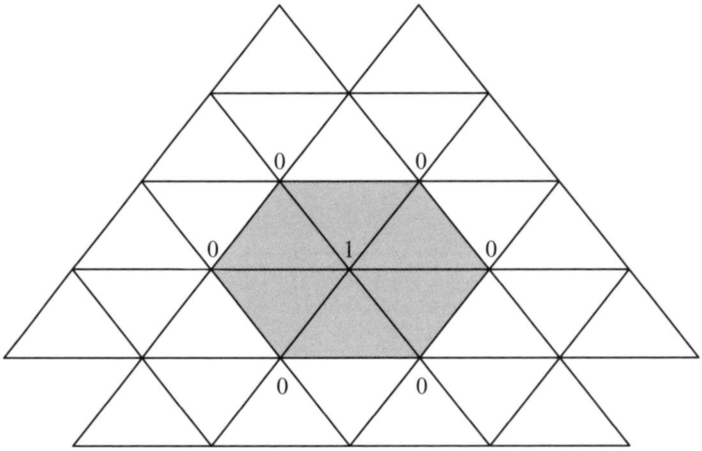

Figure 3.2: extent of a basis function on a triangular mesh

of its derivatives.

Sometimes, and it is so for quadrilaterals, the element is too complex to get a simple definition of the interpolation. So, we go back to a more regular reference element (a square in the case of a quadrilateral) in a "reference" space and define a geometric transformation between the two spaces. Interpolation is simple only in the reference space, except in the case of simplicial elements: segments, triangles and tetrahedra with linear interpolation. The use of a reference space will be exemplified in Section 7.4.1, for building finite element matrices. The finite elements to be used will be presented in detail in Chapter 7. A precise description of a large number of other elements can also be found in reference [40].

3.3 Variational principle

Having restricted our function u to a representation u_h, based on a discrete number of unknowns u_i, we now wish to minimize $L(u_h) - f$. The variational method states that the dot product $\int_\Omega (L(u_h) - f)\varphi_i d\Omega$ is zero for some functions φ_i called test functions. The choice of these functions defines the variants of the finite element method. The most classic technique, the Galerkin technique, consists of choosing test functions equal to the basis functions: $\varphi_i = \Psi_i$. In this case, if u_h is decomposed as $\sum_{i=1}^n u_i \Psi_i$, the variational method will clearly lead to a linear system of unknowns u_i.

Moreover, to each linear operator L is associated a bilinear form a such as:

$$a(u, v) = \int_\Omega (L(u) - f)\, v\, d\Omega \tag{3.10}$$

The finite element theory aims at determining the conditions on L and on a for which the problem would have a unique solution. An important condition is that, for

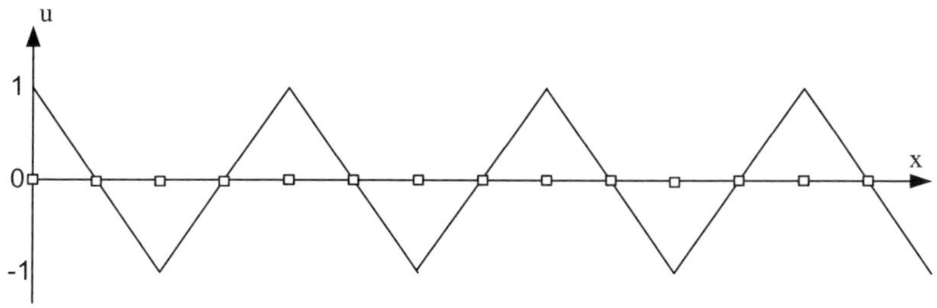

Figure 3.3: an invisible function for the gradient operator

every non-zero function u, there exists a real α positive, such that $\|L(u)\| \geq \alpha \|u\|$. If the norm L is defined by:

$$\|L\| = \sup_{u \neq 0} \left(\frac{\|L(u)\|}{\|u\|} \right)$$

this condition signifies that L should have a strictly positive norm. Otherwise, if a non-zero function u_p verifies $L(u_p) = 0$, the solution will not be unique because if u is a solution of the equation $L(u) = f$, $u + u_p$ will also be one. Within the variational method frame, the condition is written as:

$$\inf_{u \neq 0} \left(\sup_{v \neq 0} \left(\frac{a(u,v)}{\|u\| \, \|v\|} \right) \right) \geq \alpha \tag{3.11}$$

This is known as the "inf–sup condition". It prevents the occurrence of parasitic solutions which could arise from the bilinear form. For example, for a one-dimensional domain with regularly spaced degrees of freedom, if L is the gradient operator, and if the basis functions Ψ_i are linear and equal to the test functions φ_i, a function u whose values at the nodes are a succession of 1 and -1 (see Figure 3.3) would lead to $\int_\Omega L(u)\varphi_i d\Omega = 0$ for every test function φ_i and the inf–sup condition would not be verified.

This situation arises when the boundary conditions are not considered, but it can take place also far from the boundaries of a calculation domain. The bilinear forms resulting from the variational formulations of the Navier–Stokes equations do not always satisfy the inf–sup condition. A well-known condition for internal flows of incompressible fluids consists of choosing different elements to represent pressure and velocity: for example, a quadratic velocity and a linear pressure. In principle, this choice is not mandatory for compressible fluids and, by extension, for the Saint-Venant equations. However, the choice of different elements for velocity and water depth gives more regular results. This is exemplified in Plate 5, where a tide is simulated on a quarter-circle beach with a quadratic bottom elevation. Advection and all dissipative terms are removed from equations. It is clear that in this academic situation spurious oscillations occur with a linear discretization, but they disappear if the velocity is treated with a quasi-bubble discretization (see [2] for more details). Another solution, efficient with regular mesh and the finite difference method, is to have two different and staggered meshes for pressure and velocity. Finally, for elliptic equations, we can use a discretization of the Laplacian, known as "non-compatible", for which the Laplacian is not, at the discrete level, equal to the divergence of the gradient. The result of this method is detailed in Section 4.12, where the Saint-Venant equations are written in the form of a wave equation.

A more thorough examination of these questions, especially for the Navier–Stokes equations, can be found in references [29] and [56], and in the more recent work by Pironneau [156] on the methods of finite elements for fluids.

Chapter 4

Resolution of the Saint-Venant equations

Everything is vague to a degree you do not realise till you have tried to make it precise.
Bertrand Russell.

4.1 A glance at the existing methods

Presenting all the methods used to solve the Saint-Venant equations raises even more challenges. In fact, a combination of variants of the equations and of resolution methods leads to hundreds of numerical schemes.

Concerning the equations, the following should be added to the conservative and non-conservative forms of the primitive equations seen earlier:

- The "pseudo wave equation form", in which velocity is withdrawn from the continuity equation thanks to the momentum equation (see [189]).
- The "wave equation" form obtained by deriving the equations with respect to space and time, as has been done in Section 2.3.4.
- The Boltzmann equation, the starting point for "kinetic" schemes, which will be briefly discussed further.

Concerning the resolution methods, we can list the following:

- Finite differences.
- Finite volumes.
- Finite elements.
- Lagrangian or semi-Lagrangian methods.
- Spectral methods.

Hydrodynamics of Free Surface Flows: Modelling with the Finite Element Method J.-M. Hervouet

A numerical scheme will combine a choice of original equations and a choice of resolution methods. If a fractional steps method is used (see Section 4.3 below), it is possible to combine different methods, but that complicates even more taxonomy and comprehension. Thus the algorithm presented in and after Section 4.2 is founded mainly on the finite element method, and at the same time borrows from the semi-Lagrangian methods for advection terms by using the method of characteristics or from finite volumes by using fluctuation splitting schemes (Chapter 6). The choice of the computation domain is another source of diversity. In fact, finite differences could be considered as a form of finite elements where certain matrices are replaced by diagonals, but they are implemented on regular meshes where it is easier to calculate high-order derivatives, giving access to a wider choice of numerical schemes.

Before detailing our finite element scheme, we shall explain what general properties are expected from an efficient numerical scheme and also briefly describe some other methods.

4.1.1 Main properties of a (good) numerical scheme

We consider systems of partial differential equations, including time and space derivatives, advection/diffusion, flux (e.g. diffusive flux) and source terms. Non-linearity of equations results in both turbulence and discontinuous solutions, even though the initial conditions are continuous. Equations are hyperbolic and the discretization of some of their terms needs to take into account the direction of the flow. In these conditions, the numerical properties expected for a resolution scheme are as follows:

Precision

Numerical solutions should be as close as possible to the exact solutions, without diverging. For periodic solutions, or in a more general manner, for waves, a distinction is made between precision on the amplitude and precision on the phase.

Consistency

This condition requires that equations are well discretized, i.e. if the time step and mesh size tend to zero, we are back to continuous equations.

Stability

Solutions should not diverge. We can show that consistency and stability imply convergence to the correct solution (for a linear problem, well-posed, without boundary conditions and with initial values; this is the Lax theorem [123]).

Uniqueness

This property is generally satisfied by a condition known as "entropy". However, many real problems do not have just one solution (emergence of von Karman vortices by the rupture of symmetry, turbulence, Coanda effect, etc.). In the Saint-Venant equations, entropy is the integral of the head along the water depth. It is also the total energy of a column of water with a height h. It is written as:

$$E = \frac{hu^2}{2} + g\frac{h^2}{2} + ghZ_f \tag{4.1}$$

and the entropy condition is:

$$\frac{\partial E}{\partial t} + div\left[u\left(E + g\frac{h^2}{2}\right)\right] \leq 0 \tag{4.2}$$

This entropy condition becomes equality in the absence of discontinuities.

Monotonicity

An advection and diffusion scheme should not create minima or maxima other than those of the initial conditions. Schemes using low interpolation orders will be monotonic more easily, but also less precise and more diffusive (wave amplitude will be decreased artificially). Higher order schemes will be more precise but subject to spurious oscillations.

Conservation

A numerical scheme should ensure conservation of physical quantities like mass, momentum, heat, etc.

Behaviour in the presence of shocks

Either the scheme is capable of reproducing discontinuities without a specific treatment (this technique is called "shock capturing"), or shocks should be followed and they receive a special treatment ("shock fitting").

4.1.2 Finite difference schemes

Finite difference schemes, whose use is limited to regular grids, are progressively replaced by techniques applicable to non-structured meshes. Here, we are just presenting a method widely used in the past, the ADI (Alternating-Direction Implicit) method. It uses the regularity of the mesh for the treatment of the derivatives along y first, then along x. To make it clear, we apply the method to the following linearized and simplified equations:

$$\frac{\partial h}{\partial t} + h_0 div(\overrightarrow{u}) + \overrightarrow{u}_0.\overrightarrow{grad}(h) = 0 \tag{4.3}$$

$$\frac{\partial u}{\partial t} + u_0\frac{\partial u}{\partial x} + v_0\frac{\partial u}{\partial y} + g\frac{\partial h}{\partial x} = 0 \tag{4.4}$$

$$\frac{\partial v}{\partial t} + u_0\frac{\partial v}{\partial x} + v_0\frac{\partial v}{\partial y} + g\frac{\partial h}{\partial y} = 0 \tag{4.5}$$

So, friction and diffusion are ignored and it is assumed that h, u and v remain close to h_0, u_0 and v_0. The three equations can be written in a vector form:

$$\frac{\partial F}{\partial t} + A\frac{\partial F}{\partial x} + B\frac{\partial F}{\partial y} = 0 \tag{4.6}$$

with:

$$F = \begin{pmatrix} h \\ u \\ v \end{pmatrix}, \; A = \begin{pmatrix} u_0 & h_0 & 0 \\ g & u_0 & 0 \\ 0 & 0 & u_0 \end{pmatrix} \text{ and } B = \begin{pmatrix} v_0 & 0 & h_0 \\ 0 & v_0 & 0 \\ g & 0 & v_0 \end{pmatrix}$$

$\frac{\partial F}{\partial t}$ is discretized as $\frac{F^{n+1}-F^n}{\Delta t}$. Spatial discretization of the operator $\Delta t\, A\frac{\partial}{\partial x}$ gives a matrix P. Spatial discretization of the operator $\Delta t\, B\frac{\partial}{\partial y}$ gives a matrix Q. For space derivatives, a Crank–Nicholson scheme is used. That is:

$$\frac{\partial F}{\partial x} = \theta\frac{\partial F^{n+1}}{\partial x} + (1-\theta)\frac{\partial F^n}{\partial x} \tag{4.7}$$

with $\theta \geq 0.5$ for the sake of stability. Equation 4.6 is then written as:

$$(I + \theta P + \theta Q)F^{n+1} = (I - (1-\theta)P - (1-\theta)Q)F^n \tag{4.8}$$

On the basis of $\Delta F = F^{n+1} - F^n$, Beam and Warming [8] write the equation as:

$$(I + \theta P + \theta Q)\Delta F = -(P + Q)F^n \tag{4.9}$$

and factorize the left-hand side as:

$$(I + \theta P)(I + \theta Q)\Delta F$$

The approximation respects time order accuracy.

First, we resolve:

$$(I + \theta P)X = -(P + Q)F^n$$

where X is an intermediate unknown. Then:

$$(I + \theta Q)\Delta F = X$$

The linear equations obtained for each resolution, applicable only to one dimension, are tri-diagonal, as in a one-dimensional problem. Peaceman and Rachford [152] break down the problem differently by introducing an intermediate matrix $F^{n+1/2}$:

$$\left(I + \frac{P}{2}\right)F^{n+1/2} = \left(I - \frac{Q}{2}\right)F^n \tag{4.10}$$

$$\left(I + \frac{Q}{2}\right)F^{n+1} = \left(I - \frac{P}{2}\right)F^{n+1/2} \tag{4.11}$$

4.1.3 Finite volume schemes for hyperbolic equations

To simplify matters, explanations are in one dimension. Finite volume schemes deal with equations in the following conservative form:

$$\frac{\partial F}{\partial t} + \frac{\partial G(F)}{\partial x} = S(F) \tag{4.12}$$

where:

$$F = \begin{pmatrix} h \\ hu \end{pmatrix}$$

$$G(F) = \begin{pmatrix} hu \\ hu^2 + g\frac{h^2}{2} \end{pmatrix}$$

and $S(F)$ is the right-hand side containing the diffusion terms, the source terms and, for the second component, the bottom gradient: $-g\frac{\partial Z_f}{\partial x}$. The analytical solution of this system involves the Jacobian matrix of the system:

$$\frac{\partial G}{\partial F} = \begin{pmatrix} 0 & 1 \\ gh - u^2 & 2u \end{pmatrix} \tag{4.13}$$

whose eigenvalues are $u + \sqrt{gh}$ and $u - \sqrt{gh}$ and eigenvectors are:

$$\begin{pmatrix} 1 \\ u + \sqrt{gh} \end{pmatrix} \text{ and } \begin{pmatrix} 1 \\ u - \sqrt{gh} \end{pmatrix}$$

These eigenvalues make the Saint-Venant equations hyperbolic. The finite volume method looks for constant quantities over each volume or cell. These may either coincide with the elements of a non-structured mesh, or be centred on the nodes of a non-structured mesh. Basically, if we discard $S(F)$, Equation 4.12 is summed over each volume Ω_i, which gives:

$$vol(\Omega_i)\frac{\partial F}{\partial t} + \int_{\Gamma(\Omega_i)} G(F).\vec{n}\, d\Gamma = 0 \tag{4.14}$$

where $vol(\Omega_i)$ denotes the length, the surface or the volume of the element, depending on the dimension, and $\vec{n}$ the normal vector. Hence, it suffices to know $G(F)$ at the boundaries of the control cell to find the evolution in time of F. As F is constant on each cell, it is discontinuous at the cell boundaries. Finding $G(F)$ means finding the solution of the Riemann problem, with constant velocities and water depth on either side of a discontinuity. This problem, similar to a dam break on a wet bottom (see Section 4.13.4), is resolved by the theory of the characteristic curves and the Riemann invariants. Roe [166] proposed an approximation of the Riemann problem based on a linearized Jacobian matrix, which is the starting point of many schemes. However, the Roe solution should be modified in the presence of dry zones as it gives rise to non-physical hydraulic jumps which should be replaced by expansion waves. The calculation of flux should take into account the direction of the flow. This is obtained with the help of the Jacobian matrix. In two dimensions, Equation 4.12 is written as:

$$\frac{\partial F}{\partial t} + \frac{\partial G_x(F)}{\partial x} + \frac{\partial G_y(F)}{\partial y} = S(F) \tag{4.15}$$

The principle of resolution is the same. By solving the Riemann problem, the finite volume methods obtain excellent results for all systems of equations with shocks. They verify the conservation of mass and momentum properties. On the other hand, the different source terms contained in $S(F)$, mainly the bottom gradient and the friction term, are a difficulty. Friction at the bottom is dealt with in an additional fractional step, generally implicit. This affects the balance established in the first step. New schemes, like that of Bermudez *et al.* [12], for example, propose an upwind computation of the bottom gradient as well as the flux, to fulfil the properties of conservation. They also propose an implicit variant of their method, while the classic schemes are explicit, with the values of the Courant number $\frac{u\Delta t}{\Delta x}$ limited to 0.5 or 1.

4.1.4 Kinetic schemes

Kinetic schemes (in particular, see [155]) are finite volume schemes, based on an original idea, since they are supported by the Boltzmann kinetic equations for gases. In fact, the kinetic approach consists of associating a point whose water depth is h and velocity u and a density M of particles whose velocities ξ vary around an average u. This density is defined thus:

$$M(h,\xi) = \frac{h}{\sqrt{\frac{gh}{2}}}\chi\left(\frac{\xi - u}{\sqrt{\frac{gh}{2}}}\right) \tag{4.16}$$

where the function χ is a kernel with the following properties:

$$\int_R \chi(\omega)d\omega = 1$$

and:

$$\int_R \omega^2\chi(\omega)d\omega = 1$$

The Boltzmann equation obtained for a gas of N particles, when N tends to infinity, is:

$$\frac{\partial M}{\partial t} + \xi\frac{\partial M}{\partial x} - g\frac{\partial Z_f}{\partial x}\frac{\partial M}{\partial \xi} = Q(M,M) \tag{4.17}$$

Q is a "collision" term, which verifies $\int_R Q d\xi = 0$ and $\int_R \xi Q d\xi = 0$. Ultimately, it will not be necessary to calculate this unknown term. The main point is that the Saint-Venant equations are contained in Equation 4.17, called the "microscopic equation". This is illustrated by the following equations:

$$h = \int_R M(h, \xi - u)d\xi \tag{4.18}$$

$$hu = \int_R \xi M(h, \xi - u) d\xi \tag{4.19}$$

$$hu^2 + \frac{gh^2}{2} = \int_R \xi^2 M(h, \xi - u) d\xi \tag{4.20}$$

The unknowns of the Saint-Venant equations are hence replaced by the single function M, and are found on the basis of its moments. Moreover, the equation of M is linear (but M is a non-linear function of h and u). The choice of the function χ distinguishes different variants of the method. Perthame and Simeoni [155] opt for a function χ such that the function M minimizes the energy provided that its zero- and first-order moments restore h and hu. The entropy condition is thus satisfied. On a flat bottom, the Saint-Venant system is reduced to the resolution of an advection equation with source term, Equation 4.17. This equation is resolved in finite volumes. The direction of the current is taken into account in the calculation of flux terms and this is adapted to each value of velocity ξ. The macroscopic value of fluxes is then found by summing over ξ. The schemes thus conceived ensure conservation of mass and positivity of the water depth. This property is extended to the tracer equations. The condition of stability has the form:

$$\frac{\max(u, \sqrt{2gh})\Delta t}{\Delta x} \leq 1 \tag{4.21}$$

4.1.5 Finite element schemes

The classic Galerkin method, using centred derivatives even for transport terms, opposes the hyperbolic nature of equations as it does not distinguish between the upstream and downstream directions of the flow. Other specific techniques were designed to cope with this difficulty. The CDG (Characteristic–Dissipative–Galerkin, see [99]) method is based on the SUPG method which will be used and described in Section 6.2.

The Taylor–Galerkin [43] method is an explicit formulation identical to the Lax–Wendroff family of finite difference schemes. The unknowns are developed in series in relation to time. It can be shown that the method introduces an artificial diffusion to equations dependent on the time step (see [170]).

The least squares method [21] is an implicit scheme where the residual is minimized by least squares.

Hicks and Steffler [100] give a Fourier analysis of the three methods just mentioned, as well as a comparison on the test case of the Stoker solution (see Section 4.13.4). The article establishes the superiority of the CDG method whose numerical diffusion is less than Taylor–Galerkin. The least squares method is considered unacceptable for unsteady flows and gives rise to difficulties for supercritical flow.

4.2 Overall view of the Telemac-2D algorithm

An overview of the methods used will help the reader to get a general idea of the resolution scheme:

- The initial set of equations is the non-conservative Saint-Venant equations in the depth–velocity form. If the method of characteristics is chosen, these equations are solved in two stages using the fractional steps method.

- In the first step, advection terms which correspond to the transport of physical quantities u, v and, where applicable, h, k, ε and T. The hyperbolic nature of equations justifies the use of the method of characteristics, relevant to it.

- In the second step the remaining equations are considered: propagation, diffusion and source terms, plus advection terms when the method of characteristics cannot be used. This stage is resolved by the finite element technique; time discretization allows the elimination of non-linear terms from the equations. Variational formulation and discretization in space and time then transform continuous equations into a discrete linear system whose unknowns are the values of the physical quantities h, u, v at the nodes of the mesh. This system is then resolved by an iterative method of the conjugate gradient type.

4.3 Fractional steps method

This method is only applied when the method of characteristics is used.

As for the computation domain, time is discretized and the unknowns h, u, v, etc., are only determined at instants $t^0 + \Delta T$, $t^0 + 2\,\Delta T$, $t^0 + 3\,\Delta T$, and so on. Denoting $t^n = t^0 + n\,\Delta T$, the derivative in time of a function f is discretized as:

$$\frac{\partial f}{\partial t} = \frac{f^{n+1} - f^n}{\Delta t}$$

where f^{n+1} is the function f at time t^{n+1} (new time step), and f^n is the function f at time t^n (old time step). This is a first-order approximation, except if one considers that the derivative is expressed at time $t^n + \frac{\Delta t}{2}$, in which case we obtain second-order accuracy. For this reason we try to discretize the other terms of the equations at the same time. This is not always possible, particularly for advection, or in case of non-linearity.

Note: this presentation of derivatives in time is a "finite difference" approach. In finite elements, the difference $f^{n+1} - f^n$ could also be interpreted as the result of the integration by parts of a term $\int_{t^n}^{t^{n+1}} \frac{\partial f}{\partial t} dt$ which would arise from a variational formulation including time.

The general principle of resolution is as follows.

From an initial solution f^0 given at time t^0, the solution f^n is built with successive iterations (or time steps), first by finding f^1, then f^2, etc.

Within a time step, and in order to find f^{n+1} from f^n, the fractional steps method computes intermediate values by considering only a part of the terms of the equations. In our case, we shall solve:

$$\frac{\widetilde{f} - f^n}{\Delta t} + \text{advection terms} = 0 \tag{4.22}$$

then:

$$\frac{f^{n+1} - \widetilde{f}}{\Delta t} + \text{other terms} = 0 \tag{4.23}$$

where f represents h, u, v, k , ε or T.

The sum of these two equations restores the initial set; however, approximation in time will be of the first order as some terms (advection) are not discretized at time $t^n + \frac{\Delta t}{2}$.

The inconvenience of time order is compensated by the fact that each of these terms can be treated with the technique most appropriate to it.

Remark: in fact, nothing prevents us from getting second-order in time with the method of characteristics; for example, by computing two stages of advection with two different time steps (Δt and $2\Delta t$) and an advection operator equal to $2\overrightarrow{u}^n - \overrightarrow{u}^{n-1}$. The results of the two advections are denoted $\widetilde{f}^1$ and $\widetilde{f}^2$. The second order is obtained by replacing $\frac{f^{n+1}-\widetilde{f}}{\Delta t}$ by:

$$\frac{\frac{3}{2}f^{n+1} - 2\widetilde{f}^1 + \frac{1}{2}\widetilde{f}^2}{\Delta t}$$

However, this solution is quite awkward.

4.4 Advection stage using the method of characteristics

The whole or part of the following equations is solved (depending on the choice of variables to be treated with the method of characteristics), between time t^n and time t^{n+1}:

$$\frac{\widetilde{h} - h^n}{\Delta t} + \overrightarrow{u}.\overrightarrow{grad}(h) = 0 \tag{4.24}$$

$$\frac{\widetilde{u} - u^n}{\Delta t} + \overrightarrow{u}.\overrightarrow{grad}(u) = 0 \tag{4.25}$$

$$\frac{\widetilde{v} - v^n}{\Delta t} + \overrightarrow{u}.\overrightarrow{grad}(v) = 0 \tag{4.26}$$

$$\frac{\widetilde{T} - T^n}{\Delta t} + \overrightarrow{u}.\overrightarrow{grad}(T) = 0 \tag{4.27}$$

$$\frac{\widetilde{k} - k^n}{\Delta t} + \overrightarrow{u}.\overrightarrow{grad}(k) = 0 \tag{4.28}$$

$$\frac{\widetilde{\varepsilon} - \varepsilon^n}{\Delta t} + \overrightarrow{u}.\overrightarrow{grad}(\varepsilon) = 0 \tag{4.29}$$

Each of these equations simply means that the total derivative of functions h, u, v, etc., is zero along some trajectories called characteristics. The quantities transported remain unchanged along these characteristics. The calculation of characteristic curves is detailed in Chapter 6. Special attention should be given to the final remark regarding the treatment of the terms which normally appear on the right-hand side of equations.

The presence of h in the above list may appear surprising. This is because we can solve the continuity equation in two fractional steps, by splitting the term $div(h\overrightarrow{u})$ into $\overrightarrow{u}\,.\overrightarrow{grad}(h) + h\,div(\overrightarrow{u})$, where the first term can be interpreted as a transport of h.

4.5 Propagation, diffusion, source terms

When the method of characteristics is not used, this step resolves all the equations and hence also takes transport terms into account.

4.5.1 Time discretization

As explained earlier for the fractional steps method, the derivative in time of a function f takes the form:

$$\frac{\partial f}{\partial t} = \frac{f^{n+1} - f^n}{\Delta t} \tag{4.30}$$

Case of a prior advection stage

In this case we must solve:

$$\frac{h^{n+1} - \widetilde{h}}{\Delta t} + h\,div(\overrightarrow{u}) = Sce \tag{4.31}$$

$$\frac{u^{n+1} - \widetilde{u}}{\Delta t} = -g\frac{\partial Z_s}{\partial x} + F_x + \frac{1}{h}div(h\nu_e\overrightarrow{grad}(u)) \tag{4.32}$$

$$\frac{v^{n+1} - \widetilde{v}}{\Delta t} = -g\frac{\partial Z_s}{\partial y} + F_y + \frac{1}{h}div(h\nu_e\overrightarrow{grad}(v)) \tag{4.33}$$

where $\widetilde{h}$, $\widetilde{u}$ and $\widetilde{v}$ are the result of advection of h, u and v.

Strictly speaking, the terms on the right-hand side should come from integration in time along the characteristic curves (see the remark in Section 6.1).

Case without prior advection stage

$$\frac{h^{n+1} - h^n}{\Delta t} + \overrightarrow{u}\,.\overrightarrow{grad}(h) + h\,div(\overrightarrow{u}) = Sce \tag{4.34}$$

$$\frac{u^{n+1} - u^n}{\Delta t} + \overrightarrow{u}\,.\,\overrightarrow{grad}(u) = -g\frac{\partial Z_s}{\partial x} + F_x + \frac{1}{h}div(h\nu_e\overrightarrow{grad}(u)) \tag{4.35}$$

$$\frac{v^{n+1} - v^n}{\Delta t} + \overrightarrow{u}\,.\overrightarrow{grad}(v) = -g\frac{\partial Z_s}{\partial y} + F_y + \frac{1}{h}div(h\nu_e\overrightarrow{grad}(v)) \tag{4.36}$$

According to the choice of numerical schemes on the different variables, we have, in fact, an amalgam of the two earlier cases. At this point, it is not yet specified if the terms h, u, v subsisting in these equations are variables at time t^n or t^{n+1}. Strictly speaking, if we have in mind the second-order accuracy in time, one solution would be to write $h = (h^n + h^{n+1})/2$ and analogous formulae for u and v. This time discretization, called semi-implicitation, should be applied carefully. Actually, Fourier analysis of the semi-implicit scheme would give an amplification coefficient equal to 1. Such schemes, in principle perfect, tend to diverge in practice. This is because the Fourier analysis does not consider either non-linearity or boundary conditions. A safety margin is provided by writing for function f:

$$f = \theta f^{n+1} + (1-\theta) f^n$$

with $\theta > 0.5$ but very close to 0.5.

The name "semi-implicit" is retained even if θ is greater than 0.5.

Sub-iterations

This semi-implicitation does not solve all problems. Actually, the non-linear term $h\,div(\vec{u})$ brings out a product of unknowns h^{n+1} and $\vec{u}^{n+1}$. Without any other treatment, we would get, at the end of the calculation, a non-linear system that we would not know how to solve. So, the following formula is adopted:

$$h\,div(\vec{u}) = h_{prop} div(\theta_u \vec{u}^{n+1} + (1-\theta_u)\vec{u}^n) \tag{4.37}$$

where θ_u is called the "relaxation coefficient for velocity".

h_{prop} is called the "height of propagation" (this would be used to calculate the celerity of propagation of long waves). To get a linear problem, it should be equal to h^n, but we can take a value closer to $(h^n + h^{n+1})/2$ by solving in succession the same time step (these time steps are called sub-iterations) with the formula:

$$h_{prop} = \theta_h h'^{n+1} + (1-\theta_h) h^n \tag{4.38}$$

θ_h is the relaxation coefficient for water depth.

h'^{n+1} is the current estimation of h^{n+1}. It is equal to h^n, without sub-iteration, but with sub-iterations it is equal to h^{n+1}, the value resulting from the preceding sub-iteration.

Similarly, when advection terms are not processed by the method of characteristics, we write $\vec{u}\,.\,\overrightarrow{grad}(f)$ as:

$$\vec{u}\,.\,\overrightarrow{grad}(f) = \vec{u}_{conv}.\,\overrightarrow{grad}(\theta_f f^{n+1} + (1-\theta_f) f^n) \tag{4.39}$$

where $\vec{u}_{conv}$ is the advection field equal to $\theta_u \vec{u}'^{n+1} + (1-\theta_u)\vec{u}^n$, and $\vec{u}'^{n+1}$ is itself an approximation of $\vec{u}^{n+1}$ resulting from sub-iterations.

The relaxation coefficients are, in principle, always greater than 0.5 (generally 0.55 or 0.6 give good results).

Diffusion term

Semi-implicitation of the diffusion terms gives the equation (example of u):

$$\frac{1}{h} div(h\nu_e \overrightarrow{grad}(\theta_u^d u^{n+1} + (1-\theta_u^d)u^n)) \tag{4.40}$$

To gain more flexibility in the implementation of the different numerical schemes, a specific relaxation coefficient θ_u^d is used here for diffusion, which can be different from θ_u (e.g. in the case of the pseudo wave equation).

Increment formulation for water depth

The function Z_s (free surface) should be expressed as a function of the bottom elevation and water depth h. The semi-implicitation of h gives the following equation:

$$-g\,\overrightarrow{grad}(Z_s) = -g\,\overrightarrow{grad}(\theta_h h^{n+1} + (1-\theta_h)h^n + Z_f) \tag{4.41}$$

that is to say, also:

$$-g\,\overrightarrow{grad}(Z_s) = -g\theta_h\,\overrightarrow{grad}(h^{n+1} - h^n) - g\,\overrightarrow{grad}(Z_s^n) \tag{4.42}$$

where $Z_s^n = h^n + Z_f$.

Considering this expression and the continuity equation, it is more elegant to replace the unknown h^{n+1} by the increment $h^{n+1} - h^n$. In fact, in the case of a lake at rest with an irregular bottom, the term $-g\,\overrightarrow{grad}(Z_s^n)$, which reappears on the right-hand side of the final linear system, will be zero, as well as the increment $h^{n+1} - h^n$, which is an unknown of the system. Free surface elevation, in the case of a lake at rest with an irregular bottom, is thus naturally constant. This would not be the case if the bottom gradient and the water depth gradient were treated separately. Subsequently, the increment of h will be denoted as δh. Knowing δh, it is then very easy to find h^{n+1}.

It would be less natural to introduce the increments of u and v (in particular, they would increase the right-hand side of the continuity equation instead of decreasing it).

Specific treatment of the free surface gradient in the presence of uncovered beds is detailed in Section 4.11.

Note: when porosity Θ is added to the Saint-Venant equations, the term $\frac{\partial(\Theta h)}{\partial t}$ will appear when discretizing the derivative $\frac{\Theta^{n+1}h^{n+1} - \Theta^n h^n}{\Delta t}$. This term is then written as:

$$\frac{\Theta^{n+1}(h^{n+1} - h^n)}{\Delta t} + \frac{\Theta^{n+1}h^n - \Theta^n h^n}{\Delta t}$$

where the unknown $h^{n+1} - h^n$ reappears in the first term (Θ^{n+1} is assumed to be known for solving the equations between t^n and t^{n+1}). The second term is entirely known and is added to the right-hand side of the equation.

Time discretization of friction terms

Friction terms are non-linear (except of course with the linear law). They have to be treated in an implicit way if we want to avoid undesirable numerical effects, like a totally artificial inversion of velocities under the effect of friction. In order to conserve a linear form, $u\sqrt{u^2+v^2}$ and $v\sqrt{u^2+v^2}$ are discretized as follows:

$$u\sqrt{u^2+v^2} \approx u^{n+1}\sqrt{(u^n)^2+(v^n)^2} \tag{4.43}$$

$$v\sqrt{u^2+v^2} \approx v^{n+1}\sqrt{(u^n)^2+(v^n)^2} \tag{4.44}$$

We thus obtain an entirely implicit treatment of friction terms corresponding to a somewhat semi-implicitation in the form $u^2 = u^n u^{n+1}$. An evident imperfection (but not very disturbing in practice) is that a time step with initial conditions of zero velocity is not subject to friction.

4.5.2 Space discretization

We look for function values at the discretization points, by projecting these functions onto the bases:

$$f = \sum_{i=1}^{n} f^i \Psi_i$$

n here is the number of discretization points, f^i is the value of the function f at point i and Ψ_i the basis function associated to the same point. Each of the bases is thus associated to a degree of freedom. As we have seen in Chapter 3, Ψ_i is a function whose value is 1 at this point and 0 at all the others.

Within a given element, a function is interpolated from the values at the discretization nodes.

Interpolation functions are simple only on a reference element. Each actual element is the image of this reference element by simple geometric transformation (there is one for each element). See Section 7.1 for a description of basis functions.

Important note: to avoid a possible resurgence of the inf–sup condition (see Chapter 3) spatial discretization is not always the same for variables h, u and v. Two sets of basis functions are therefore necessary in variational formulation, presented next.

4.5.3 Variational formulation

A variational formulation was established in Goutal's thesis [58], with the variables u and h. The water depth and the velocity should belong to $H_0^1(\Omega)$ and $L^2(\Omega)$ respectively. $L^2(\Omega)$ is the set of square-integrable functions defined on the domain and $H_0^1(\Omega)$ is the subset of $L^2(\Omega)$ consisting of the functions whose derivatives are square integrable and zero along the boundary of Ω. Demonstrations established in reference [58] will not be repeated here. They suppose Dirichlet-type boundary conditions for velocity u. They should be extended to other types of boundary conditions, such as the Dirichlet condition for water depth h, and also to the case of impermeability boundary condition in a weak form, which will be examined shortly.

Convergence of our algorithm with low Reynolds number has also been established by Bernardi and Pironneau [13].

In practice, the variational formulation consists of choosing a set of functions, called "test functions", denoted φ_j and each associated to a point of discretization (degree of freedom of number j), then replacing an equation with the form $E = 0$ by $\int_\Omega E\varphi_j d\Omega = 0$.

In this chapter we assume that the test functions are different from the basis functions and that discretization of the water depth can be different from that of velocity. Hence, we denote:

Ψ_i^h: basis at point i to express the water depth.

φ_i^h: test function at point i for the continuity equation.

Ψ_i^u: basis at point i to express the components of velocity.

φ_i^u: test function at point i for the two momentum equations.

The number of bases related to water depth is denoted as nph.

The number of bases related to velocity is denoted as npu.

npu and nph are not necessarily equal (the case of a linear water depth and a quadratic velocity, for example). As the existence of a prior step of advection only suppresses terms, we shall deal with the general case where advection terms still appear. Put simply, these last terms are treated like the others, which would correspond to centred schemes of poor quality. What is really done is explained in detail in Chapter 6. Hence, the initial equations, after time discretization, are:

Continuity:

$$\frac{h^{n+1} - h^n}{\Delta t} + \vec{u}_{conv} \,.\, \overrightarrow{grad}(h) + h_{prop} div(\vec{u}) = Sce \tag{4.45}$$

Momentum:

$$\begin{aligned} &\frac{u^{n+1} - u^n}{\Delta t} + \vec{u}_{conv} \,.\, \overrightarrow{grad}(u) \\ = \; & -\theta_h \, g \, \frac{\partial(h^{n+1} - h^n)}{\partial x} - g\frac{\partial Z_s^n}{\partial x} + F_x + \frac{1}{h} div(h\nu_e \overrightarrow{grad}(u)) \end{aligned} \tag{4.46}$$

$$\begin{aligned} &\frac{v^{n+1} - v^n}{\Delta t} + \vec{u}_{conv} \,.\, \overrightarrow{grad}(v) \\ = \; & -\theta_h \, g \, \frac{\partial(h^{n+1} - h^n)}{\partial y} - g\frac{\partial Z_s^n}{\partial y} + F_y + \frac{1}{h} div(h\nu_e \overrightarrow{grad}(v)) \end{aligned} \tag{4.47}$$

When h, u or v are mentioned without a time index n or $n+1$, these are semi-implicit values which have not been developed for the sake of conciseness. For the definition of $\vec{u}_{conv}$ and h_{prop} see Section 4.5.1.

On the basis of these equations, the operations are separated into two stages: writing the integral on the domain with multiplication by each test function, then projection of the unknown functions onto the bases.

Integration on the domain and multiplication by the test functions

Multiplication by the test functions and integration on the domain give the following equations:

$$\int_\Omega \left(\frac{h^{n+1}-h^n}{\Delta t}\right) \varphi_i^h \, d\Omega + \int_\Omega (\vec{u}_{conv} \cdot \overrightarrow{grad}(h)) \, \varphi_i^h \, d\Omega + \int_\Omega h_{prop} div(\vec{u}) \, \varphi_i^h \, d\Omega$$

$$= \int_\Omega Sce \varphi_i^h \, d\Omega \tag{4.48}$$

for every degree of freedom of water depth denoted i, between 1 and nph, and:

$$\int_\Omega \left(\frac{u^{n+1}-u^n}{\Delta t}\right) \varphi_i^u \, d\Omega + \int_\Omega \vec{u}_{conv} \cdot \overrightarrow{grad}(u) \varphi_i^u \, d\Omega$$

$$= - \int_\Omega \theta_h \, g \, \frac{\partial (h^{n+1}-h^n)}{\partial x} \varphi_i^u \, d\Omega - \int_\Omega g \frac{\partial Z_s^n}{\partial x} \varphi_i^u \, d\Omega$$

$$+ \int_\Omega F_x \varphi_i^u \, d\Omega + \int_\Omega \frac{1}{h} div(h \nu_e \overrightarrow{grad}(u)) \varphi_i^u \, d\Omega \tag{4.49}$$

$$\int_\Omega \left(\frac{v^{n+1}-v^n}{\Delta t}\right) \varphi_i^u \, d\Omega + \int_\Omega \vec{u}_{conv} \cdot \overrightarrow{grad}(v) \varphi_i^u \, d\Omega$$

$$= - \int_\Omega \theta_h \, g \, \frac{\partial (h^{n+1}-h^n)}{\partial y} \varphi_i^u \, d\Omega - \int_\Omega g \frac{\partial Z_s^n}{\partial y} \varphi_i^u \, d\Omega$$

$$+ \int_\Omega F_y \varphi_i^u \, d\Omega + \int_\Omega \frac{1}{h} div(h \nu_e \overrightarrow{grad}(v)) \varphi_i^u \, d\Omega \tag{4.50}$$

for every i degree of freedom of velocity, between 1 and npu.

In order to make the impermeability boundary condition appear in "weak form", the term:

$$\int_\Omega h_{prop} div(\vec{u}) \, \varphi_i^h \, d\Omega$$

of the continuity equation is integrated by parts to give:

$$\int_\Omega h_{prop} div(\vec{u}) \, \varphi_i^h \, d\Omega = \int_\Gamma h_{prop} \, \vec{u} \, . \, \vec{n} \, \varphi_i^h d\Gamma - \int_\Omega \vec{u} \, . \, \overrightarrow{grad}(h_{prop} \, \varphi_i^h) \, d\Omega \tag{4.51}$$

With the classic resolution methods of the Saint-Venant or Navier–Stokes equations or in the formulation by Goutal (see [58]), impermeability is generally treated in a "strong form" due to Dirichlet conditions in the conservation of momentum equation. In this case the conservation of momentum equation is no longer treated along the boundaries.

As a continuation of variational formulation, the diffusion terms like:

$$\int_\Omega \frac{1}{h} div(h\nu_e \overrightarrow{grad}(u))\, \varphi_i^u \, d\Omega$$

are simplified and treated as:

$$\int_\Omega div(\nu_e \overrightarrow{grad}(u))\, \varphi_i^u \, d\Omega$$

this term also should be integrated by parts as it contains second derivatives not relevant to the spaces chosen for our functions. Boundary terms also appear:

$$\int_\Omega div(\nu_e \overrightarrow{grad}(u))\, \varphi_i^u \, d\Omega$$

$$= \int_\Gamma \varphi_i^u \nu_e \overrightarrow{grad}(u) \,.\, \vec{n} \, d\Gamma - \int_\Omega \nu_e \overrightarrow{grad}(u) . \overrightarrow{grad}(\varphi_i^u) \, d\Omega \tag{4.52}$$

The boundary terms will provide the boundary conditions for friction on the walls. Actually, $\nu_e \overrightarrow{grad}(u) \,.\, \vec{n}$ is a shear stress.

Note: a stricter treatment of diffusion with the term $1/h$ gives rise to the problem of representing $1/h$ in the space of approximation. One solution consists of writing:

$$\int_\Omega \frac{1}{h} div(h\nu_e \overrightarrow{grad}(u))\, \varphi_i^u \, d\Omega \approx \frac{1}{h_i} \int_\Omega div(h\nu_e \overrightarrow{grad}(u))\, \varphi_i^u \, d\Omega \tag{4.53}$$

which allows a satisfactory demonstration of conservation of momentum (see Section 4.10.1 for a similar demonstration for tracer mass). This solution also brings out problems of division by zero in the presence of uncovered beds.

Projection of functions onto the bases

As indicated earlier, we now write:

$$h = \sum_{j=1}^{nph} h_j \Psi_j^h, \quad u = \sum_{j=1}^{npu} u_j \Psi_j^u, \quad v = \sum_{j=1}^{npu} v_j \Psi_j^u, \text{ etc.}$$

After this projection, the equations become:
Continuity:

$$\sum_{j=1}^{nph} \left(\frac{h_j^{n+1} - h_j^n}{\Delta t} \right) \int_\Omega \Psi_j^h \, \varphi_i^h \, d\Omega + \sum_{j=1}^{nph} h_j \int_\Omega \vec{u}_{conv} \,.\, \overrightarrow{grad}(\Psi_j^h)\, \varphi_i^h \, d\Omega$$

$$+ \int_\Gamma h_{prop} \vec{u} \,.\, \vec{n} \, \varphi_i^h d\Gamma - \sum_{j=1}^{npu} \vec{u}_j \int_\Omega \Psi_j^u \, \overrightarrow{grad}(h_{prop}\, \varphi_i^h) \, d\Omega$$

$$= \sum_{j=1}^{nph} Sce_j \int_\Omega \Psi_j^h \varphi_i^h \, d\Omega \tag{4.54}$$

Momentum:

$$\sum_{j=1}^{npu}\left(\frac{u_j^{n+1}-u_j^n}{\Delta t}\right)\int_\Omega \Psi_j^u\varphi_i^u\,d\Omega+\sum_{j=1}^{npu}u_j\int_\Omega \vec{u}_{conv}\,.\,\overrightarrow{grad}(\Psi_j^u)\varphi_i^u\,d\Omega$$

$$=-\sum_{j=1}^{nph}(h_j^{n+1}-h_j^n)\int_\Omega \theta_h\,g\,\frac{\partial\Psi_j^h}{\partial x}\varphi_i^u\,d\Omega-\int_\Omega g\frac{\partial Z_s^n}{\partial x}\varphi_i^u\,d\Omega$$

$$+\int_\Omega F_x\varphi_i^u\,d\Omega+\int_\Gamma \varphi_i^u\nu_e\overrightarrow{grad}(u)\,.\,\vec{n}\,d\Gamma-\sum_{j=1}^{npu}u_j\int_\Omega \nu_e\overrightarrow{grad}(\Psi_j^u)\,.\,\overrightarrow{grad}(\varphi_i^u)\,d\Omega \quad (4.55)$$

$$\sum_{j=1}^{npu}\left(\frac{v_j^{n+1}-v_j^n}{\Delta t}\right)\int_\Omega \Psi_j^u\varphi_i^u\,d\Omega+\sum_{j=1}^{npu}v_j\int_\Omega \vec{u}_{conv}\,.\,\overrightarrow{grad}(\Psi_j^u)\varphi_i^u\,d\Omega$$

$$=-\sum_{j=1}^{nph}(h_j^{n+1}-h_j^n)\int_\Omega \theta_h\,g\,\frac{\partial\Psi_j^h}{\partial y}\varphi_i^u\,d\Omega-\int_\Omega g\frac{\partial Z_s^n}{\partial y}\varphi_i^u\,d\Omega$$

$$+\int_\Omega F_y\varphi_i^u\,d\Omega+\int_\Gamma \varphi_i^u\nu_e\overrightarrow{grad}(v)\,.\,\vec{n}\,d\Gamma-\sum_{j=1}^{npu}v_j\int_\Omega \nu_e\overrightarrow{grad}(\Psi_j^u)\,.\overrightarrow{grad}(\varphi_i^u)\,d\Omega \quad (4.56)$$

Before changing these equations to a linear system, the boundary integrals obtained are described more precisely.

4.5.4 Natural boundary conditions

The natural boundary conditions appear in the preceding equations as boundary integrals.

Mass conservation

In the mass conservation equation, these boundary integrals are:

$$\int_\Gamma h_{prop}\,\vec{u}\,.\,\vec{n}\,\varphi_i^h d\Gamma$$

$$=\theta_u\int_\Gamma h_{prop}\,\vec{u}^{\,n+1}.\,\vec{n}\,\varphi_i^h d\Gamma+(1-\theta_u)\int_\Gamma h_{prop}\,\vec{u}^{\,n}.\,\vec{n}\,\varphi_i^h d\Gamma \quad (4.57)$$

These terms are expressions of flux through walls. Two cases appear:

- In the case of a rigid wall, they are simply cancelled. This method is referred to as “impermeability boundary condition in weak form”.
- In the case of a liquid boundary, the integrals have to be calculated. A problem appears for the term:

$$\theta_u \int_\Gamma h_{prop} \overrightarrow{u}^{n+1} . \overrightarrow{n} \varphi_i^h d\Gamma$$

which contains the unknown $\overrightarrow{u}^{n+1}$.

Tests of treatment of this term with $\overrightarrow{u}^{n+1}$ being considered as an unknown are still not conclusive (the case of free output where parasitic reflections occur); $\overrightarrow{u}^{n+1}$ is thus replaced by $\overrightarrow{u}^n$ except if the components of velocity are given at time t^{n+1} (point with Dirichlet-type condition for velocities).

The impermeability boundary condition in weak form was described for the first time in 1991 in refereence [73]. Westerink *et al.* ([190] and [115]) used it to solve the Saint-Venant equations in a wave equation form. In this framework they showed that this condition suppresses parasitic modes observed earlier. Plate 6 represents von Karman eddies, namely vortex streets behind bridge piers. Details of the velocity field around the upper pier are given at the bottom of the plate. It results from a combination of a slip condition and an impermeability condition in weak form. The condition $\overrightarrow{u} . \overrightarrow{n} = 0$ is fulfilled without any subsequent correction.

Momentum

In each momentum equation a single boundary term appears:

$$\int_\Gamma \varphi_i^u \nu_e \overrightarrow{grad}(u) . \overrightarrow{n} \, d\Gamma$$

or:

$$\int_\Gamma \varphi_i^u \nu_e \overrightarrow{grad}(v) . \overrightarrow{n} \, d\Gamma$$

Now, by definition:

$$\overrightarrow{grad}(f) . \overrightarrow{n} = \frac{\partial f}{\partial n}$$

So the boundary conditions on friction are used (e.g. the logarithmic law) to calculate the boundary terms of the conservation of momentum equation, on the basis of a value of the shear stress $\nu_e \frac{\partial u_t}{\partial n}$. For a classic condition of friction we get, for example:

$$\nu_e \frac{\partial u_t}{\partial n} = -(u^*)^2$$

However, it is preferable to use an implicit treatment for friction, such as:

$$\nu_e \frac{\partial u_t}{\partial n} = a u_t$$

which offers more stability. This can be obtained numerically by choosing:

$$a = -\frac{(u^*)^2}{u_t^n}$$

where u_t^n is the tangential velocity of the previous time step.

The final form of the boundary term is hence written as:

$$\int_\Gamma \varphi_i^u \nu_e \overrightarrow{grad}(u) \,.\, \vec{n} \; d\Gamma = \int_\Gamma \varphi_i^u \, a \, u \, d\Gamma \tag{4.58}$$

As opposed to the continuity equation, this term is treated implicitly, so it will appear in the matrices and not in the right-hand terms (see the following subsection).

When such a friction condition is applied, the velocity on the solid boundaries is not zero, and u^* depends on the local flow and is *a priori* an unknown. In fact we have to assume that the velocity is not really taken at the boundary but at a small distance δ from the boundary, and that we have a logarithmic profile as given in formulae 2.269 and 2.271 in Section 2.4.3. For example, in a hydraulically rough flow we have:

$$u = \frac{u_*}{\kappa} \ln\left(\frac{33\,\delta}{k_s}\right) \tag{4.59}$$

which allows us to evaluate u_*:

$$u_* = \frac{\kappa u}{\ln(\frac{33\,\delta}{k_s})} \tag{4.60}$$

If u is taken at the boundary, δ is either an arbitrarily small distance or a fraction of the local mesh size, but in both cases it is a new parameter that can influence the results. Another solution to evaluate u_* consists of choosing for u the nearest point inside the domain, and for δ the distance between this point and the boundary. The assumption is thus that this inner point is still in the logarithmic profile. With a hydraulically smooth flow, an extra difficulty is that formula 2.269 giving u_* is implicit, so that an iterative process is necessary to get the friction velocity.

4.5.5 Sources and sinks

It has been seen that in this case the continuity equation is written as:

$$\frac{\partial h}{\partial t} + div(\vec{Q}) = Sce \tag{4.61}$$

without specifying the value of the source term Sce. By stating that the function Sce should only refer to the source point, whose number will be $isce$, we write $Sce = X\Psi_{isce}$, Ψ_{isce} being the basis of the point $isce$ and X an unknown coefficient. We aim at imposing an inflow Q_{sce} at the source point. In the absence of flux at the boundaries and by integrating the equation on the entire calculation domain, we get, with the usual discretization of the time derivative:

$$\int_\Omega \frac{(h^{n+1} - h^n)}{\Delta t} \, d\Omega = \int_\Omega X \Psi_{isce} \, d\Omega$$

But then, we should have:

$$\int_\Omega \frac{(h^{n+1} - h^n)}{\Delta t} \, d\Omega = Q_{sce}$$

which then imposes:

$$X = \frac{Q_{sce}}{\int_\Omega \Psi_{isce}\, d\Omega}$$

And we finally get:

$$Sce = \frac{Q_{sce}\Psi_{isce}}{\int_\Omega \Psi_{isce}\, d\Omega}$$

So, this concept of sources and sinks introduces, at the continuum level, terms which depend on the discretization of the domain. This is due to the initial hypothesis $Sce = X\Psi_{isce}$ which supposes that the source is not reduced to a single point but spread out on the domain of definition of the base Ψ_{isce}. Another solution consists of writing that $Sce = Q_{sce}\delta(isce)$, where $\delta(isce)$ is the Dirac function relative to the basis Ψ_{isce}, which is equivalent to directly adding the term Q_{sce} to the right-hand side of the continuity equation at the point $isce$, admitting that $\int_\Omega Sce\Psi_i\, d\Omega$ is equal to Q_{sce} if $i = isce$, and 0 otherwise.

4.5.6 Matrix form of the system

The set of equations elaborated in Section 4.5.3 only contains linear combinations of unknowns. Hence, they constitute a linear system that can have the form $AX = B$. In this system, the points with the Dirichlet condition (i.e. imposed values) are kept formally as unknowns (see Section 7.4.7). X is the vector of unknowns composed of three vectors with respective dimensions nph, npu and npu, and containing the values δh, u and v. These vectors will be called δH, U and V:

$$X = \begin{pmatrix} \delta H \\ U \\ V \end{pmatrix} \tag{4.62}$$

B is also composed of three vectors:

$$B = \begin{pmatrix} CV1 \\ CV2 \\ CV3 \end{pmatrix} \tag{4.63}$$

A is a matrix initially composed of nine square or rectangular matrices. Among these nine matrices some of them are zero due to the absence of coupling between the two components of velocity. This is the consequence of treating in weak form the impermeability condition at the walls. A is written as:

$$A = \begin{pmatrix} AM1 & BM1 & BM2 \\ CM1 & AM2 & 0 \\ CM2 & 0 & AM3 \end{pmatrix} \tag{4.64}$$

Before giving the expression of matrices composing A, a few others which appear in the linear system of Section 4.5.3 will be defined:

The mass matrices:

$$M^h_{ij} = \int_\Omega \Psi^h_j\, \varphi^h_i \tag{4.65}$$

$$M_{ij}^{u} = \int_{\Omega} \Psi_j^u \, \varphi_i^u \tag{4.66}$$

The advection matrices:

$$T_{x\ ij}^{h}(u_{conv}) = \int_{\Omega} u_{conv} \, \frac{\partial \Psi_j^h}{\partial x} \, \varphi_i^h \, d\Omega \tag{4.67}$$

$$T_{y\ ij}^{h}(v_{conv}) = \int_{\Omega} v_{conv} \, \frac{\partial \Psi_j^h}{\partial y} \, \varphi_i^h \, d\Omega \tag{4.68}$$

$$T_{x\ ij}^{u}(u_{conv}) = \int_{\Omega} u_{conv} \, \frac{\partial \Psi_j^u}{\partial x} \, \varphi_i^u \, d\Omega \tag{4.69}$$

$$T_{y\ ij}^{u}(v_{conv}) = \int_{\Omega} v_{conv} \, \frac{\partial \Psi_j^u}{\partial y} \, \varphi_i^u \, d\Omega \tag{4.70}$$

The diffusion matrix:

$$D_{ij}^{u} = \int_{\Omega} \nu_e \overrightarrow{grad}(\Psi_j^u) \, . \, \overrightarrow{grad}(\varphi_i^u) \, d\Omega \tag{4.71}$$

The “gradient” or “divergence” type matrices are as follows.
In the continuity equation:

$$B_{x\,ij}^{uh}(h_{prop}) = -\int_{\Omega} \Psi_j^u \, \frac{\partial}{\partial x}(h_{prop} \, \varphi_i^h) \, d\Omega \tag{4.72}$$

$$B_{y\,ij}^{uh}(h_{prop}) = -\int_{\Omega} \Psi_j^u \, \frac{\partial}{\partial y}(h_{prop} \, \varphi_i^h) \, d\Omega \tag{4.73}$$

In fact, the two preceding matrices are not truly classic gradient or divergence matrices due to the presence of the term h_{prop} in the integrals.

We also have in the momentum equation:

$$C_{x\,ij}^{uh} = \int_{\Omega} \frac{\partial \Psi_j^h}{\partial x} \varphi_i^u \, d\Omega \tag{4.74}$$

$$C_{y\,ij}^{uh} = \int_{\Omega} \frac{\partial \Psi_j^h}{\partial y} \varphi_i^u \, d\Omega \tag{4.75}$$

The boundary matrix:

$$G_{ij}^{u} = \int_{\Gamma} \varphi_i^u \nu_e \, a \, \varphi_i^u \Psi_j^u \, d\Gamma \tag{4.76}$$

The matrices of friction are as follows.

With the time discretization chosen, the value of friction terms is (example of component u):

$$-\int_{\Omega} \frac{1}{\cos(\alpha)} \frac{C_f}{2h^n} \, u^{n+1} \sqrt{(u^n)^2 + (v^n)^2} \varphi_i^u d\Omega \tag{4.77}$$

These terms are treated in the following simplified form:

$$-u_i^{n+1} \frac{1}{\cos(\alpha)} \frac{C_f}{2h_i^n} \sqrt{(u_i^n)^2 + (v_i^n)^2} \int_\Omega \varphi_i^u d\Omega \tag{4.78}$$

which gives:

$$F_{ii}^u = -\frac{1}{\cos(\alpha)} \frac{C_f}{2h_i^n} \sqrt{(u_i^n)^2 + (v_i^n)^2} \int_\Omega \varphi_i^u d\Omega \tag{4.79}$$

This approximation (which is a mass lumping) has the advantage of giving diagonal matrices. An entirely strict treatment is not possible on account of square roots, for example. However, more complete matrices could be kept (u^{n+1} should not be removed from the integral as u_i^{n+1}) but tests show that the results do not change significantly. On the other hand, the conditioning of the system degrades slightly. We notice that the equations in u and in v give the same matrix. If C_f is variable in space, to calculate F_{ii}^u the value of C_f is taken at point i.

Blocks of matrices in matrix A:

The matrices contained in A are then:

$$AM1 = \frac{M^h}{\Delta t} + \theta_h T^h_{x\ ij}(u_{conv}) + \theta_h T^h_{y\ ij}(v_{conv}) \tag{4.80}$$

The matrices T^h disappear if one executes a prior fractional step with the method of characteristics.

The mass matrix can be "lumped", if necessary. This signifies that it is replaced by a diagonal D where each coefficient D_i is equal to the sum of the terms of the corresponding line of M. If r is the rank of the matrix, we thus get: $D_i = \sum_{j=1}^r M_{ij}$. Mass lumping can be partial (in which case, only a percentage of $\sum_{j=1}^r M_{ij}$ is carried over to the diagonal of M). The effect of mass lumping is to reinforce the diagonal of $AM1$ and hence improve the conditioning of the final linear system. On the other hand there is a slight fall in precision (we tend towards a formulation similar to finite differences, although all the properties of mass conservation remain unchanged). We go on with the matrices in A:

$$AM2 = AM3 = \frac{M^u}{\Delta t} + \theta_u T_x^u(u_{conv}) + \theta_u T_y^u(v_{conv}) + D^u - F^u - G^u \tag{4.81}$$

where the matrices T^u disappear if one executes a prior fractional step with the method of characteristics.

$$BM1 = \theta_u B_x^{uh}(h_{prop}) \quad BM2 = \theta_u B_y^{uh}(h_{prop}) \tag{4.82}$$

$$CM1 = g\theta_h C_x^{uh} \quad CM2 = g\theta_h C_y^{uh} \tag{4.83}$$

The right-hand sides in vector B are:

$$CV1 = (\theta_h - 1)(T_x^h + T_y^h)H^n + (\theta_u - 1)(B_x^{uh}U^n + B_y^{uh}V^n) + M^h SCE + TB1 \tag{4.84}$$

where $TB1$ represents the term $-\int_\Gamma h_{prop}\vec{u}.\vec{n}\varphi_i^h d\Gamma$.

$$CV2 = \frac{M^u}{\Delta t}U^n + (\theta_u - 1)(T_x^u + T_y^u)U^n - gC_x^{uh}Z_s^n + M^u F_x \tag{4.85}$$

$$CV3 = \frac{M^u}{\Delta t}V^n + (\theta_u - 1)(T_x^u + T_y^u)V^n - gC_y^{uh}Z_s^n + M^u F_y \tag{4.86}$$

Note: with the method of characteristics, we should replace:
in the expression for $CV1$: $(\theta_h - 1)(T_x^h + T_y^h)H^n$ by $\frac{M^h}{\Delta t}(\widetilde{H} - H^n)$
in the expression for $CV2$: $\frac{M^u}{\Delta t}U^n + (\theta_u - 1)(T_x^u + T_y^u)U^n$ by $\frac{M^u}{\Delta t}\widetilde{U}$
in the expression for $CV3$: $\frac{M^u}{\Delta t}V^n + (\theta_u - 1)(T_x^u + T_y^u)V^n$ by $\frac{M^u}{\Delta t}\widetilde{V}$
Remark: the majority of the matrices of the system are dependent on time.

Once the matrices and the right-hand sides of the system are built, the Dirichlet-type boundary conditions (imposed values) have to be taken into account and the linear system has to be resolved. These operations are independent of the type of equation to be solved and are described in Sections 7.4.7 and 7.5.

4.6 Radiation conditions

The existence of artificial boundaries in the calculation domains, e.g. in the open sea, often raises the question of boundary conditions and leads to ill-posed problems. In the open sea, generally only the free surface elevation is known. It is tempting to adopt a condition with imposed elevation and free velocity. When the current is directed inwards on these boundaries, the problem is ill-posed and the solution could diverge. The theory of characteristics, which would help to emphasize the required conditions, cannot be applied strictly in two dimensions. The characteristic curves highlighted in one dimension are transformed into cones (see [33]). The "radiation" boundary conditions use the theory of characteristics with a few assumptions. This is particularly so for the Thompson method [182]. In practice, the following approach is used:

- All the data available at the open boundaries (elevation and velocities) are provided and declared as imposed values.

- The local conditions at the boundaries are studied in the light of the theory of characteristics and the Riemann invariants, to determine which information is really necessary and which should be kept aside. The new elevations and velocities to be imposed are deduced. These are not necessarily given by the user. The Thompson method adapted to the Saint-Venant equations is given below. The uniqueness of the solutions thus obtained has been studied by Lich and Muat [127].

Let us start with the conservative form of the Saint-Venant equations using the notations of Section 4.1.3 and Equation 4.15 recalled here:

$$\frac{\partial F}{\partial t} + \frac{\partial G_x(F)}{\partial x} + \frac{\partial G_y(F)}{\partial y} = S(F)$$

Taking into account all the source terms, and the possibility of discharge sources within the domain, the right-hand side is:

$$S(F) = \begin{pmatrix} Sce \\ -gh\frac{\partial Z_s}{\partial x} + hF_x \\ -gh\frac{\partial Z_s}{\partial y} + hF_y \end{pmatrix}$$

We consider an open boundary with a local normal vector $\overrightarrow{n}$ and a local tangent vector $\overrightarrow{t}$. By a rotation, we introduce derivatives with respect to η and ζ, where η is the normal and ζ the tangent at the boundary:

$$\frac{\partial F}{\partial t} + \frac{\partial G_\eta^*(F)}{\partial \eta} + \frac{\partial G_\zeta^*(F)}{\partial \zeta} = S(F) \tag{4.87}$$

The determinant of the Jacobian matrix:

$$J = \begin{pmatrix} \frac{\partial \eta}{\partial x} & \frac{\partial \eta}{\partial y} \\ \frac{\partial \zeta}{\partial x} & \frac{\partial \zeta}{\partial y} \end{pmatrix}$$

is equal to 1. Thus we get:

$$G_\eta^* = \frac{\partial \eta}{\partial x}G_x + \frac{\partial \eta}{\partial y}G_y \text{ and } G_\zeta^* = \frac{\partial \zeta}{\partial x}G_x + \frac{\partial \zeta}{\partial y}G_y \tag{4.88}$$

The system 4.87 is then written as:

$$\frac{\partial F}{\partial t} + A\frac{\partial F}{\partial \eta} + B\frac{\partial F}{\partial \zeta} = S(F) \tag{4.89}$$

with:

$$A = \begin{pmatrix} 0 & \frac{\partial \eta}{\partial x} & \frac{\partial \eta}{\partial y} \\ \frac{\partial \eta}{\partial x}(c^2 - u^2) - \frac{\partial \eta}{\partial y}uv & 2u\frac{\partial \eta}{\partial x} + v\frac{\partial \eta}{\partial y} & u\frac{\partial \eta}{\partial y} \\ \frac{\partial \eta}{\partial y}(c^2 - u^2) - \frac{\partial \eta}{\partial x}uv & v\frac{\partial \eta}{\partial x} & u\frac{\partial \eta}{\partial x} + 2v\frac{\partial \eta}{\partial y} \end{pmatrix} \tag{4.90}$$

or even, by denoting u_η as the normal component of velocity and u_ζ the tangential component:

$$A = \begin{pmatrix} 0 & \frac{\partial \eta}{\partial x} & \frac{\partial \eta}{\partial y} \\ \frac{\partial \eta}{\partial x}c^2 - uu_\eta & u\frac{\partial \eta}{\partial x} + u_\eta & u\frac{\partial \eta}{\partial y} \\ \frac{\partial \eta}{\partial y}c^2 - vu_\eta & v\frac{\partial \eta}{\partial x} & u_\eta + v\frac{\partial \eta}{\partial y} \end{pmatrix} \tag{4.91}$$

Subsequently, we ignore the variations along the tangent at the boundary and try to solve the system:

$$\frac{\partial F}{\partial t} + A\frac{\partial F}{\partial \eta} = S_\eta(F) \tag{4.92}$$

where:

$$S_\eta(F) = \begin{pmatrix} Sce \\ -gh\frac{\partial \eta}{\partial x}\frac{\partial Z_s}{\partial \eta} + hF_x \\ -gh\frac{\partial \eta}{\partial y}\frac{\partial Z_s}{\partial \eta} + hF_y \end{pmatrix} \tag{4.93}$$

A is diagonalized as $A = L^{-1}\Lambda L$ with:

$$L = \begin{pmatrix} u_\zeta & \frac{\partial \eta}{\partial y} & -\frac{\partial \eta}{\partial x} \\ c - u_\eta & \frac{\partial \eta}{\partial x} & \frac{\partial \eta}{\partial y} \\ c + u_\eta & -\frac{\partial \eta}{\partial x} & -\frac{\partial \eta}{\partial y} \end{pmatrix}$$

and:

$$\Lambda = \begin{pmatrix} u_\eta & 0 & 0 \\ 0 & u_\eta + c & 0 \\ 0 & 0 & u_\eta - c \end{pmatrix}$$

By stating that $dW = LdF$, we then get back to the diagonalized system:

$$\frac{\partial W}{\partial t} + \Lambda \frac{\partial W}{\partial \eta} = LS_\eta \tag{4.94}$$

each of whose lines is a simple transport equation with source term. Thompson proposes to consider that L is constant in the vicinity of a boundary point, and to write $W = \overline{L}F$, where:

$$\overline{L} = \begin{pmatrix} \overline{u_\zeta} & \frac{\partial \eta}{\partial y} & -\frac{\partial \eta}{\partial x} \\ \overline{c} - \overline{u_\eta} & \frac{\partial \eta}{\partial x} & \frac{\partial \eta}{\partial y} \\ \overline{c} + \overline{u_\eta} & -\frac{\partial \eta}{\partial x} & -\frac{\partial \eta}{\partial y} \end{pmatrix} \tag{4.95}$$

the overbar values being considered as constant. The Riemann invariants of the vector W are thus:

- $W_1 = -h(u_\zeta - \overline{u_\zeta})$ (advection with velocity u_η).
- $W_2 = h(\overline{c} + u_\eta - \overline{u_\eta})$ (advection with velocity $u_\eta + c$).
- $W_3 = h(-\overline{c} + u_\eta - \overline{u_\eta})$ (advection with velocity $u_\eta - c$).

and to which can be added, if a tracer T also has to be considered:

- $W_4 = h(T - \overline{T})$ (advection with velocity u_η).

Pure advection is treated with the method of characteristics. Source terms, if any, are considered in a fractional step. Once the Riemann invariants are known, the primitive variables are restored by the following formulae:

$$h = \frac{W_1 - W_3}{2\overline{c}} \tag{4.96}$$

$$u = \frac{\frac{\partial \eta}{\partial y} W_1 + \frac{\partial \eta}{\partial x} W_2 - h\overline{c}\frac{\partial \eta}{\partial x}}{h} + \overline{u} \tag{4.97}$$

$$v = \frac{\frac{\partial \eta}{\partial y} W_2 - \frac{\partial \eta}{\partial x} W_1 - h\overline{c}\frac{\partial \eta}{\partial y}}{h} + \overline{v} \tag{4.98}$$

$$T = -\frac{W_4}{h} + \overline{T} \tag{4.99}$$

The ill-posed nature of a problem at the boundary appears with this method when moving along the characteristic curves in the upstream direction, if a characteristic leaves the domain without the corresponding information being given by the user. On the other hand, a redundant condition will simply be put aside. In principle, the method is not time consuming as moving along the characteristic in the upstream direction is required only for points situated on an open boundary. However, three (or four, with a tracer) characteristics per point have to be calculated, and, each time, the corresponding advection field prepared. This is done for the entire calculation domain. In fact, certain tide simulations are executed with Courant number values $\frac{c\Delta t}{\Delta x}$ around 50. This prohibits restriction to a calculation of the local advection field at the boundary.

Plate 7 shows an application of the Thompson radiation conditions for letting a circular wave, propagating in a square reservoir, leave the domain (see also Section 4.13.3). In this case no information was given at the boundaries. However, we observe that the wave goes outwards with only a slight reflection at the corners, due to the fact that tangent terms at the boundary were neglected in our approach.

4.7 Resolution of the Boussinesq equations

Very few models have been proposed for solving the Boussinesq equations in finite elements. The main difficulty is the higher order of derivatives in the term:

$$-\frac{H_0^2}{6}\overrightarrow{grad}\left(div\left(\frac{\partial \overrightarrow{u}}{\partial t}\right)\right) + \frac{H_0}{2}\overrightarrow{grad}\left(div\left(H_0\frac{\partial \overrightarrow{u}}{\partial t}\right)\right) \tag{4.100}$$

which is added to the Saint-Venant momentum equation. Do Carmo and Seabra-Santos [41] have proposed one of the first models. With an assumption of curl-free flow, these authors show that the supplementary term leads to the following right-hand terms in the conservation of momentum equations.

In the u equation:

$$\frac{H_0^2}{3}\left(\frac{\partial^3 u}{\partial x^2 \partial t} + \frac{\partial^3 u}{\partial y^2 \partial t}\right) + \frac{\partial F_u}{\partial t} \tag{4.101}$$

In the v equation:

$$\frac{H_0^2}{3}\left(\frac{\partial^3 v}{\partial x^2 \partial t} + \frac{\partial^3 v}{\partial y^2 \partial t}\right) + \frac{\partial F_v}{\partial t} \tag{4.102}$$

with:

$$F_u = \frac{H_0}{2}\left(u\frac{\partial^2 H_0}{\partial x^2} + \frac{\partial H_0}{\partial x}\frac{\partial v}{\partial y} + \frac{\partial H_0}{\partial y}\frac{\partial v}{\partial x} + v\frac{\partial^2 H_0}{\partial x \partial y} + 2\frac{\partial H_0}{\partial x}\frac{\partial u}{\partial x}\right) \tag{4.103}$$

$$F_v = \frac{H_0}{2}\left(v\frac{\partial^2 H_0}{\partial y^2} + \frac{\partial H_0}{\partial x}\frac{\partial u}{\partial y} + \frac{\partial H_0}{\partial y}\frac{\partial u}{\partial x} + u\frac{\partial^2 H_0}{\partial x \partial y} + 2\frac{\partial H_0}{\partial y}\frac{\partial v}{\partial y}\right) \tag{4.104}$$

The idea of the algorithm is then to add two auxiliary variables r and s, such as:

$$r = u - \frac{H_0^2}{3}\left(\frac{\partial^2 u}{\partial x^2} + \frac{\partial^2 u}{\partial y^2}\right) - F_u \tag{4.105}$$

$$s = v - \frac{H_0^2}{3}\left(\frac{\partial^2 v}{\partial x^2} + \frac{\partial^2 v}{\partial y^2}\right) - F_v \tag{4.106}$$

The Saint-Venant momentum equations take into account the Boussinesq term if the derivatives $\frac{\partial u}{\partial t}$ and $\frac{\partial v}{\partial t}$ are changed into $\frac{\partial r}{\partial t}$ and $\frac{\partial s}{\partial t}$. The new equations can be solved in three steps:

1) Calculation of h with the continuity equation.
2) Calculation of r and s with the momentum equation.
3) Deduction of u and v from r and s.

The most time-consuming stage is the calculation of u and v on the basis of r and s. The matrix conditioning after discretization of the equation:

$$u - \frac{H_0^2}{3}\left(\frac{\partial^2 u}{\partial x^2} + \frac{\partial^2 u}{\partial y^2}\right) - F_u = r \tag{4.107}$$

is quite poor. Such conditioning is common to all the resolution methods of the Boussinesq equations.

The earlier algorithm is satisfactory for reproducing the propagation of waves in a rectilinear canal. However, the assumption of curl free flow proved to be the cause of instabilities with complex flows. A more robust algorithm is obtained by directly adding the supplementary Boussinesq term to the algorithm elaborated for the Saint-Venant equations.

The discretization of the term 4.100 contains an implicit term:

$$\frac{1}{\Delta t}\frac{H_0^2}{6}\overrightarrow{grad}(div(\overrightarrow{u}^{n+1})) - \frac{1}{\Delta t}\frac{H_0}{2}\overrightarrow{grad}(div(H_0\overrightarrow{u}^{n+1})) \tag{4.108}$$

in the left-hand side of the momentum equation and an explicit term:

$$\frac{1}{\Delta t}\frac{H_0^2}{6}\overrightarrow{grad}(div(\overrightarrow{u}^{n})) - \frac{1}{\Delta t}\frac{H_0}{2}\overrightarrow{grad}(div(H_0\overrightarrow{u}^{n})) \tag{4.109}$$

in the right-hand side.

This second term is easily deduced from the matrices to be built for the left-hand term. In the variational formulation, this term becomes:

$$\int_\Omega \left(\frac{1}{\Delta t}\frac{H_0^2}{6}\overrightarrow{grad}(div(\overrightarrow{u}^{n+1}))\right)\Psi_i d\Omega - \int_\Omega \left(\frac{1}{\Delta t}\frac{H_0}{2}\overrightarrow{grad}(div(H_0\overrightarrow{u}^{n+1}))\right)\Psi_i d\Omega \tag{4.110}$$

or even, by integrating by parts (and ignoring the boundary terms!):

$$-\frac{1}{\Delta t}\int_\Omega div(\overrightarrow{u}^{n+1})\,\overrightarrow{grad}\left(\frac{H_0^2}{6}\Psi_i\right) d\Omega + \frac{1}{\Delta t}\int_\Omega div(H_0\overrightarrow{u}^{n+1})\,\overrightarrow{grad}\left(\frac{H_0}{2}\Psi_i\right) d\Omega \tag{4.111}$$

which finally gives:

$$\frac{1}{3\Delta t}\int_\Omega H_0^2\, div(\overrightarrow{u}^{n+1})\,\overrightarrow{grad}(\Psi_i)\,d\Omega$$

$$+\frac{1}{6\Delta t}\int_\Omega H_0\,\overrightarrow{grad}(H_0)\,div(\overrightarrow{u}^{n+1})\Psi_i\,d\Omega$$

$$+\frac{1}{2\Delta t}\int_\Omega H_0\,\overrightarrow{u}^{n+1}.\overrightarrow{grad}(H_0)\;\overrightarrow{grad}(\Psi_i)\,d\Omega$$

$$+\frac{1}{2\Delta t}\int_\Omega \overrightarrow{u}^{n+1}.\overrightarrow{grad}(H_0)\,\Psi_i\;\overrightarrow{grad}(H_0)\,d\Omega \tag{4.112}$$

This term is a vector with two components. If we project $\overrightarrow{u}^{n+1}$ onto the bases Ψ_j and consider the component along x and the component along y of the vector, we get, along x, a term:

$$A_{UU}\,U^{n+1} + A_{UV}V^{n+1}$$

and along y, a term:

$$A_{VU}\,U^{n+1} + A_{VV}V^{n+1}$$

where A_{UU} is, for example, the matrix:

$$A_{UU} = \frac{1}{3\Delta t}\int_\Omega H_0\,H_0\,\frac{\partial\Psi_j}{\partial x}\,\frac{\partial\Psi_i}{\partial x}\,d\Omega + \frac{1}{6\Delta t}\int_\Omega H_0\,\frac{\partial H_0}{\partial x}\,\frac{\partial\Psi_j}{\partial x}\,\Psi_i\,d\Omega$$

$$+\frac{1}{2\Delta t}\int_\Omega H_0\,\frac{\partial H_0}{\partial x}\,\Psi_j\,\frac{\partial\Psi_i}{\partial x}\,d\Omega + \frac{1}{2\Delta t}\int_\Omega \frac{\partial H_0}{\partial x}\,\frac{\partial H_0}{\partial x}\,\Psi_j\,\Psi_i\,d\Omega \tag{4.113}$$

So a whole class of matrices appears as:

$$\int_\Omega f_1(H_0)\,f_2(H_0)\;f_3(\Psi_j)\;f_4(\Psi_i)\,d\Omega$$

where the functions f_1 to f_4 can be the identity, the derivative with respect to x or the derivative with respect to y. A terminology is now proposed, where identity is represented by "0", the derivative with respect to x by "X" and the derivative with respect to y by "Y". So we denote M_{0X0Y} as the matrix:

$$\int_\Omega H_0\,\frac{\partial H_0}{\partial x}\,\Psi_j\,\frac{\partial\Psi_i}{\partial y}\,d\Omega$$

Let us point out interesting properties of these matrices, such as $M_{0X0Y} = M_{X00Y}$ and $M_{0XY0} = {}^tM_{0X0Y}$.

We can now write A_{UU}, A_{UV}, A_{VU} and A_{VV}, with a limited number of new matrices to be built:

$$A_{UU} = \frac{1}{\Delta t}\left(\frac{M_{0XX0}}{6} + \frac{M_{XX00}}{2} + \frac{{}^tM_{0XX0}}{2} + \frac{M_{00XX}}{3}\right) \tag{4.114}$$

$$A_{UV} = \frac{1}{\Delta t}\left(\frac{M_{0Y0X}}{2} + \frac{M_{XY00}}{2} + \frac{{}^tM_{00XY}}{3} + \frac{{}^tM_{0X0Y}}{6}\right) \tag{4.115}$$

$$A_{VU} = \frac{1}{\Delta t}\left(\frac{M_{0X0Y}}{2} + \frac{M_{00XY}}{3} + \frac{M_{XY00}}{2} + \frac{{}^tM_{0Y0X}}{6}\right) \tag{4.116}$$

$$A_{VV} = \frac{1}{\Delta t}\left(\frac{M_{0YY0}}{6} + \frac{M_{YY00}}{2} + \frac{{}^tM_{0YY0}}{2} + \frac{M_{00YY}}{3}\right) \tag{4.117}$$

The main extra cost is a coupling of u^{n+1} and v^{n+1}. The matrix:

$$A = \begin{pmatrix} AM1 & BM1 & BM2 \\ CM1 & AM2 & 0 \\ CM2 & 0 & AM3 \end{pmatrix}$$

becomes:

$$A' = \begin{pmatrix} AM1 & BM1 & BM2 \\ CM1 & AM2' & A_{UV} \\ CM2 & A_{VU} & AM3' \end{pmatrix}$$

with $AM2' = AM2 + A_{UU}$ and $AM3' = AM3 + A_{VV}$.

On the right-hand side of the equation for u^{n+1}, $A_{UU}U^n + A_{UV}V^n$ should be added.

On the right-hand side of the equation for v^{n+1}, $A_{VU}U^n + A_{VV}V^n$ should be added.

The new system to be solved has the same poor conditioning as the Helmholtz equations found in the first method, but the algorithm is much more robust.

4.8 Resolution of k–ε model equations in 2D

4.8.1 Advection step

For the resolution of k–ε model equations, the fractional steps method is also used. Consequently, the advection stage of turbulent quantities is treated separately and is solved at the same time as hydraulic quantities, if the method of characteristics is required.

4.8.2 Production, diffusion, source terms

Time discretization

As in the propagation stage, and knowing $\widetilde{k}$ and $\widetilde{\varepsilon}$, the results of the advection step, we need to solve:

$$\frac{k^{n+1} - \widetilde{k}}{\Delta t} - div\left(\frac{\nu_t}{\sigma_k}\overrightarrow{grad}(k)\right) = G + P_{kv} - \varepsilon \tag{4.118}$$

$$\frac{\varepsilon^{n+1} - \widetilde{\varepsilon}}{\Delta t} - div\left(\frac{\nu_t}{\sigma_\varepsilon}\overrightarrow{grad}(\varepsilon)\right) = C_{1\varepsilon}\frac{\varepsilon}{k}G + P_{\varepsilon v} - C_{2\varepsilon}\frac{\varepsilon^2}{k} \tag{4.119}$$

Linearization is executed according to the principle of semi-implicitation explained earlier, but here with the choice of coefficients which separate the variables k and ε. This leads to the system:

$$\frac{k^{n+1}}{\Delta t} - div\left(\frac{\nu_t}{\sigma_k}\overrightarrow{grad}(k^{n+1})\right) + \frac{\varepsilon^n}{k^n}k^{n+1} = \frac{\widetilde{k}}{\Delta t} + G^n + P^n_{kv} \tag{4.120}$$

$$\frac{\varepsilon^{n+1}}{\Delta t} - div\left(\frac{\nu_t}{\sigma_\varepsilon}\overrightarrow{grad}(\varepsilon^{n+1})\right) + C_{2\varepsilon}\frac{\varepsilon^n}{k^n}\varepsilon^{n+1} = \frac{\widetilde{\varepsilon}}{\Delta t} + C_{1\varepsilon}\frac{\varepsilon_n}{k^n}G^n + P^n_{\varepsilon v} \tag{4.121}$$

Let us point out the specific treatment of the term ε on the right-hand side of the equation for k, written as:

$$\varepsilon \approx \frac{\varepsilon^n}{k^n}k^{n+1}$$

This substitution is indispensable to separate the variables and for the numerical stability of the model. ε^n/k^n is the inverse of a time. An examination of Equations 4.120 and 4.121 shows that the choice of a variable ε/k in place of ε would perhaps be more appropriate.

Spatial discretization

Variational formulation, as well as spatial discretization, are executed exactly as in the hydrodynamic system, after integration by parts of diffusion terms for discretization in space. The boundary terms obtained during this integration by parts are assumed to be zero. This does not affect the entry points of the domain or the walls since k and ε are imposed. This conforms to an ordinary variational formulation where the space of test functions is limited to zero functions on the parts of the boundary where the unknown is given (Dirichlet condition). For a liquid exit, this is the same as assuming that the gradients of k and along the normal at the exit are zero.

If the same notations are kept as before for the matrices (mass, diffusion), the final system can take the form:

$$\begin{pmatrix} A1 & 0 \\ 0 & A2 \end{pmatrix}\begin{pmatrix} K \\ \varepsilon \end{pmatrix} = \begin{pmatrix} C1 \\ C2 \end{pmatrix} \tag{4.122}$$

with:

$$A1 = \left(\frac{1}{\Delta t} + \frac{\varepsilon^n}{k^n}\right)M + \frac{D}{\sigma_k} \tag{4.123}$$

$$A2 = \left(\frac{1}{\Delta t} + C_{2\varepsilon}\frac{\varepsilon^n}{k^n}\right)M + \frac{D}{\sigma_k} \tag{4.124}$$

$$C1 = M\left(\frac{\widetilde{K}}{\Delta t} + G^n + P^n_{kv}\right) \tag{4.125}$$

$$C2 = M\left(\frac{\widetilde{\varepsilon}}{\Delta t} + C_{1\varepsilon}\frac{\varepsilon^n}{k^n}G^n + P^n_{kv}\right) \tag{4.126}$$

Since the equations for k and ε are uncoupled, they are solved separately.

4.9 Solving the tracer equation in 2D

Paradoxically, this equation is more complex to solve in 2D than in 3D, because the value of the tracer is an average over the vertical.

The equation resolved for the tracer at a degree of freedom i is as follows:

$$\int_\Omega \frac{(T^{n+1} - T^n)}{\Delta t}\varphi_i \, d\Omega + \int_\Omega \vec{u}.\overrightarrow{grad}(T)\varphi_i \, d\Omega - \frac{1}{h_i}\int_\Omega div(h\nu_t\overrightarrow{grad}(T^{n+1}))\varphi_i \, d\Omega$$

$$= \frac{1}{h_i}\int_\Omega (T_{sce} - T)Sce\varphi_i \, d\Omega \tag{4.127}$$

In this equation:

T_{sce} is the value of the tracer at the source.

Sce is the source of the tracer (in m/s). It is connected to the discharge of the sources Q_{sce} (in m^3/s).

$T = \theta_T T^{n+1} + (1 - \theta_T)T^n$ where θ_T is the relaxation coefficient for the tracer.

h is the water depth and, in theory, should be equal to $(1 - \theta_T)h^{n+1} + \theta_T h^n$ (see Section 4.10.2 below).

h_i is the water depth h defined above and taken at point i.

It is impossible to work out the terms in $1/h$ in the variational formulation since the function $1/h$ is not in the approximation space. The term $1/h$ has been removed from the integral on Ω and replaced by $1/h_i$. This choice has again been dictated by considerations on the conservation of tracer quantity (see Section 4.10.2).

Moreover, diffusion terms are entirely implicit.

The different operations for the resolution of the tracer equation are identical to those described for the k–ε model. The only difference comes from the boundary term arising from the integration of the diffusion term, which is retained at this point. However, its treatment is similar to its counterparts in the hydrodynamic system. Actually, at the walls, the boundary condition for the tracer is the data of dispersive flux in the form:

$$\nu_t \frac{\partial T}{\partial n} = aT + b$$

as in 3D.

For a free exit, and when the method of characteristics is used for advection, the result of the advection step is imposed in the diffusion step. The effect of diffusion is thus overlooked, in this case. This process is preferred to imposing an artificial wall boundary condition on a liquid boundary, something that would happen if a point in a liquid exit was treated as a degree of freedom in the diffusion step.

The final linear system is resolved by iterative methods described in Section 7.5.

4.10 Laws of conservation in 2D

Water alone is eternal. Yun Son-Do (1587–1671, Korea)

Controlling the mass in numerical simulations emphasizes a certain number of difficulties and demonstrations of mass conservation are a guide for building algorithms; they sometimes question them (see for example the treatment of the Mercator projection). The proof of mass conservation of fluid and the proof of mass conservation of a tracer will be examined successively and, subsequently, the head, momentum and energy conservation.

4.10.1 Mass conservation of the fluid

Let us now consider the case of the continuity equation without a source term:

$$\frac{\partial h}{\partial t} + div(h\vec{u}) = 0$$

which can be written as:

$$\frac{\partial h}{\partial t} + div(h_{prop}\vec{u}) = 0$$

as before.

After time discretization, variational formulation, integration by parts and resolution of the final linear system, we shall resolve for each point i where the continuity equation is treated:

$$\int_{\Omega} \frac{(h^{n+1} - h^n)}{\Delta t} \varphi_i \, d\Omega + \int_{\Gamma} \varphi_i \, h_{prop} \vec{u} \, . \vec{n} \, d\Gamma - \int_{\Omega} h_{prop} \vec{u} \, . \overrightarrow{grad}(\varphi_i) \, d\Omega \tag{4.128}$$

Time discretization of $\vec{u}$ is not specified. This does not change the demonstration in any way.

To show overall mass conservation, we add the equations of all the points. Taking into account that $\sum_{i=1}^{npoin} \varphi_i = 1$, we get $\sum_{i=1}^{npoin} \overrightarrow{grad}(\varphi_i) = 0$. The sum of all the equations then gives:

$$\int_{\Omega} (h^{n+1} - h^n) \, d\Omega = -\Delta t \int_{\Gamma} h_{prop} \vec{u} \, . \vec{n} \, d\Gamma \tag{4.129}$$

This equation expresses that the evolution of mass in the domain Ω is simply due to the flux through the boundaries of Ω. This is the proof of mass conservation.

This formula also shows how to calculate fluxes at the boundaries, by strictly following the finite element formulation. We see, in particular, that the choice of h_{prop} remains free.

The demonstration fails when:

- Mass conservation has not been resolved for all the points. Unfortunately, this happens when water depth conditions are imposed; moreover, in this last case the outflow is evaluated just at those points where mass conservation has not been resolved. This explains the persistence of a residual error at liquid boundaries even when really conservative schemes are used.

- Discretized equations cannot be reduced to a conservative form. This is so if advection of h is done with the method of characteristics, or when discretization in a non-conservative form of $\vec{u}.\overrightarrow{grad}(h)$ and $h\, div(\vec{u})$ in the continuity equation is not compatible; that is, they cannot be reduced to a divergence operator. The loss of compatibility is generally due to time discretization and linearization which eliminates the products of unknown functions. This is why the use of sub-iterations, where non-linear terms are actualized, restores mass conservation.

4.10.2 Conservation of the tracer

On the basis of what was mentioned for the fluid, everything becomes complicated to such a point that building non-conservative schemes for a non-conservative equation of the tracer remains a partially unresolved problem.

Let us assume that the equation resolved for the tracer is as follows:

$$\int_\Omega \frac{(T^{n+1}-T^n)}{\Delta t}\varphi_i\, d\Omega + \int_\Omega \vec{u}.\overrightarrow{grad}(T)\varphi_i\, d\Omega - \frac{1}{h_i}\int_\Omega div(h\nu_t \overrightarrow{grad}(T^{n+1}))\varphi_i\, d\Omega$$

$$= \frac{1}{h_i}\int_\Omega T_{sce} Sce \varphi_i\, d\Omega - \frac{1}{h_i}\int_\Omega T\left(\frac{(h^{n+1}-h^n)}{\Delta t} + div(h\vec{u})\right)\varphi_i\, d\Omega \tag{4.130}$$

This equation has been seen in Section 4.9, with the same notations, but by replacing the term:

$$\frac{1}{h_i}\int_\Omega -T\ Sce\varphi_i\, d\Omega$$

by:

$$-\frac{1}{h_i}\int_\Omega T\left(\frac{(h^{n+1}-h^n)}{\Delta t} + div(h\vec{u})\right)\varphi_i\, d\Omega$$

that is, by replacing the right-hand side of the continuity equation by its left-hand side. Writing this is indispensable for the demonstration and will bring out an additional constraint. Demonstration of the tracer conservation uses the demonstration of the mass of fluid conservation. There should be strict compatibility between the continuity equation and the tracer equation. This compatibility cannot be assured if the tracer is treated with a different time step. This is due to the processing of non-linear terms: a mass conservation equation resolved in the discretized form from the instant t^n to t^{n+1}, then from t^{n+1} to t^{n+2}, does not give the same result as a resolution in a single step from t^n to t^{n+2}.

The demonstration for conservation of the tracer follows.

Each tracer equation at point i is multiplied by h_i (we now understand why $\frac{1}{h_i}$ has been put outside the integral of the diffusion term), then all the equations are added. By using the decomposition of h on the test functions (assuming that the test functions of the tracer are the basis functions of the water depth), we then get:

$$\int_\Omega \frac{(T^{n+1}-T^n)}{\Delta t}h\, d\Omega + \int_\Omega h\vec{u}.\overrightarrow{grad}(T)d\Omega - \int_\Omega div(h\nu_T\, \overrightarrow{grad}(T^{n+1}))d\Omega$$

$$= \int_{\Omega} T_{sce}\ Sce\, d\Omega - \int_{\Omega} T \left(\frac{(h^{n+1} - h^n)}{\Delta t} + div(h\overrightarrow{u}) \right) d\Omega \tag{4.131}$$

In the preceding equation we assumed that:

$$T = (1 - \theta_T)T^{n+1} + \theta_T T^n \text{ and } h = (1 - \theta_h)h^{n+1} + \theta_h h^n \tag{4.132}$$

At this point, we realize that the condition $\theta_h + \theta_T = 1$ is necessary such that the terms coming from time derivatives combine to give the following time derivative:

$$\frac{h^{n+1}T^{n+1} - h^n T^n}{\Delta t}$$

On this condition, we finally get:

$$\int_{\Omega} \frac{(h^{n+1}T^{n+1} - h^n T^n)}{\Delta t} d\Omega + \int_{\Omega} div(hT\overrightarrow{u})d\Omega - \int_{\Omega} div(h\nu_T \overrightarrow{grad}(T^{n+1}))d\Omega$$

$$= \int_{\Omega} T_{sce}\ Sce\, d\Omega \tag{4.133}$$

This equation expresses that the variation of the tracer quantity in the domain is due to the flux at the boundaries (by including a diffusive flux) and at the rate of creation. It also shows that the tracer quantity in the domain should be calculated as $\int_{\Omega} hT\, d\Omega$, which can be interpreted as the dot product (MH, T) where M is the mass matrix and H and T the vectors containing the water depth and tracer values. The compatible form of fluxes with their specific discretization of h is also given.

4.10.3 Head and momentum conservation

Our initial choice of the non-conservative Saint-Venant equations, and the "shock capturing" approach in which there is no real treatment of shocks, raises a fundamental question: what would be the numerical solution when the real theoretical solution has discontinuities? The reasoning will take place in two stages. First, we examine the nature of discontinuities which could appear in solutions of the non-conservative equations. This first examination will tend to prove that the non-conservative equations lead to incorrect solutions. In the second stage, we accept that this approach is not valid if the numerical method used does not allow discontinuities, as is the case with finite elements with linear interpolation.

Discontinuities with the non-conservative equations

Let us recall the conditions of Section 2.3.6 on hydraulic jumps: 1D, source terms, diffusion and friction being neglected, flat bottom and steady flow. The momentum equation in conservative form is reduced to:

$$\frac{\partial}{\partial x}\left(hu^2 + g\frac{h^2}{2} \right) = 0 \tag{4.134}$$

while the same equation in non-conservative form gives:

$$\frac{\partial}{\partial x}\left(\frac{u^2}{2}+gh\right)=0 \tag{4.135}$$

In the non-conservative form of the Saint-Venant momentum equation, the physical quantity conserved on either side of the hydraulic jump is no longer:

$$hu^2+g\frac{h^2}{2}$$

but the head:

$$h+\frac{u^2}{2g}$$

So we find different conjugate depths and velocities relations:

$$\frac{4h_1^2h_2^2}{h_1+h_2}=h_c^3$$

$$u_1u_2(u_1+u_2)=2gQ$$

and:

$$F_1^{2/3}F_2^{2/3}=\frac{2}{F_1^{2/3}+F_2^{2/3}}$$

which are compared with the real ones in Tables 4.1 and 4.2. The curves $F_1=f(F_2)$ obtained with the conservation of momentum on the one hand, and the conservation of head on the other hand, are very close in the vicinity of $F=1$. They do not differ by more than 1% up to a Froude number of 1.16 and not more than 10% up to 1.72. These "non-conservative" relations are effectively observed on both sides of a transcritical flow over a bump, a case where the head is conserved while the momentum is not.

Table 4.1: conservative and non-conservative conjugate relations in a hydraulic jump

	Conjugate depths	Conjugate velocities
Conservative form of momentum equation	$h_1h_2(h_1+h_2)=\frac{2Q^2}{g}$	$\frac{u_1^2u_2^2}{u_1+u_2}=\frac{gQ}{2}$
Non-conservative form of momentum equation	$\frac{h_1^2h_2^2}{h_1+h_2}=\frac{Q^2}{2g}$	$u_1u_2(u_1+u_2)=2gQ$

To illustrate this, in Table 4.3 there is a comparison of the variations in head, momentum and total energy in a hydraulic jump with conservative and non-conservative forms. The table gives the difference between the upstream and downstream values. Thus, we observe that with both formulations the total energy of a column of water increases during a hydraulic jump, as does entropy in a shock. However, there is definitely a loss of energy in both cases, since its flux decreases.

Table 4.2: conjugate relations between Froude numbers in a hydraulic jump

	Conjugate Froude numbers
Conservative form of momentum equation	$F_1^{2/3} F_2^{2/3} = \frac{1}{2}(\frac{1}{F_1^{2/3}} + \frac{1}{F_2^{2/3}})$
Non-conservative form of momentum equation	$F_1^{2/3} F_2^{2/3} = \frac{2}{F_1^{2/3}+F_2^{2/3}}$

Table 4.3: variations of the physical quantities in a hydraulic jump

	Head	Momentum	Total energy
Conservative form of momentum equation	$\frac{(h_2-h_1)^3}{4h_1h_2}$	0	$\frac{g}{4}(h_1^2 - h_2^2)$
Non-conservative form of momentum equation	0	$\frac{g(h_1-h_2)^3}{2(h_1+h_2)}$	$\frac{g(h_1-h_2)(h_1^2+h_2^2)}{2(h_1+h_2)}$

A paradox

The question at this point is to know whether the use of a non-conservative equation will lead the numerical simulations to the erroneous relations of the hydraulic jump we have just seen. A consistent scheme, together with the time and space steps tending to the zero limit, should give the "exact" solution corresponding to the initial equations. In fact, every "finite element" solution without special treatment of shocks is continuous. Hence the two formulations are equivalent and should lead to the same solution if consistent numerical schemes are used. This solution will conserve momentum and head and will hence be false in both cases. In practice, we observe that this solution is a good approximation of the conservative discontinuous solution, as is shown in the result of Plate 8 (see Section 4.13.2, flow over a weir with a hydraulic jump). Besides, we cannot rule out that local numerical diffusion just appears and uses up all or a part of the energy whose disappearance is foreseen by the theory.

4.10.4 Conservation of energy

In the absence of friction, diffusion, source terms and discontinuities, the total energy of a flow should be conserved. We also assume a flat bottom, to simplify matters. The non-conservative Saint-Venant equations are then written as:

$$\frac{\partial h}{\partial t} + div(h\overrightarrow{u}) = 0 \tag{4.136}$$

$$\frac{\partial u}{\partial t} + \overrightarrow{u}.\overrightarrow{grad}(u) + g\frac{\partial h}{\partial x} = 0 \tag{4.137}$$

$$\frac{\partial v}{\partial t} + \overrightarrow{u}.\overrightarrow{grad}(v) + g\frac{\partial h}{\partial y} = 0 \tag{4.138}$$

By multiplying Equation 4.136 by $gh + \frac{u^2+v^2}{2}$, Equation 4.137 by hu and Equation 4.138 by hv, the sum of the three resultant equations then gives:

$$\frac{\partial}{\partial t}\left(\frac{gh^2}{2} + h\frac{u^2+v^2}{2}\right) + div\left[\left(gh^2 + h\frac{u^2+v^2}{2}\right)\overrightarrow{u}\right] = 0 \tag{4.139}$$

which expresses the conservation of the following quantity:

$$\frac{gh^2}{2} + h\frac{u^2+v^2}{2}$$

Integrated on the 2D domain, this will provide the total energy, except for one factor, ρ. In fact this quantity is the sum of potential energy and kinetic energy of a column of water whose height is h, velocity u and cross-section equal to unity. We again find the condition of entropy of Equation 4.2, in the absence of discontinuity.

Remark: the absence of a factor $\frac{1}{2}$ in front of the term gh^2 within the divergence can look like an error and might suggest that only half the energy entering at the boundary is available inside the domain. In fact $\frac{gh^2}{2}$ is not potential energy but the integral of potential energy along a column of water, which is not an extensive quantity. By multiplying Equation 4.136 by h, we show that:

$$\frac{\partial(h^2/2)}{\partial t} + div(h^2\overrightarrow{u}) - \overrightarrow{u}.\overrightarrow{grad}(h^2/2) = 0 \tag{4.140}$$

If we consider the simplified case of the slow filling of a reservoir, with a depth which remains constant in space so as to cancel the term $\overrightarrow{u}.\overrightarrow{grad}(h^2/2)$, we observe that the flux h^2 at the boundaries causes a variation of $h^2/2$ in the domain.

Let us examine the conservation of energy at the discrete level. We ought to know whether it is possible to obtain a conservation of energy equation as:

$$\frac{(\frac{gh^2}{2} + h\frac{u^2+v^2}{2})^{n+1} - (\frac{gh^2}{2} + h\frac{u^2+v^2}{2})^{n}}{\Delta t} + div\left[\left(gh^2 + h\frac{u^2+v^2}{2}\right)\overrightarrow{u}\right] = 0 \tag{4.141}$$

Our degrees of freedom are:

- Choice of discretization of quantities $gh + \frac{u^2+v^2}{2}$, hu and hv by which Equations 4.136, 4.137 and 4.138 are multiplied.
- Choice of discretization of the same three equations.

In order to form the derivatives in time at the discrete level, it is necessary: for $gh + \frac{u^2+v^2}{2}$ to be written as:

$$g\left(\frac{h^n + h^{n+1}}{2}\right) + \frac{(u^n)^2 + \left(u^{n+1}\right)^2 + (v^n)^2 + \left(v^{n+1}\right)^2}{4}$$

for hu to be written as:

$$hu = \left(\frac{h^n + h^{n+1}}{2}\right)\left(\frac{u^n + u^{n+1}}{2}\right) \tag{4.142}$$

for hv to be written as:

$$hv = \left(\frac{h^n + h^{n+1}}{2}\right)\left(\frac{v^n + v^{n+1}}{2}\right) \tag{4.143}$$

With this choice, we can verify that $gh\frac{\partial h}{\partial t}$ is actually discretized as:

$$gh\frac{\partial h}{\partial t} = \frac{\frac{g\left(h^{n+1}\right)^2}{2} - \frac{g(h^n)^2}{2}}{\Delta t} \tag{4.144}$$

In the same way $hu\frac{\partial u}{\partial t} + \frac{u^2}{2}\frac{\partial h}{\partial t}$ becomes:

$$hu\frac{\partial u}{\partial t} + \frac{u^2}{2}\frac{\partial h}{\partial t} = \frac{\frac{h^{n+1}\left(u^{n+1}\right)^2}{2} - \frac{h^n(u^n)^2}{2}}{\Delta t} \tag{4.145}$$

In order that the term:

$$gh\, div(h\overrightarrow{u}) + h\overrightarrow{u}.\overrightarrow{grad}(gh)$$

actually gives $div(gh^2\overrightarrow{u})$, it then becomes necessary that $div(h\overrightarrow{u})$ is discretized as:

$$div(h\overrightarrow{u}) = div\left[\left(\frac{h^n + h^{n+1}}{2}\right)\left(\frac{\overrightarrow{u}^n + \overrightarrow{u}^{n+1}}{2}\right)\right] \tag{4.146}$$

and $\overrightarrow{grad}(gh)$ as:

$$\overrightarrow{grad}(gh) = \overrightarrow{grad}\left[g\left(\frac{h^n + h^{n+1}}{2}\right)\right] \tag{4.147}$$

Finally, the discretization of:

$$\frac{u^2+v^2}{2} div(h\overrightarrow{u}) + hu\,\overrightarrow{u}\,.\,\overrightarrow{grad}(u) + hv\,\overrightarrow{u}\,.\,\overrightarrow{grad}(v)$$

should give a discretization of $div(h\frac{u^2+v^2}{2}\overrightarrow{u})$, the only degrees of freedom left being the discretization of $\overrightarrow{u}\,.\,\overrightarrow{grad}(u)$ and $\overrightarrow{u}\,.\,\overrightarrow{grad}(v)$. This brings us back to imposing that the term:

$$\frac{(u^n)^2 + \left(u^{n+1}\right)^2 + (v^n)^2 + \left(v^{n+1}\right)^2}{4} div\left[\left(\frac{h^n + h^{n+1}}{2}\right)\left(\frac{\overrightarrow{u}^n + \overrightarrow{u}^{n+1}}{2}\right)\right]$$

$$+\left(\frac{h^n + h^{n+1}}{2}\right)\left(\frac{u^n + u^{n+1}}{2}\right)\overrightarrow{u}\,.\,\overrightarrow{grad}(u) + \left(\frac{h^n + h^{n+1}}{2}\right)\left(\frac{u^n + u^{n+1}}{2}\right)\overrightarrow{u}\,.\,\overrightarrow{grad}(v)$$

becomes equal to:

$$div\left[\left(\frac{h^n+h^{n+1}}{2}\right)\frac{(u^n)^2+\left(u^{n+1}\right)^2+(v^n)^2+\left(v^{n+1}\right)^2}{4}\left(\frac{\vec{u}^{\,n}+\vec{u}^{\,n+1}}{2}\right)\right]$$

This is possible only by discretizing:

$$\vec{u}\,.\,\overrightarrow{grad}(u) \text{ as } \frac{\vec{u}^{\,n}.\,\overrightarrow{grad}(u^n)+\vec{u}^{\,n+1}.\,\overrightarrow{grad}(u^{n+1})}{2}$$

$$\vec{u}\,.\,\overrightarrow{grad}(v) \text{ as } \frac{\vec{u}^{\,n}.\,\overrightarrow{grad}(v^n)+\vec{u}^{\,n+1}.\,\overrightarrow{grad}(v^{n+1})}{2}$$

In conclusion we find that:

- For an unsteady flow, only a numerical scheme perfectly centred in time, and hence non-linear in this case, can conserve energy.
- The initial choice of non-conservative equations does not appear as a handicap for conservation of energy.
- As all the conditions found only concern time discretization, the conservation of energy is assured for steady flow.

What has just been done is, however, valid only within the framework of a strict conformity with the Galerkin method. The use of the method of characteristics or the SUPG method, for example, adds terms which cannot be put together in a divergence, and hence ruins the demonstration.

4.11 The treatment of uncovered beds

How awful is a river without water! (A saying from ancient Egypt)

The solution usually adopted for dealing with uncovered beds or dry zones is to limit the value of the water depth to a threshold. This solution has several disadvantages: first, any arbitrary intervention on the water depth ruins the conservation of mass; then the conservation of momentum equation applied to a fine film of water creates a laminar flow whose velocities are not negligible and act as parasitic values which are transmitted to zones under water. Two other solutions have been elaborated and fine tuned. The first one consists of treating the equations in the whole domain and in

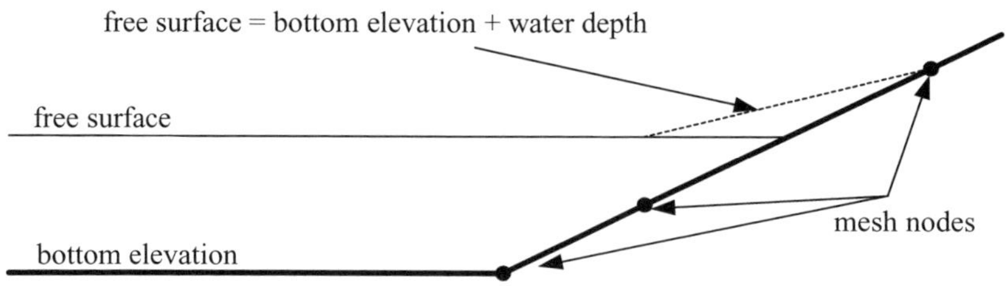

Figure 4.1: parasitic free surface on a shoreline

their totality, simply by correcting the terms that are evidently rendered false by the absence of water, like the free surface gradient. In fact, in an uncovered area, the free surface gradient becomes the bottom gradient and creates parasitic driving terms as is shown in 1D in Figure 4.1.

Considering the entire free surface gradient for a lake at rest with banks partially covered with water highlights the presence of parasitic driving terms due to the erroneous introduction in the equations of dry bottom gradient terms.

The second option consists of removing from the calculation all elements which are not completely wet.

4.11.1 Option 1: correction of the free surface gradient

The half-dry, half-wet elements are characterized by a discontinuity in the free surface gradient which passes from a value close to 0 where there is water to a value close to $\overrightarrow{grad}(Z_f)$ where there is none. So, linear interpolation is ill-adapted and a specific treatment of the term $-g\,\overrightarrow{grad}(Z_s^n)$ should take place.

A second fundamental idea is to abandon the minimum threshold criterion to determine if an element is dry or wet. We decide that elements requiring a special calculation are those whose bottom elevation of one node (referred to as "high") is higher than the elevation of the free surface of another one (referred to as "low"). Finally, in these contentious elements, the free surface gradient is not brutally cancelled. One simply reduces the free surface elevation of the "high" node by the difference between its bottom elevation and the free surface elevation of the "low" node (this is done only to calculate the free surface gradient). The problem of a lake at rest is solved by this method which, moreover, conserves real free surface gradients, e.g. when a triangle contains two wet nodes and one dry node.

The only limitation to the method comes from the criterion deciding whether an element is dry or wet. In certain cases (large elements along the slope and low water depth) this criterion is false and declares as dry elements which are entirely wet. This limitation should be taken into account while building the mesh. It should be ensured that the difference in level on an element is less than the depth.

Plate 9 illustrates the capacity of this approach to simulate drying on a beach, to zero depth. In this test case, the elevation of the free surface at the left boundary of the calculation domain (a 46 m long canal whose cross-section is shown in the plate) is lowered according to a sinusoidal law, as during a tide. The beach empties and water retained in a pool forms a puddle which gradually goes back to rest. The problem of a "lake at rest" is also brought out in this configuration, at the bottom of the plate. The situation shown is stable; the post-processor, by linear interpolation, draws a sloping free surface at the point where the free surface reaches the beach. The algorithm dealing with exposed beds does not consider this parasitic gradient and does not generate any movement.

4.11.2 Option 2: masking of exposed elements

In this option the criterion used to decide whether an element is partially dry depends on a comparison between an average value of the bottom per element and an average value of the free surface. Moreover, all elements that are not entirely wet are eliminated from the mesh. For this purpose, each element is associated with a mask equal to 1 for a wet element and equal to 0 for an exposed element (even partially exposed). All contributions of the elements in the calculation of vectors and matrices are multiplied by this mask, which leads us to a modification of the mesh by the removal of elements.

This technique is not as good as the correction of the free surface gradient, and in particular poses problems in the dynamics of wetting and drying, as well as in mass conservation. These problems and a few solutions will be given in detail together with the treatment of exposed beds in 3D for which the method has been more widely applied.

4.12 Pseudo wave equation

The pseudo wave equation is obtained from the Saint-Venant equations by eliminating the velocity from the continuity equation, using a value of this velocity given by the momentum equation. To obtain the wave equation, velocity is eliminated in the same way, but by considering the partial time derivative of the continuity equation. The numerical problems generally encountered in the wave equation have led to a variant, the generalized wave equation, where a penalty is added to ensure good mass conservation. However, the penalty coefficient has to be adjusted, as the case may be, to eliminate spurious oscillations, or ensure the continuity equation. On their side, supporters of the generalized wave equation affirm that the solutions of primitive equations themselves suffer from parasitic oscillations. Reference [85], a communication presented at the Congress Computational Methods in Water Resources, shows that these oscillations, which seem to be due to a resurgence of the inf–sup condition, can be eliminated, either by sufficient precision, or by the adoption of a different discretization for the water depth and for velocity by using, for example, the element known as the "quasi-bubble" for the velocity.

By establishing a bridge between primitive equations and the pseudo wave equation, we shall show here that there is no real difference between both approaches except the use in the latter case of a "non-compatible" Laplacian to discretize the divergence of a gradient in finite elements.

The idea comes from an attempt to simplify the linear system combining the water depth and velocity of Section 4.5.6, by eliminating velocity from the continuity equation at the discrete level. To simply draw an expression of velocity from the momentum equation, we look for a set of options for which the matrices $AM2$ and $AM3$ are diagonal. The form of our linear system is:

$$\begin{pmatrix} AM1 & BM1 & BM2 \\ CM1 & AM2 & 0 \\ CM2 & 0 & AM3 \end{pmatrix} \begin{pmatrix} \delta H \\ U^{n+1} \\ V^{n+1} \end{pmatrix} = \begin{pmatrix} CV1 \\ CV2 \\ CV3 \end{pmatrix} \tag{4.148}$$

$AM2$ and $AM3$ contain matrices arising from time derivatives, of diffusion, friction and, in the case of the SUPG method, of advection. The friction matrix is diagonal; the mass matrix resulting from the time derivative can be made diagonal by mass lumping; and diffusion can be carried over to the right-hand side if it is written in an explicit way. Finally, the use, for transport terms, of the method of characteristics or of any other formally explicit method (PSI distributive scheme, for example) also enables us to shift them to the right-hand side. Put together, these conditions give diagonal matrices $AM2$ and $AM3$, and we get the following expressions for U and V:

$$U = AM2^{-1}CV2 - AM2^{-1}CM1\,\delta H \tag{4.149}$$

$$V = AM3^{-1}CV3 - AM3^{-1}CM2\,\delta H \tag{4.150}$$

By carrying over these values to the discretized continuity equation we get:

$$(AM1 - BM1\,AM2^{-1}CM1 - BM2\,AM3^{-1}CM2)\,\delta H$$

$$= CV1 - BM1\,AM2^{-1}CV2 - BM2\,AM3^{-1}CV3 \tag{4.151}$$

where we can replace:

$$BM1\,AM2^{-1}\,CM1 + BM2\,AM3^{-1}\,CM2$$

by the notation G.

Thus we can write that the new equation system is:

$$\begin{pmatrix} AM1 - G & 0 & 0 \\ 0 & AM2 & 0 \\ 0 & 0 & AM3 \end{pmatrix} \begin{pmatrix} \delta H \\ U^{n+1} \\ V^{n+1} \end{pmatrix}$$

$$= \begin{pmatrix} CV1 - BM1\,AM2^{-1}CV2 - BM2\,AM3^{-1}CV3 \\ CV2 - CM1\,\delta H \\ CV3 - CM2\,\delta H \end{pmatrix} \tag{4.152}$$

In this system the three equations are uncoupled, so δH can be computed first, then U^{n+1} and V^{n+1}. Depth and velocity may have different discretization, so we will be able to couple the pseudo wave equation and quasi-bubble discretization of the velocity.

This system is resolved as such by Casulli [22] in finite differences. In fact, we can build without approximation, in finite differences, a matrix like $BM1\,AM2^{-1}\,CM1$.

In 1D for example, $AM1$ would be tri-diagonal and $BM1\,AM2^{-1}CM1$ penta-diagonal. As we do not wish to deal with complex matrices in finite elements, the matrix G will be approximated. In order to find out the nature of G, we resume elimination of velocity at the continuum level, on the basis of equations discretized only in time, as follows:

Continuity:

$$\frac{\delta h}{\Delta t} + \theta_u div(h^n \overrightarrow{u}^{n+1}) + (1-\theta_u)\, div(h^n \overrightarrow{u}^n) \tag{4.153}$$

Recall that $\delta h = h^{n+1} - h^n$.

Momentum:

$$\frac{\overrightarrow{u}^{n+1} - \overrightarrow{u}^n}{\Delta t} + \overrightarrow{u}.\nabla(\overrightarrow{u}) = -g\,\overrightarrow{grad}(Z_s) - fric\,\overrightarrow{u}^{n+1} + \frac{1}{h} div(h\nu_e \overrightarrow{grad}(\overrightarrow{u}^n)) \tag{4.154}$$

by denoting for the sake of simplification $fric = \frac{g\|\overrightarrow{u}^n\|}{K^2 h^{4/3}}$ (the example of Strickler's friction formula).

With the same assumptions adopted at the discrete level, this last equation becomes:

$$\frac{\overrightarrow{u}^{n+1} - \overrightarrow{\tilde{u}}}{\Delta t} = -g\theta_h \overrightarrow{grad}(\delta h) - g\,\overrightarrow{grad}(Z_s^n) - fric\,\overrightarrow{u}^{n+1} + \frac{1}{h} div(h\nu_e \overrightarrow{grad}(\overrightarrow{u}^n)) \tag{4.155}$$

We derive:

$$\overrightarrow{u}^{n+1} = \frac{\frac{\overrightarrow{\tilde{u}}}{\Delta t} - g\,\overrightarrow{grad}(Z_s^n) + \frac{1}{h} div(h\nu_e \overrightarrow{grad}(\overrightarrow{u}^n)) - g\theta_h \overrightarrow{grad}(\delta h)}{\frac{1}{\Delta t} + fric} \tag{4.156}$$

By carrying over this value $\overrightarrow{u}^{n+1}$ to the continuity equation, we find the pseudo wave equation:

$$\frac{\delta h}{\Delta t} + div\left(\frac{-\theta_h\theta_u g h^n}{\frac{1}{\Delta t} + fric} \overrightarrow{grad}(\delta h)\right) = -(1-\theta_u) div(h^n \overrightarrow{u}^n)$$

$$-\theta_u div\left[\frac{h^n}{\frac{1}{\Delta t} + fric}\left(\frac{\overrightarrow{\tilde{u}}}{\Delta t} - g\,\overrightarrow{grad}(Z_s^n) + \frac{1}{h} div(h\nu_e \overrightarrow{grad}(\overrightarrow{u}^n))\right)\right] \tag{4.157}$$

Each of these different terms can be compared with its discrete equivalent:

$$\frac{\delta h}{\Delta t} \text{ becomes: } AM1\,\delta H$$

$$-(1-\theta_u) div(h^n \overrightarrow{u}^n) \text{ becomes: } CV1$$

$$-\theta_u div\left[\frac{h^n}{\frac{1}{\Delta t}+fric}\left(\frac{\vec{\tilde{u}}}{\Delta t}-g\,\overrightarrow{grad}(Z_s^n)+\frac{1}{h}div(h\nu_e\overrightarrow{grad}(\vec{u}^n))\right)\right]$$

becomes:

$$-BM1\,AM2^{-1}CV2-BM2\,AM3^{-1}CV3$$

and:

$$div\left(\frac{-\theta_h\theta_u g h^n}{\frac{1}{\Delta t}+fric}\overrightarrow{grad}(\delta h)\right)$$

becomes:

$$BM1\,AM2^{-1}\,CM1+BM2\,AM3^{-1}\,CM2=-G$$

We recognize in:

$$div\left(\frac{-\theta_h\theta_u g h^n}{\frac{1}{\Delta t}+fric}\overrightarrow{grad}(\delta h)\right)$$

a diffusion term, with a diffusion coefficient equal to:

$$\frac{\theta_h\theta_u g h^n}{\frac{1}{\Delta t}+fric}$$

whose dimension is m^2/s. This term can be written in a classic manner by integration by parts giving a diffusion matrix which will be an approximation of $-G$. For this, we write:

$$\int_\Omega \Psi_i^h\, div\left(\frac{-\theta_h\theta_u g h^n}{\frac{1}{\Delta t}+fric}\overrightarrow{grad}(\delta h)\right)d\Omega$$

$$=\int_\Gamma \Psi_i^h\,\frac{-\theta_h\theta_u g h^n}{\frac{1}{\Delta t}+fric}\overrightarrow{grad}(\delta h).\vec{n}\;d\Omega+\int_\Omega \frac{\theta_h\theta_u g h^n}{\frac{1}{\Delta t}+fric}\overrightarrow{grad}(\delta h)\;.\;\overrightarrow{grad}(\Psi_i^h)\;d\Omega \quad (4.158)$$

An approximation of G is thus the matrix given by the term:

$$-\int_\Omega \frac{\theta_h\theta_u g h^n}{\frac{1}{\Delta t}+fric}\overrightarrow{grad}(\Psi_j^h)\;.\;\overrightarrow{grad}(\Psi_i^h)\;d\Omega$$

This value of G is deliberately kept in this form if the velocity is quasi-bubble.

In 1D, our equation can have two interpretations: either diffusion with coefficient:

$$\nu=\frac{\theta_h\theta_u g h^n}{\frac{1}{\Delta t}+fric}$$

or, by dividing it by Δt, wave propagation with celerity:

$$c=\sqrt{\frac{\theta_h\theta_u g h^n}{1+\Delta t\,fric}}$$

To get a physical celerity in the absence of friction, it is necessary that $\theta_h = \theta_u = 1$.

The approximate calculation of G is in fact a Laplacian calculation called "non-compatible". That is, we do not verify at the discrete level that:

$$div(\overrightarrow{grad}(f)) = \Delta(f)$$

The use of the non-compatible Laplacian is one of the well-known solutions for removing spurious oscillations arising from a violation of the inf–sup condition, and resolution with the pseudo wave equation may benefit from this effect.

We have just shown that a specific configuration in our resolution of primitive equations, such as:

- choice of the method of characteristics for the advection of velocities
- mass lumping of the mass matrix of the momentum equation
- explicit diffusion
- use of a non-compatible Laplacian

could be interpreted as a resolution of a pseudo wave equation. We can compare primitive equations and pseudo wave equations for an unsteady case, then for a steady case. Plate 10 shows the results of four different resolutions for flow around a bridge pier:

- Primitive equations with linear discretization.
- Primitive equations with "quasi-bubble" velocity.
- Pseudo wave equation with linear discretization.
- Pseudo wave equation with "quasi-bubble" velocity.

For all the calculations, the mass matrix of velocity is "lumped" and diffusion is made explicit (a configuration which, in fact, is not the best for primitive equations). It is clear that primitive equations and pseudo wave equation are very close. The main difference in the four results appears to be due to the discretization of velocity. With linear interpolation some "trembling" of the free surface is visible, these spurious oscillations are removed by a quasi-bubble velocity. The CPU times obtained with a 750 MHz Hewlett Packard C3700 are also given in Plate 10: our new "pseudo wave equation" algorithm is always faster by a factor of 2 or even more; it is generally more robust and seems more appropriate for steady flow. In fact, in this case, mass lumping of the time derivative and explicitation of diffusion have no effect on the final solution; the only difference is then the treatment of the Laplacian for writing $div(\overrightarrow{grad}(h))$. This difference, although perhaps important for stability, is minimal in the results, as is shown by the superposition of curves in Plate 8, relative to the flow over a weir with a hydraulic jump (see Section 4.13.2 below).

4.13 Some validation test cases

We have mentioned, during the presentation of the algorithms, some tests by showing the main particulars. Here is a recapitulation of the main points for the validation of a code resolving the Saint-Venant equations, as well as a few new cases of validation.

4.13.1 Test of a lake at rest

In a lake at rest with a variable bottom, the numerical solution should not generate any movement. This evidence is not guaranteed if the bottom gradient and the depth gradient are not treated in a totally compatible manner (hence the necessity of the same upwinding in finite volumes). In our algorithm, this condition is obtained because the unknown "water depth" is in fact the increment $h^{n+1} - h^n$. The driving term in the conservation of momentum equation is $grad(Z_s^n)$, which is actually zero if one starts from a flat free surface. The test becomes complicated if dry zones are added in the calculation domain; see the treatment of exposed banks (Section 4.11) and Plate 9.

4.13.2 Rapid flow over a weir with a hydraulic jump downstream

This test case refers to a flow in a canal which is 2 m wide, about 21 m long, with a flat bottom whose elevation is -0.2 m, except between the abscissae x=7.12 m and x=12.88 m, where there is a parabolic weir culminating at x=10 m at elevation 0, according to the formula:

$$Z_f(x) = \max(-0.2, -0.0246875(x-10)^2) \tag{4.159}$$

The mesh is made up of squares whose sides measure 16 cm, cut into triangles. A constant discharge of $Q = 2$ m^3/s enters the canal. A depth of 60 cm is maintained downstream. Strickler's friction law is used with a coefficient of 40. Viscosity is zero. Assuming a permanent flow and a constant width for the canal, the momentum equation becomes:

$$u\frac{\partial u}{\partial x} = -g\frac{\partial Z_s}{\partial x} - g\frac{u^2}{Kh^{4/3}} \tag{4.160}$$

By replacing u by $\frac{Q}{hL}$ and writing that $Z_s = h + Z_f$ we find:

$$\left(1 - \frac{Q^2}{gL^2h^3}\right)\frac{\partial h}{\partial x} = -\frac{\partial Z_f}{\partial x} - \frac{Q^2}{L^2K^2h^{10/3}} \tag{4.161}$$

The critical water depth is:

$$h_c = \left(\frac{Q^2}{gL^2}\right)^{1/3} = 0.46714 \text{ m}$$

In the absence of friction this critical water depth is observed at the top of the weir. Here, it is attained if:

$$\frac{\partial Z_f}{\partial x} = -\frac{Q^2}{L^2K^2h_c^{10/3}} = -\frac{g}{K^2h_c^{1/3}} \tag{4.162}$$

The form of the weir was chosen such that the critical depth is attained one element (16 cm) after the top of the weir (the slope is then equal to −0.0079). This helps us to build an "exact" solution by making use of the differential equation:

$$\frac{\partial h}{\partial x} = \frac{-\frac{\partial Z_f}{\partial x} - \frac{Q^2}{L^2 K^2 h^{10/3}}}{1 - \frac{Q^2}{g L^2 h^3}} \tag{4.163}$$

with a Runge–Kutta method to second-order accuracy. The solution upstream of the critical section is calculated with 1000 negative space steps starting from this section. We compute downstream, still with 1000 steps, starting from the same section. This gives a first solution which is supercritical. A downstream subcritical solution is then calculated beginning from the right of the canal and starting from the boundary condition. The passage from the supercritical solution to the subcritical solution is achieved by a hydraulic jump, when the jump conditions are fulfilled; that is, when $h_1 h_2 (h_1 + h_2) = 2h_c^3$, as shown in Section 2.2. In Plate 8, the "exact" solution is compared with the solutions obtained with non-conservative primitive equations and with the pseudo wave equation. The use of a non-conservative form proves to be not very detrimental and no parasitic oscillation is observed on either side of the hydraulic jump. The false "exact" solution obtained with a non-conservative form (dashed line in Plate 8) differs by the position of the hydraulic jump. In fact, the critical weir upstream and the water depth imposed downstream decide the solution, but the place where the hydraulic jump relation triggers the transition from the supercritical flow imposed by the upstream condition, and the subcritical flow imposed at the downstream end, is staggered. We observe, as indicated in Section 4.10.3, that the numerical calculation gives a very good approximation of the "real" conservative solution. A more advanced theoretical study would be necessary to understand this surprising result better.

This case can also be simulated with the non-hydrostatic 3D Navier–Stokes equations (see Plate 11). The numerical parameters are the same as those of the Saint-Venant equations; however, a viscosity of 0.01 m^2/s along the vertical is used (a zero diffusion would not transmit the shear stress of friction on the bottom). The dynamic pressure field, at the bottom of Plate 11 (basically, this is the variable $\frac{p\Delta t}{\rho}$), reveals that the flow is hydrostatic everywhere except at the location of the hydraulic jump. The hydraulic jump is less marked and is followed by an undulation, which is more in conformity with the observations.

4.13.3 Wave propagation

Two points have to be verified:

- Propagation of waves with the correct celerity. This generally is done in a canal with a rectangular section and constant depth. It can be verified with the Boussinesq equations in Plate 12 that the wave crosses 434 m in 40 s, corresponding to the theoretical celerity $\sqrt{g(h+A)}$ with g=9.81 m/s^2, h=10 m and $A = 2$ m.

- The absence of polarization, i.e. a propagation of identical waves in all directions. This is the aim of the test-case presented in Plate 13 whose initial condition is a Gaussian-shaped free surface with an elevation equal to

$2.4(1 + e^{-(x^2+y^2)/4})$. We can clearly see that a circular wave is propagated and then reflected by the walls of the square domain. Shown at the top of Plate 7 is the result of the reflection by the walls.

4.13.4 Ritter and Stoker solutions

These 1D solutions reproduce a dam break (with a reservoir at constant depth h_g) in a river with a constant rectangular cross-section. In the case of Ritter [164], the bottom on the downstream side is dry. In the case of Stoker [178], the depth on the downstream side is constant and velocity is zero. If the river flows along the x axis and the dam is at x=0, the free surface is a parabola in the Ritter solution, limited upstream by the abscissa $x = -\sqrt{gh_g}t$ and downstream by the abscissa $x = 2\sqrt{gh_g}t$. By considering the Riemann invariants and the theory of characteristics, and by denoting $c_g = \sqrt{gh_g}$, we find that:

$$\begin{aligned} c = \sqrt{gh} &= \tfrac{2}{3}(c_g - \tfrac{x}{2t}) \\ u &= \tfrac{2}{3}(\tfrac{x}{t} + c_g) \end{aligned} \tag{4.164}$$

In the Stoker solution with a depth of h_d downstream, to which is associated a celerity $c_d = \sqrt{gh_d}$, the Ritter expansion wave is limited on the downstream side by a zone with constant depth h_m and constant velocity u_m, itself connected to the river at rest by a hydraulic jump which advances at a speed of:

$$W = \frac{h_m u_m}{h_m - h_d}$$

a formula established in Section 2.2 (with $u_d = 0$ here). In the case of Stoker, there are four distinct zones of solutions, separated by three points C, D and E (see Figure 4.2). The abscissae of the points C, D and E are, denoting $c_m = \sqrt{gh_m}$:

$$x_C = -c_g\, t$$

$$x_D = (u_m - c_m)\, t$$

$$x_E = W\, t$$

The four zones then have the following solutions:

- Between the upstream position and the point C: reservoir at rest.
- Between C and D: the Ritter solution.
- Between D and E: $h = h_m$ and $u = u_m$.
- Downstream of E: river at rest, with a depth of h_d.

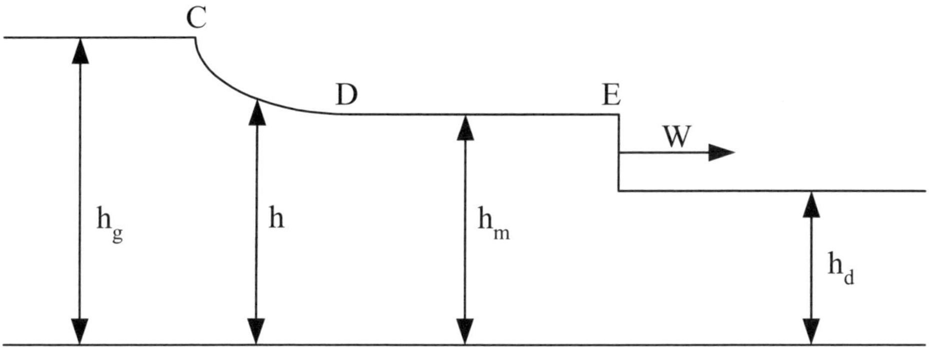

Figure 4.2: dam break on a wet bottom, Stoker's solution

The velocity u_m can be found at point D which belongs to the expansion wave for which $u + 2c = 2c_g$.

So $u_m = 2(c_g - c_m)$. The last unknown c_m is found by writing the conservation of momentum across the hydraulic jump, in a relative referential attached to the jump:

$$h_m(u_m - W)^2 + g\frac{h_m^2}{2} = h_d W^2 + g\frac{h_d^2}{2} \tag{4.165}$$

This is expressed by an equation of the sixth degree (resolved by iterations):

$$c_m^6 - 9c_d^2 c_m^4 + 16c_g c_d^2 c_m^3 - c_d^2(c_d^2 + 8c_g^2)c_m^2 + c_d^6 = 0 \tag{4.166}$$

The case $c_d = 0$ restores the Ritter solution. In 1999, Wu *et al.* [191] published a new solution to this problem concerning a river with a trapezoidal section.

Plate 14 shows a numerical simulation of the Stoker solution on a 24 km long reach with a mesh made of 300 m long elements. Here h_g=6 m and h_d=2 m. The time step is 5 s and we display the solution at t=1200 s. In this calculation, the velocity is entirely implicit and discretized using quasi-bubble elements. The advection of velocity is done with the distributive PSI scheme (Section 6.4.1). The continuity equation is resolved in the conservative form with a SUPG upwinding (Section 6.3). The hydraulic jump is delayed by one element and the transition takes place in one element with slight oscillations upstream and downstream. This shows the quality of the scheme, which does not have any specific treatment for shocks.

Chapter 5

Resolution of the Navier–Stokes equations

I believe there is no philosophical high-road in Science, with epistemological signposts. No, we are in a jungle and find our way by trial and error, building our road behind us as we proceed. Max Born (1882–1970)

5.1 Introduction

The first numerical three-dimensional models for free surface flow date from the 1970s [36]. A well-established practice consists of assuming hydrostatic pressure [149], leading to what is known as the three-dimensional Saint-Venant equations. Several examples of these models exist in finite differences:

- The Princeton Ocean Model (POM), see [15]. In this model the gradients of baroclinic and barotropic pressures are separated into two distinct modes.
- The Davies and Lawrence model [35], applied to calculations of tides in the Irish Sea.
- The Trim3D model (see [24], [25] and [23]) which combines a semi-implicit formulation and a Lagrangian–Eulerian model for transport equations, quite close to what we shall be doing in finite elements. The pressure and velocity nodes are staggered. With a mesh made of one layer, the hydrostatic pressure hypothesis restores the Saint-Venant equations of the Trim2D software. This model has been applied to calculations of salinity in the bay area of San Francisco [62].
- The Delft3D-Flow model [39] functions with a curvilinear mesh (formed with hexahedra) which has the advantage of being able to follow the boundary lines to a certain extent. The inconvenience in relation to finite elements is that local refining propagates along the mesh lines. A local geometric transformation is applied to do the calculations in a regular mesh. This introduces a metric

Hydrodynamics of Free Surface Flows: Modelling with the Finite Element Method J.-M. Hervouet

which complicates the spatial derivatives. This metric can be simplified if the two families of the mesh lines are perpendicular.

All the models mentioned so far use a sigma transform. Another solution proposed in finite differences, and compared to the sigma transform [14], is called "geopotential z-coordinate". It consists simply of working with fixed size elements, but varying in number along the vertical to follow the evolution of the surface. The bottom and the free surface have a staircase shape and this leads to errors in the evaluation of stress or transport terms (see [52]).

In 1981, Frederiksen and Watts [53] proposed a space and time finite element to account for variation of the free surface.

Also in finite elements, Luettich and colleagues [132] compiled the software ADCIRC (ADvanced CIRCulation model for oceanic, coastal and estuarine waters), dedicated to oceanic circulation and estuaries, and used by the U.S. Army Corps of Engineers. This software also solves the Saint-Venant equations and has been used on parallel computer architecture with the MPI library (see Chapter 8). For more information refer to the Internet website http://www.marine.unc.edu.

The non-hydrostatic models are not so common. The idea of decomposing the pressure into hydrostatic and dynamic parts, which will be discussed later, was formulated in 1995 by Casulli and Stelling [26]. In 1996, Mahadevan *et al.* [135] introduced an oceanic non-hydrostatic model to conveniently reproduce upwelling and downwelling phenomena. They also showed that the non-hydrostatic problem was well-posed.

One of the main difficulties of three-dimensional models is the reproduction of the free surface. Several new techniques have been developed to try to overcome this problem:

- The ALE method (Arbitrary Lagrangian–Eulerian, [42]). The idea is to assign a specific velocity to the mesh in order to follow the movements of the domain. The originality of the method is that this velocity is not the actual velocity of particles of water.

- The MAC method (Marker And Cell, [65]) uses fictitious particles without mass to mark the element full of water. The principal advantage of this method is that it helps to deal with breaking waves or with any flow whose free surface is not a univocal function. The inconvenience is the time cost for additional calculation to follow the particles, in addition to the resolution of equations governing the movement of the fluid.

- The VOF method (Volume Of Fluids, [102]) resorts to a quantity c, the volume occupation ratio, which indicates if a volume is full of liquid (c=1) or empty (c=0). The evolution of c is governed by a conservation equation:

$$\frac{\partial c}{\partial t} + div(c\overrightarrow{u}) = 0 \tag{5.1}$$

The resolution of this equation requires perfectly conservative and monotonic numerical schemes. In its original version, the VOF method only solves equations of the liquid phase. A pocket of gas enclosed by a breaking wave will not be subject to any effect of buoyancy. The often spectacular results of this method lack realism, however.

In more recent versions (refer for example to [111]) this error is corrected by solving the equations of the two phases, which are separated by the interface whose equation is c=0.5. So, it is necessary to plan a specific processing of volumes containing the interface, each of their nodes being attributed physical quantities for each of the two phases, if necessary by a simple extrapolation, to avoid basing the numerical scheme on non-defined values. Another solution, published by Lafaurie *et al.* [119] in 1994 and taken up more recently by Muzaferija and Peric [145], consists of considering only one fluid, the change of phase being only a discontinuity of density.

Another method, SPH (Smoothed Particles Hydrodynamics), was first introduced in astrophysics [143], then applied to free surface flows [144]. This technique proves to be more precise and more promising than the MAC and VOF methods, and deserves more explanations. It is a completely Lagrangian approach, without a mesh, with particles moving due to the influence of pressure forces, viscosity and gravity. Each particle possesses density, pressure and velocity. The originality of SPH lies in the way in which a continuous definition of the fields is given on the basis of nodal values of particles. Interpolation is inspired by the equality:

$$A(\overrightarrow{r}) = \int_{\Omega} A(\overrightarrow{r'})\, \delta(\overrightarrow{r} - \overrightarrow{r'})\, d\Omega \tag{5.2}$$

where δ is the Dirac distribution. With SPH, we write:

$$A(\overrightarrow{r}) = \int_{\Omega} A(\overrightarrow{r'}) w_h(\overrightarrow{r} - \overrightarrow{r'})\, d\Omega + O(h) \tag{5.3}$$

where w_h is an approximation of δ, a spherical kernel having a compact support with a diameter $2h$, h being called the attenuation length. The kernel depends only on the distance r to its centre. Here is an example based on a spline of order 3 in two dimensions.

We denote $q = r/h$:

$$w_h(q) = \frac{10}{7\pi h^2} \left\{ \begin{array}{c} 1 - \frac{3}{2}q^2 + \frac{3}{4}q^3 \text{ if } 0 \leq q \leq 1 \\ \\ \frac{1}{4}(2-q)^3 \text{ if } 1 \leq q \leq 2 \end{array} \right\} \tag{5.4}$$

Once the kernel w_h is defined, Equation 5.3 is rewritten as a sum over all the particles:

$$A(\overrightarrow{r}) = \sum_b \frac{m_b}{\rho_b} A_b w_h(\overrightarrow{r} - \overrightarrow{r_b}) \tag{5.5}$$

where m_b is the mass of the particle b, ρ_b its density. m_b/ρ_b is the volume of the particle and replaces the volume element $d\Omega$ in formula 5.3. The interest in this form of interpolation is that the calculation of derivatives is simple. Thus the gradient of a field A is simply:

$$\nabla A(\overrightarrow{r}) = \sum_b \frac{m_b}{\rho_b} A_b \nabla w_h(\overrightarrow{r} - \overrightarrow{r_b}) \tag{5.6}$$

Assuming the flow to be compressible, the equations describing the movement of particles are:

- A mass conservation equation written as:

$$\frac{d\rho}{dt} = \vec{U}.\overrightarrow{grad}(\rho) - div(\rho\vec{U}) \tag{5.7}$$

- A momentum equation as:

$$\frac{d\vec{U}}{dt} = -\frac{1}{\rho}\overrightarrow{grad}(p) + \nu\Delta\vec{U} + \vec{F}_{ext} \tag{5.8}$$

- An equation of state:

$$p = \frac{\rho_0 c_0^2}{\gamma}\left[\left(\frac{\rho}{\rho_0}\right)^{\gamma} - 1\right] \tag{5.9}$$

c_0 is the speed of sound in the fluid and γ the ratio of specific heats at constant pressure and constant volume. The method is strictly applied to a compressible fluid. Moreover, the mass conservation equation only conserves mass, not volume.

The equations are resolved in an explicit manner with a limitation of the time step by the value:

$$\Delta t = \min\left(0.25\sqrt{\frac{h}{F_b}}, 0.4\frac{h}{c_0}\right) \tag{5.10}$$

where F_b is the maximum force applied to a particle. In practice, in order to gain stability, the values γ and c_0 adopted are different from the real values for water, giving a slightly more compressible fluid.

Another original feature of the SPH method is the treatment of boundary conditions, the walls themselves being modelled by particles to which the "Lennard-Jones" type of repulsion forces are associated.

In the future, SPH could become the method which takes over finite elements for complex cases, particularly for mixed flow, wave breaking and all the cases where free surface is not univocal.

After this rapid overview, we shall study in detail the resolution of the Navier–Stokes equations in finite elements, first with the hydrostatic hypothesis, by using the fractional steps method, then with an algorithm adapted to the non-hydrostatic equations.

5.2 Equations with hydrostatic hypothesis

5.2.1 Fractional steps method

The basis algorithm consists of three fractional steps. We write:

$$\frac{\partial f}{\partial t} = \frac{f^{n+1} - f_D + f_D - f_C + f_C - f^n}{\Delta t} \tag{5.11}$$

This leads to a successive resolution of three steps:

1) An advection step:

$$\frac{f_C - f^n}{\Delta t} + \text{advection terms} = 0 \quad (f = U, V, K, \varepsilon \text{ or } T) \tag{5.12}$$

2) A diffusion step:

$$\frac{f_D - f_C}{\Delta t} + \text{diffusion terms} = \text{source terms} \quad (f = U, V, K, \varepsilon \text{ or } T) \tag{5.13}$$

For a tracer T, the source terms not only are due to sources or sinks but may also contain a production or destruction term.

3) A pressure-continuity step:

$$\frac{f^{n+1} - f_D}{\Delta t} + \text{pressure terms} = 0 \tag{5.14}$$

This last step includes also the resolution of the continuity equation $div(\overrightarrow{U}) = 0$ which helps to deduce the water depth h and the vertical component of velocity W.

Each of these steps will be resolved in detail one after the other. Prior to this, however, discretization in space and building up the mesh have to be specified.

5.2.2 Spatial discretization

The spatial discretization chosen is a finite element discretization using prisms (see Section 7.1.4). Prisms have six nodes. Their quadrangular sides are vertical. The prism has been retained because its two-dimensional horizontal projection constitutes one of the finite elements (triangle with linear interpolation) used to resolve the Saint-Venant equations in two dimensions. It is thus possible to build a three-dimensional mesh from a two-dimensional mesh. It only requires to mesh the two-dimensional domain Ω with triangles, then to duplicate this mesh vertically as sketched in Figure 5.1. Calculation in two and three dimensions with the same mesh is thus possible.

5.2.3 Building the three-dimensional mesh

To build the three-dimensional domain with prisms, it suffices to build the triangulation of the two-dimensional domain, then to repeat it along the vertical in superimposed layers (called "planes"). These "planes" are in fact surfaces with variable elevations. The only condition is that the elevation of the points belonging to a vertical should increase from the bottom to the free surface. Several simple formulae guarantee this condition, as in the following four methods.

Method 1: planes evenly spaced along the vertical

This first option, the most classic, consists of starting with the coordinate nodes (x, y) of the 2D mesh, then defining the elevation for each point of the 3D mesh with coordinates (x, y, z), according to the formula:

$$z(x, y, ip) = Z_f\,(x, y) + \frac{ip - 1}{np - 1}\,(Z_s\,(x, y) - Z_f\,(x, y))$$

Z_f always being the bottom elevation, Z_s the free surface elevation and ip is the rank of the plane under consideration (planes range from 1 to np if we go from the bottom to the free surface). This is the simplest method based on the sigma transform. In this case the vertical coordinate z^* of each plane in the transformed mesh is simply $\frac{ip-1}{np-1}$.

Method 2: data of a function θ constant for each plane

The coordinate z^* of plane ip in the transformed mesh is no longer $\frac{ip-1}{np-1}$ but a given number θ_{ip} such that $0 \leq \theta_{ip} \leq 1$ and, for intermediary planes, $\theta_{ip-1} < \theta_{ip} < \theta_{ip+1}$. We now have:

$$z(x, y, ip) = Z_f\,(x, y) + \theta_{ip}\,(Z_s\,(x, y) - Z_f\,(x, y))$$

Figure 5.1 shows an example of a 3D mesh built with the following values: $\theta_1 = 0$, $\theta_2 = \frac{1}{6}$, $\theta_3 = \frac{1}{3}$, $\theta_4 = \frac{2}{3}$, $\theta_5 = 1$. The choice of values of θ_{ip} is free, the only constraints being a value of 0 for the bottom ($ip = 1$), a value of 1 for the free surface ($ip = np$) and a series of intermediate values in a strictly increasing order.

Method 3: double sigma transform

Here we prescribe a constant elevation for one plane, then the planes between the bottom and this plane are evenly spaced, and the same is done between this plane and the free surface.

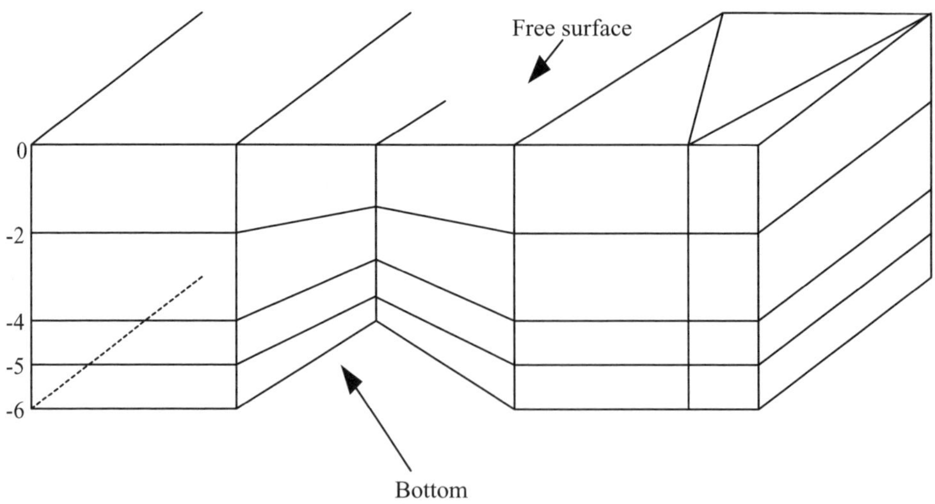

Figure 5.1: mesh built with a constant proportionality coefficient per plane

Method 4: horizontal planes

In order to represent the temperature stratification zones (thermoclines) better, it is sometimes worth imposing that a maximum number of intermediate "planes" be horizontal. In this case it is possible to indicate the desired elevation for each plane, which will be denoted as z_{ip}. This elevation is not always accessible because it could be situated below the bottom or above the free surface. In this case, the elevation of the point of the plane ip will be maintained slightly above the bottom, or slightly below the surface. This can be executed with the help of a "MinMod" limiter by the formula:

$$z = \min\left\{\max\left[Z_f(x,y) + \frac{ip-1}{np-1}d_{\min}, z_{ip}\right], Z_s(x,y,t) - \frac{np-ip}{np-1}d_{\min}\right\} \qquad (5.15)$$

where d_{min} is an imposed minimum distance. We impose that plane 1 (the bottom) is at a minimum distance $d_{\min}$ from the free surface, and that the plane np (the free surface) is at a minimum distance d_{min} from the bottom.

Figure 5.2 shows an example of a 3D mesh built with this method and the following elevations z_{ip}: $z_1 = -6$, $z_2 = -5$, $z_3 = -4$, $z_4 = -2$, $z_5 = 0$.

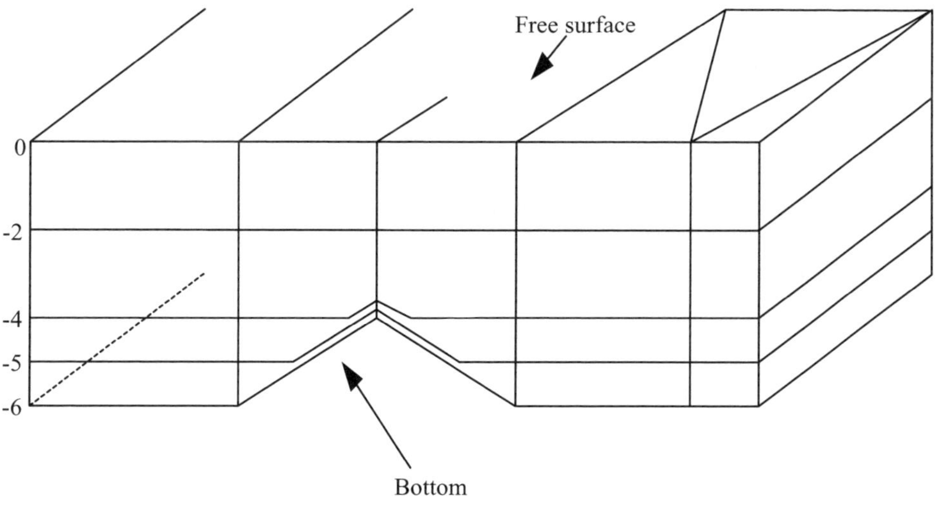

Figure 5.2: mesh made of horizontal planes

In conformity with the earlier formula, when the bottom rises to the level of plane number 2 normally situated at elevation −5 m, this plane number 2 then adopts the form of the bottom, at a distance $d_{min}/4$.

The advantage of this distribution is the elimination of truncation errors in the calculation of buoyancy terms. These errors can generate unacceptable instability when these terms are dominant. Actually, when the planes are not horizontal, it is impossible to represent correctly a function f that only depends on z. Now, the density of the fluid should tend to this type of function corresponding to a position of equilibrium without horizontal pressure gradients.

With this method, non-horizontal mesh planes are inevitable, but only at a distance less than d_{min} from the bottom or from the surface. In these zones with quasi-zero volume, it is suggested that the forces arising from the parasitic horizontal pressure gradients (called hydrostatic inconsistencies) be cancelled by the intermediary of a filter which acts in places where the prisms are, such that a node of the lower base (number 1, 2 or 3) has an elevation higher than a node of the upper base (number 4, 5 or 6). This filter also cancels the different diffusion coefficients on these elements.

5.2.4 Sigma transform and generalized sigma transform

As the free surface evolves with time, the elevations z of the 3D mesh vary from one time step to another. The variables f^{n+1}, f_D, f_C and f^n ($f = u$, v or T) are each defined on different meshes. However, it is still possible to execute a change of variables so as to fix the mesh during a time step. As we have already seen in Section 2.2.5, this can be particularly interesting in the advection step, which has a definite influence on the evolution of the free surface. The change of variables currently adopted is the sigma transform which consists of passing from the coordinate z to the coordinate z^* independent of time.

For the classical sigma transform, we recall the formula used:

$$z^* = \frac{z - Z_f}{Z_s - Z_f} \tag{5.16}$$

When a plane has a prescribed constant elevation, a double sigma transform is preferred, with a function z^* varying between -1 and $+1$. If $Z_{prescribed}$ is the required constant elevation, we have below the fixed plane:

$$z^* = \frac{z - Z_{prescribed}}{Z_{prescribed} - Z_f}$$

and above:

$$z^* = \frac{z - Z_{prescribed}}{Z_s - Z_{prescribed}}$$

For a generalized sigma transform, it is convenient to use the following function z^* between planes ip and $ip + 1$:

$$z^* = ip - 1 + \frac{z - z_{ip}}{z_{ip+1} - z_{ip}} \tag{5.17}$$

where z_{ip} is the elevation of plane ip, and z_{ip+1} the elevation of plane $ip + 1$. This function z^* now varies between 0 and $np - 1$.

These forms of transformation are necessary when dealing with the overall mesh, e.g. during an advection step with the method of characteristics. The definition of z^*

is thus univocal over the entire vertical. For local computations within a single finite element, for example, a local definition will be simpler, and we may write that:

$$z^* = \frac{z - z_{ip}}{z_{ip+1} - z_{ip}} \tag{5.18}$$

i.e. we have a simple sigma transform between two planes.

It is interesting to notice that the four methods given above split into two different families. The first two are such that $\frac{\partial z}{\partial z^*} = h$, they thus rely on the sigma transform. For the last two, which correspond to generalized sigma transforms:

$$\frac{\partial z}{\partial z^*} = z_{ip+1} - z_{ip} = \Delta z \tag{5.19}$$

i.e. the width between two planes which we denote Δz. It is only piecewise constant vertically.

5.2.5 Advection step

The resolution of the advection equations is described in Chapter 6. This step is executed in a fixed transformed mesh (by the sigma transform), which makes it necessary to calculate a vertical velocity at each time step:

$$W^* = \frac{dz^*}{dt} = \frac{\partial z^*}{\partial t} + U\frac{\partial z^*}{\partial x} + V\frac{\partial z^*}{\partial y} + W\frac{\partial z^*}{\partial z} \tag{5.20}$$

on the basis of results of the earlier time step.

Use of the method of characteristics on the fixed mesh arising from the sigma transform is not a problem but it may be with other techniques. In order to show the complications arising from the sigma transform, here is an example of advection with the SUPG method (described in Chapter 6). This method consists of upwinding the basis functions in the direction of the flow:

$$\Psi_i^{'} = \Psi_i + \frac{\Delta t}{2}\vec{U}.\overrightarrow{grad}(\Psi_i)$$

So, the new basis functions $\Psi_i^{'}$ remain linear but become discontinuous. The variational formulation is applied to the following equation:

$$\frac{f^C - f^n}{\Delta t} + \vec{U}^*.\overrightarrow{grad}^*\left(\theta f^C + (1-\theta)f^n\right) = 0 \tag{5.21}$$

where θ is the relaxation coefficient, $\vec{U}^*$ the velocity in the fixed mesh and $\overrightarrow{grad}^*$ the gradient operator in the same mesh. The earlier equation is multiplied by the deformed basis function:

$$\Psi_i^{'*} = \Psi_i^* + \frac{\Delta t}{2}\vec{U}^*.\overrightarrow{grad}(\Psi_i^*) \tag{5.22}$$

and by using the Einstein notation $f^C = f_j^C \Psi_j^*$, we get:

$$\frac{f_j^C \Psi_j^* \Psi_i^{\prime *}}{\Delta t} + \theta \Psi_i^{'*} \vec{U}^*.\overrightarrow{grad}^*\left(f_j^C \Psi_j^*\right)$$

$$= \frac{f_j^n \Psi_j^* \Psi_i'^*}{\Delta t} + (\theta - 1)\ \Psi_i'^* \vec{U}^* . \overrightarrow{grad}^* \left(f_j^n \Psi_j^*\right) \tag{5.23}$$

Strictly speaking, the deformed basis function should be used for all the terms in the equation. In practice, only the advection term is transformed, which does not alter the conservation property. After dividing by θ, we get:

$$f_j^C \left(\frac{\Psi_j^* \Psi_i^*}{\theta \Delta t} + \Psi_i^* \vec{U}^* . \overrightarrow{grad}^* \left(\Psi_j^*\right)\right)$$

$$+f_j^C \left(\frac{\Delta t}{2} \vec{U}^* . \overrightarrow{grad}^* \left(\Psi_i^*\right) \vec{U}^* . \overrightarrow{grad}^* \left(\Psi_j^*\right)\right) = f_j^n \left(\frac{\Psi_j^* \Psi_i^*}{\theta \Delta t}\right)$$

$$+f_j^n \left(\frac{\theta - 1}{\theta} \left(\Psi_i^* \vec{U}^* . \overrightarrow{grad}^* \left(\Psi_j^*\right) + \frac{\Delta t}{2} \vec{U}^* . \overrightarrow{grad}^* \left(\Psi_i^*\right) \vec{U}^* . \overrightarrow{grad}^* \left(\Psi_j^*\right)\right)\right) \tag{5.24}$$

By multiplying by h and by integrating on the total fixed domain we finally get:

$$f_j^C \left(\frac{m_{ij}}{\theta \Delta t} + vgr_{ij} + ugug_{ij}\right) = f_j^n \left(\frac{m_{ij}}{\theta \Delta t} + \frac{\theta - 1}{\theta} \left(vgr_{ij} + ugug_{ij}\right)\right) \tag{5.25}$$

with:

$$m_{ij} = \int_{\Omega^*} h \Psi_i^* \Psi_j^* d\Omega^* = \int_{\Omega} \Psi_i \Psi_j \, d\Omega \tag{5.26}$$

M, whose general term is m_{ij}, is the mass matrix. It may be estimated either in the real mesh or in the fixed mesh.

$$vgr_{ij} = \int_{\Omega^*} h \varphi_i^* \vec{u}^* . \overrightarrow{grad}^* \left(\varphi_j^*\right) d\Omega^* \tag{5.27}$$

VGR, whose general term is vgr_{ij}, is the matrix obtained for classic advection by finite elements expressed in fixed mesh.

$$ugug_{ij} = \frac{\Delta t}{2} \int_{\Omega^*} h \vec{U}^* . \overrightarrow{grad}^* \left(\Psi_i^*\right) \vec{U}^* . \overrightarrow{grad}^* \left(\Psi_j^*\right) d\Omega^* \tag{5.28}$$

$UGUG$, whose general term is $ugug_{ij}$, is the additional matrix from SUPG. We show [93] that $ugug_{ij}$ is equivalent to an artificial diffusion which, in a dimension greater than one, is only produced in the direction of the current. In 2D, the calculation of the earlier matrices is relatively short for triangular elements. For prisms with a triangular base, the number of operations necessary to form the matrices then clearly increases. In order to simplify the calculation of their coefficients and observing that, in theory, it does not harm the conservation property, we can limit $\vec{U}^*$ in formula 5.22 to the horizontal components of velocity. In the case of essentially horizontal flow, this approximation is not a problem. Only the generic coefficient of the $UGUG$ matrix is modified in this way. It becomes:

$$ugug_{ij} = \frac{\Delta t}{2} \int_{\Omega^*} h \vec{U}_h^* . \overrightarrow{grad}^* \left(\Psi_i^*\right) \vec{U}^* . \overrightarrow{grad}^* \left(\Psi_j^*\right) d\Omega^* \tag{5.29}$$

with $\vec{U}_h^* = (U, V, 0)$. When this approximation is made the matrix $UGUG$ is no longer symmetric.

When advection of a variable is done with the help of this method, the steps of advection and diffusion are in fact treated at the same time.

5.2.6 Diffusion step

For the conditions of existence and uniqueness of the solution for this type of parabolic problem, the reader can refer to Raviart and Thomas [162] or to the book by Pironneau [156] (in French).

Given the complexity of a diffusion stage with the sigma transform, and the numerical artefacts which are its consequence, this stage is executed in the real mesh. However, the Jacobian of the isoparametric transformation for any prism is not a constant. Now, this Jacobian appears in the denominator in the calculation of a diffusion matrix, therefore we cannot get an exact polynomial expression. The necessary approximations themselves induce disturbing numerical errors. This problem does not exist with tetrahedra and a linear interpolation, the Jacobian then being a constant. Now, a prism can be broken down into three tetrahedra (see Figure 5.3, where the tetrahedra are formed by the sets of four points (3,1,2,6), (4,6,5,1) and (5,2,1,6)). It suffices to consider that interpolation on the prism is linear by pieces, exactly as in the quasi-bubble triangle. We choose to execute the diffusion stage with a new type of element. As no degree of freedom is added, the topology of a prismatic element matrix can be formally retained. After calculating the element matrices on the tetrahedra, a pre-assembly restores the element matrices prism by prism. Writing this pre-assembly shows that all the properties of a diffusion matrix (sum of the terms of each line equal to zero, sum of the terms of each column equal to zero) are actually transferred from the tetrahedra to the prisms. Another possibility is to find an approximation of the diffusion matrix in prisms, e.g. by considering a constant average Jacobian on every prism to get a polynomial expression (see Section 7.4.1).

Buoyancy terms

The source terms of buoyancy are one of the strategic points of 3D equations. Their correct treatment conditions the representation of stratifications, a current stumbling block of modelling. These source terms are, if we do not consider atmospheric pressure (see Section 2.2.4):

$$F_x = g\frac{\Delta\rho}{\rho_o}\left(\frac{\partial Z_s}{\partial x}\right)_{y,t} - g\frac{\partial}{\partial x}\left(\int_z^{Z_s}\frac{\Delta\rho}{\rho_o}\,dz\right)_{y,z,t} \quad (5.30)$$

$$F_y = g\frac{\Delta\rho}{\rho_o}\left(\frac{\partial Z_s}{\partial y}\right)_{x,t} - g\frac{\partial}{\partial y}\left(\int_z^{Z_s}\frac{\Delta\rho}{\rho_o}\,dz\right)_{x,z,t} \quad (5.31)$$

The terms F_x and F_y are calculated on the mesh at time t^n on the basis of differences in density $\Delta\rho$. These differences in density themselves have been evaluated earlier with the help of a state function in which the different concentrations of active tracers appear.

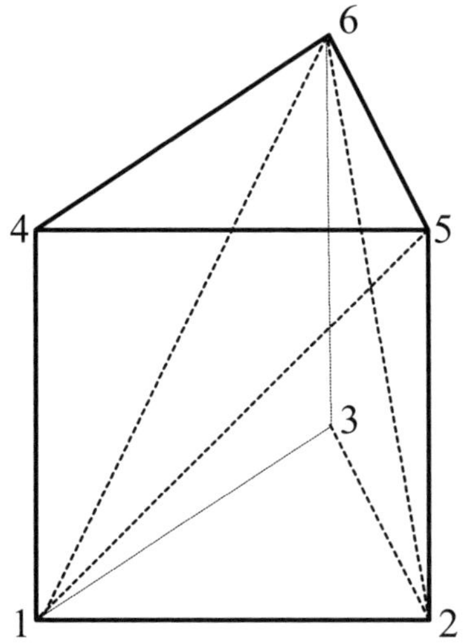

Figure 5.3: decomposition of a prism into three tetrahedra

Bijvelds [14] compares the treatment of these terms in sigma transform formulation, in different software. Given below are two modes of calculation, limited to the term F_x, the first in the real mesh and the second in the transformed mesh.

Calculation in real mesh: the second term of F_x cannot be directly expressed in the form of vectors in finite elements. It has to be transformed with the Leibnitz formula to obtain:

$$-g\frac{\partial}{\partial x}\left(\int_z^{Z_s}\frac{\Delta\rho}{\rho_o}dz\right)_{y,z,t} = -g\left[\int_z^{Z_s}\frac{\partial}{\partial x}\left(\frac{\Delta\rho}{\rho_o}\right)dz + \frac{\Delta\rho(Zs)}{\rho_o}\frac{\partial Z_s}{\partial x}\right] \tag{5.32}$$

We finally get:

$$F_x = -g\int_z^{Z_s}\frac{\partial}{\partial x}\left(\frac{\Delta\rho}{\rho_o}\right)dz + g\frac{\partial Z_s}{\partial x}\left(\frac{\Delta\rho(z)}{\rho_o} - \frac{\Delta\rho(Zs)}{\rho_o}\right) \tag{5.33}$$

This formula is simply calculated by using vertical lines of the mesh for the integral along z and to find the value of $\Delta\rho(Zs)$ above one point of elevation z.

Calculation in transformed mesh: in the transformed mesh, we have:

$$F_x = g\frac{\Delta\rho}{\rho_o}\left(\frac{\partial Z_s}{\partial x}\right)_{y,t} - g\left[\frac{\partial}{\partial x}\left(\int_z^{Z_s}\frac{\Delta\rho}{\rho_o}dz\right)_{y,z^*,t} + \left(\frac{\partial z}{\partial x}\right)_{y,z^*,t}\frac{\Delta\rho}{\rho_o}\right] \tag{5.34}$$

The variable $zs = z - Z_s$ is introduced; F_x is then simplified as:

$$F_x = -g\left[\frac{\partial}{\partial x}\left(\int_{zs}^{0}\frac{\Delta\rho}{\rho_o}\,dz\right)_{y,z^*,t} + \left(\frac{\partial zs}{\partial x}\right)_{y,z^*,t}\frac{\Delta\rho}{\rho_o}\right] \tag{5.35}$$

At this stage of reasoning, it is worth pursuing the calculation of F_x at the discrete level because the expression formulated above can be further simplified. The variation of this term will be calculated between two nodes of the mesh denoted as "inf" and "sup", one located just above the other, with the knowledge that on the surface F_x is zero:

$$F_x^{\mathrm{inf}} - F_x^{\mathrm{sup}} = -g\left[\frac{\partial}{\partial x}\left(\int_{zs^{\mathrm{inf}}}^{zs^{\mathrm{sup}}}\frac{\Delta\rho}{\rho_o}\,dz\right)_{y,z^*,t}\right]$$

$$-g\left[\left(\frac{\partial zs^{\mathrm{inf}}}{\partial x}\right)_{y,z^*,t}\frac{\Delta\rho^{\mathrm{inf}}}{\rho_o} - \left(\frac{\partial zs^{\mathrm{sup}}}{\partial x}\right)_{y,z^*,t}\frac{\Delta\rho^{\mathrm{sup}}}{\rho_o}\right] \tag{5.36}$$

With linear functions along z, this gives:

$$F_x^{\mathrm{inf}} - F_x^{\mathrm{sup}} = -g\frac{\partial}{\partial x}\left[\frac{1}{2}\left(zs^{\mathrm{sup}} - zs^{\mathrm{inf}}\right)\left(\frac{\Delta\rho^{\mathrm{sup}}}{\rho_o} + \frac{\Delta\rho^{\mathrm{inf}}}{\rho_o}\right)\right]_{y,z^*,t}$$

$$-g\left[\left(\frac{\partial zs^{\mathrm{inf}}}{\partial x}\right)_{y,z^*,t}\frac{\Delta\rho^{\mathrm{inf}}}{\rho_o} - \left(\frac{\partial zs^{\mathrm{sup}}}{\partial x}\right)_{y,z^*,t}\frac{\Delta\rho^{\mathrm{sup}}}{\rho_o}\right] \tag{5.37}$$

and we finally get:

$$F_x^{\mathrm{inf}} - F_x^{\mathrm{sup}} = \frac{g}{2}\left[\left(\frac{\Delta\rho^{\mathrm{sup}}}{\rho_o} - \frac{\Delta\rho^{\mathrm{inf}}}{\rho_o}\right)\frac{\partial}{\partial x}\left(zs^{\mathrm{sup}} + zs^{\mathrm{inf}}\right)_{y,z^*,t}\right]$$

$$-\frac{g}{2}\left[\left(zs^{\mathrm{sup}} - zs^{\mathrm{inf}}\right)\frac{\partial}{\partial x}\left(\frac{\Delta\rho^{\mathrm{sup}}}{\rho_o} + \frac{\Delta\rho^{\mathrm{inf}}}{\rho_o}\right)_{y,z^*,t}\right] \tag{5.38}$$

The two derivatives along x are discontinuous functions at the nodes of the mesh, and are calculated as follows:

- For every node M, we can define the set $ELEM(M)$ of elements of the mesh having the node M among their vertices.
- For each of the elements i of $ELEM(M)$ we can calculate the $\left(\frac{\partial f}{\partial x}\right)_i$ derivative of f at M on the side of the element i.

The derivative of f at M is then estimated by:

$$\left(\frac{\partial f}{\partial x}\right)(M) = \frac{\sum\limits_{i\,\in\,ELEM(M)}\int\limits_{\Omega}\left(\frac{\partial f}{\partial x}\right)_i\Psi_M d\Omega}{\int\limits_{\Omega}\Psi_M d\Omega} \tag{5.39}$$

where Ψ_M is the basis function associated to node M.

Friction terms

Compatibility between friction terms in 2D and 3D has already been discussed in Section 2.3.11; however, this is only valid for formulae relying on a depth-integrated velocity, such as those of Strickler or Chézy. If we want to use a local roughness factor and assume a logarithmic profile near the bottom, we are rather in the case of Section 4.5.4, and the friction velocity U_* has to be evaluated. We can use the fact that the mesh of prisms is made of piled-up meshes of triangles and consider the first plane above the bottom. If we assume that this first plane is still in the logarithmic profile, we can write (the example of a hydraulically rough flow):

$$U_* = \frac{\kappa U_{plane\,1}}{\ln(\frac{33\Delta z}{k_s})}$$

where $U_{plane\,1}$ is the velocity at the first point above the bottom and Δz the altitude of this point above the bottom.

Other source terms

The other source terms included in the terms F_x and F_y are also treated in the diffusion stage.

5.2.7 Pressure–continuity step

Associated with the hypothesis of a hydrostatic pressure, integration along the vertical of the continuity and momentum equations (for U and V) helps to obtain equations similar to the Saint-Venant equations. Integration of the momentum equations concerning U and V as they are written takes place after treating advection and diffusion. So, advection or diffusion or even friction terms at the bottom do not appear in the integrated equations. By resolving these equations between the instants t^n and t^{n+1}, we obtain the free surface and the average horizontal velocities at the instant t^{n+1}. From this we deduce the mesh at t^{n+1}, the variables U and V, then W by resolving the continuity equation.

Resolution of equations averaged over the vertical

After the advection and diffusion steps, the continuity and momentum equations related to U and V are written as:

$$\frac{\partial U}{\partial x} + \frac{\partial V}{\partial y} + \frac{\partial W}{\partial z} = 0 \tag{5.40}$$

$$\frac{U^{n+1} - U_D}{\Delta t} = -\frac{1}{\rho}\frac{\partial p_h}{\partial x} \tag{5.41}$$

$$\frac{V^{n+1} - V_D}{\Delta t} = -\frac{1}{\rho}\frac{\partial p_h}{\partial y} \tag{5.42}$$

where p_h is the hydrostatic pressure.

By integration along the vertical:

$$\overline{f}(x,y,t) = \frac{1}{h}\int_{Z_f}^{Z_s} f(x,y,z,t)\,dz$$

we obtain a system corresponding to the Saint-Venant equations without advection or diffusion:

$$\left\{\begin{array}{l} \frac{\partial h}{\partial t} + \frac{\partial(uh)}{\partial x} + \frac{\partial(vh)}{\partial y} = 0 \\[2ex] \frac{u^{n+1}-\overline{u}_D}{\Delta t} = -g\frac{\partial Z_s}{\partial x} \\[2ex] \frac{v^{n+1}-\overline{v}_D}{\Delta t} = -g\frac{\partial Z_s}{\partial y} \end{array}\right. \tag{5.43}$$

The terms averaged over the vertical are calculated on the basis of the 3D field f by the trapezoidal rule. That is:

$$\overline{f} = \frac{1}{Z_s - Z_f}\sum_{ip=1}^{np-1}\left(\frac{f_{ip}+f_{ip+1}}{2}\right)(z_{ip+1}-z_{ip}) \tag{5.44}$$

where np is the number of planes. In the case of a classical sigma transform this can be simplified to:

$$\overline{f} = \sum_{ip=1}^{np-1}\left(\frac{f_{ip}+f_{ip+1}}{2}\right)(z^*_{ip+1}-z^*_{ip}) \tag{5.45}$$

The Saint-Venant type of equations which have just been obtained are resolved by the methods already given in Chapter 4. After the resolution, we have:

$$h^{n+1},\ \overrightarrow{u}^{n+1} \text{ and } \overrightarrow{v}^{n+1}$$

The new mesh at time t^{n+1} is based on the new depth which gives the new free surface. What remains is the evaluation of the components of 3D velocity, namely U, V and W at instant t^{n+1}.

Calculation of U and V

The remaining part of the Navier–Stokes equations to be solved for the calculation of U and V is:

$$\left\{\begin{array}{l} \frac{U^{n+1}-U_D}{\Delta t} = -g\frac{\partial Z_s}{\partial x} \\[2ex] \frac{V^{n+1}-V_D}{\Delta t} = -g\frac{\partial Z_s}{\partial y} \end{array}\right. \tag{5.46}$$

This system combined with the system integrated along the vertical becomes:

$$\left\{\begin{array}{l} \frac{U^{n+1}-U_D}{\Delta t} = \frac{\overline{u}^{n+1}-\overline{U}_D}{\Delta t} \\[2ex] \frac{V^{n+1}-V_D}{\Delta t} = \frac{\overline{v}^{n+1}-\overline{V}_D}{\Delta t} \end{array}\right. \tag{5.47}$$

The calculation of U^{n+1} and V^{n+1}, henceforth, is reduced to a simple combination of vectors.

Remarks:

- The impermeability condition on the solid walls is taken into consideration in the resolution of the Saint-Venant equations. It is again taken into account in the calculation of velocities U^{n+1} and V^{n+1} where $\frac{1}{h}\int_{Z_f}^{Z_s} \overrightarrow{U}.\vec{n}\, dz = \vec{u}.\vec{n}$ is imposed; that is, the condition imposed on 3D velocities is consistent with the one imposed on 2D velocities. We assume that the impermeability condition is already verified for $\vec{u}.\vec{n}$ and we limit ourselves to project $\overrightarrow{U} - \vec{u}$ onto the tangent at the wall.

- The system 5.43 can be resolved by starting not from the result of diffusion, but from the values of time t^n averaged along the vertical. The diffusion, advection and source terms should then be added, all averaged along the vertical. We obtain the complete set of Saint-Venant equations, hence more physical. In normal flow where the effect of the free surface slope is balanced by friction, these two terms are balanced, it is thus important to treat them in the same fractional step.

Computing the vertical velocity

The vertical velocity W is calculated on the basis of the mass conservation equation. As pressure is assumed hydrostatic, W is required only for the advection step at the next time step. As this advection is executed in the fixed transformed mesh, it is W^* that we ought to know. It is more judicious to solve the 3D mass conservation equation in a fixed mesh. Thus we avoid the to and fro movement between the real and fixed mesh which makes the calculation unwieldy.

In fact, except in the case of advection by the method of characteristics, where W^* is really used, it is the $\Delta z W^*$ grouping which appears often and which is homogeneous to a velocity (we recall that Δz is the depth between two planes, see Section 5.2.4). The formula for passing from W^* to W, i.e. $W = \Delta z W^*$ + other terms (see Equation 5.65 below, we treat here the generalized sigma transform case), shows that $\Delta z W^*$ should have the same discretization as W. It appears that in a fixed mesh, the real unknown should be $\Delta z W^*$ and not W^*. We shall now examine a method for calculating $\Delta z W^*$, then how W is found on the basis of $\Delta z W^*$. We first have to establish the form of the continuity equation in the transformed mesh.

Continuity equation in the transformed mesh: we recall here the continuity equation $div(\overrightarrow{U}) = 0$ written in the transformed mesh (Equation 2.99):

$$\frac{1}{\Delta z}\left[\frac{\partial \Delta z}{\partial t} + \left(\frac{\partial\left(\Delta z U\right)}{\partial x}\right)_{y,z^*,t} + \left(\frac{\partial\left(\Delta z V\right)}{\partial y}\right)_{x,z^*,t} + \left(\frac{\partial \Delta z W^*}{\partial z^*}\right)_{x,y,t}\right] = 0 \quad (5.48)$$

For each degree of freedom i we should thus solve:

$$\int_{\Omega^*}\left[\frac{\partial \Delta z}{\partial t}+\left(\frac{\partial \Delta z U}{\partial x}\right)_{y,z^*,t}+\left(\frac{\partial \Delta z V}{\partial y}\right)_{x,z^*,t}+\left(\frac{\partial \Delta z W^*}{\partial z^*}\right)_{x,y,t}\right]\Psi_i^*\, d\Omega^*=0 \tag{5.49}$$

With a simple sigma transform we would in fact get:

$$\int_{\Omega^*}\left[\left(\frac{\partial h}{\partial t}\right)_{x,y}+\left(\frac{\partial hU}{\partial x}\right)_{y,z^*,t}+\left(\frac{\partial hV}{\partial y}\right)_{x,z^*,t}+\left(\frac{\partial hW^*}{\partial z^*}\right)_{x,y,t}\right]\Psi_i^*\, d\Omega^*=0 \tag{5.50}$$

where we see the underlying depth-integrated continuity equation.

It is actually this form of the mass conservation equation that is found while demonstrating the mass conservation of the tracer (see Section 5.6.2). Discretized as such, this equation whose unknown is W^* or rather $\Delta z W^*$ leads to a system that is very ill-conditioned, or even to a forced system. The boundary conditions are $W^*=0$ at the bottom and the free surface, if they are impermeable. The term:

$$\int_{\Omega^*}\left(\frac{\partial \Delta z W^*}{\partial z^*}\right)_{x,y,t}\Psi_i^*\, d\Omega^*$$

with the unknown $\Delta z W^*$ leads to a matrix A with general term:

$$A_{ij}=\int_{\Omega^*}\frac{\partial \Psi_j^*}{\partial z^*}\Psi_i^*\, d\Omega^* \tag{5.51}$$

Outside the boundaries, the diagonal terms of this matrix are zero. From the boundary conditions at the bottom and the free surface, we arrive at disconnected solutions between two successive planes (parasitic oscillations), or an impossibility. This is because the only unknown retained for solving mass conservation is the vertical velocity. Figure 5.4 shows the case of a mesh reduced to two planes, where no degree of freedom is available on the vertical i to find a vertical velocity for re-establishing a divergence-free flow.

A vertical velocity defined on a staggered mesh would be necessary, as in Figure 5.5.

To circumvent this problem, a solution consists of changing the unknown or, to be more precise, of giving a specific definition of the unknown $\Delta z W^*$. This definition, which will be given in the next paragraph, has been inspired by the necessity to get an exact mass conservation of tracers with distributive schemes. As a matter of fact, it will in due course appear that this definition of $\Delta z W^*$ is entirely compatible with these distributive schemes.

Calculation of $\Delta z W^*$: here we strictly solve, for every point i of the mesh:

$$\int_{\Omega^*}\left(\Delta z^{n+1}-\Delta z^{n}\right)\Psi_i^* d\Omega^*=\Delta t\int_{\Omega^*}\Delta z\vec{U}\,.\,\overrightarrow{grad}(\Psi_i^*)\, d\Omega^*-\Delta t\int_{\Gamma^*}\Delta z\vec{U}.\vec{n}\,\Psi_i^* d\Gamma^* \tag{5.52}$$

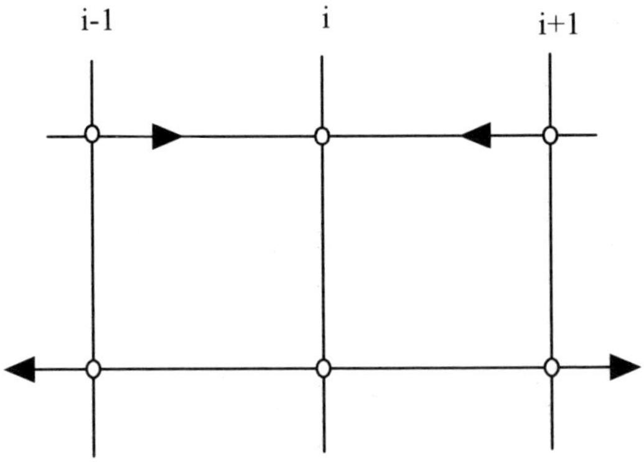

Figure 5.4: example of a forced problem

Note: on the right-hand side discretization of Δz is not specified. Actually it must be consistent with the 2D continuity equation, as will be explained later, which generally imposes that Δz is taken at time t^n.

We shall now show that the problem is well-posed if we consider that the unknowns are the average of $\Delta z W^*$ along the vertical of each prism.

If the horizontal and vertical gradients are differentiated, the continuity equation for a point i can be written as:

$$\int_{\Omega^*} \Delta z W^* \frac{\partial \Psi_i^*}{\partial z^*} d\Omega^* = \frac{1}{\Delta t} \int_{\Omega^*} \left(\Delta z^{n+1} - \Delta z^n\right) \Psi_i^* d\Omega^*$$
$$- \int_{\Omega^*} \Delta z \, \vec{U} \, . \overrightarrow{grad}_{2D}(\Psi_i^*) \, d\Omega^* + \int_{\Gamma^*} \Delta z \vec{U} . \vec{n} \, \Psi_i^* d\Gamma^* \tag{5.53}$$

In this form, the right-hand side contains only known terms (assuming that the flux boundary conditions are given at the boundaries). So, we concentrate on the left-hand side which contains unknowns. For this, it should be considered that we work with prisms in the transformed fixed mesh, hence with horizontal base and top. In these conditions, prisms are homothetic to the reference prism and the 3D bases are products of one function of x and y and one function of z (let us recall that for example $\Psi_1 = (1 - \alpha - \beta)\frac{(1-\gamma)}{2})$). The function of x and y is the same as the basis functions of triangles with linear interpolation. This allows us to break down each base as $\Psi_i^* = \Psi_i^h \Psi_i^v$. Ψ_i^h depends only on x and y, while Ψ_i^v depends only on z. Thus:

$$\frac{\partial \Psi_i^*}{\partial z^*} = \Psi_i^h \frac{\partial \Psi_i^v}{\partial z^*}$$

The function $\Delta z W^*$ can be defined on the whole domain as follows:

$$\Delta z W^* = \sum_{j=1}^{npoin2} \left[\Delta z W^*(z^*)\right]_j \Psi_i^h(x,y) \tag{5.54}$$

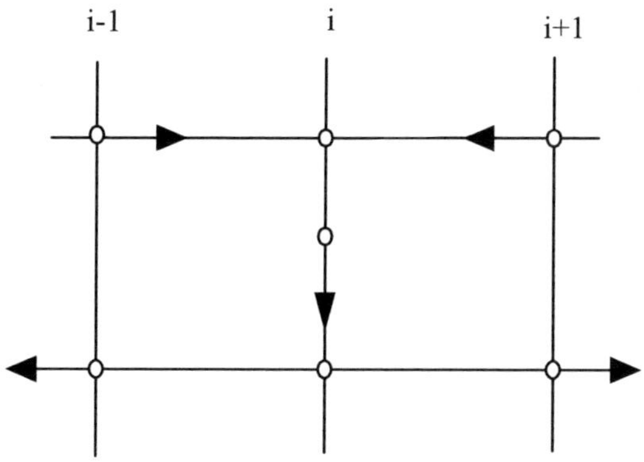

Figure 5.5: the corresponding vertical velocity

where *npoin*2 is the number of degrees of freedom in the 2D mesh. $\Delta z W^*(z^*)$ is a function defined between 0 and 1 for each vertical line of nodes of the transformed mesh.

The left-hand side of Equation 5.53 then becomes:

$$\sum_{j=1}^{npoin2} \left\{ \int_{\Omega_{2D}} \Psi_i^h \Psi_j^h d\Omega_{2D} \int_0^1 \left[\Delta z W^*(z^*)\right]_j \frac{\partial \Psi_i^v}{\partial z^*} dz^* \right\} \tag{5.55}$$

The mass matrix of the mesh of triangles appears in $\int_{\Omega_{2D}} \Psi_i^h \Psi_j^h \, d\Omega_{2D}$.

According to the position of the point i, the value of $\frac{\partial \Psi_i^v}{\partial z^*}$ in a prism is equal to $\pm\frac{1}{\Delta z^*}$, where Δz^* is the height of the prism (these bases are linear; their value is 1 at the top and 0 at the bottom or vice versa).

For a basis situated on the plane *ip*, a plane different from the bottom and the free surface, we get:

$$\int_0^1 \left[\Delta z W^*(z^*)\right]_j \frac{\partial \Psi_i^v}{\partial z^*} dz^* = \frac{1}{(z^*_{ip} - z^*_{ip-1})} \int_{z^*_{ip-1}}^{z^*_{ip}} \left[\Delta z W^*(z^*)\right]_j dz^*$$

$$-\frac{1}{(z^*_{ip+1} - z^*_{ip})} \int_{z^*_{ip}}^{z^*_{ip+1}} \left[\Delta z W^*(z^*)\right]_j dz^* \tag{5.56}$$

The quantities retrieved here are the ones which help to calculate the coefficients of distributive schemes (see Section 6.4.2).

Let us denote:

$$\overline{(\Delta z w^*)}^j_{ip-\frac{1}{2}} = \frac{1}{(z^*_{ip} - z^*_{ip-1})} \int_{z^*_{ip-1}}^{z^*_{ip}} \left[\Delta z W^*(z^*)\right]_j dz^* \tag{5.57}$$

We finally get:

$$\int_0^1 [\Delta z W^*(z^*)]_j \frac{\partial \Psi_i^v}{\partial z^*} dz^* = \overline{(\Delta z w^*)}^j_{ip-\frac{1}{2}} - \overline{(\Delta z w^*)}^j_{ip+\frac{1}{2}} \tag{5.58}$$

For a point at the bottom, we have:

$$\int_0^1 [\Delta z W^*(z^*)]_j \frac{\partial \Psi_i^v}{\partial z^*} dz^* = +\overline{(\Delta z w^*)}^j_{1+\frac{1}{2}} \tag{5.59}$$

and at the surface:

$$\int_0^1 [\Delta z W^*(z^*)]_j \frac{\partial \Psi_i^v}{\partial z^*} dz^* = -\overline{(\Delta z w^*)}^j_{np-\frac{1}{2}} \tag{5.60}$$

We arrive at a series of linear systems (one per plane), whose form is:

$$\sum_{j=1}^{npoin2} \left\{ \int_{\Omega_{2D}} \Psi_i^h \Psi_j^h d\Omega_{2D} \left[\overline{(\Delta z w^*)}^j_{ip+\frac{1}{2}} - \overline{(\Delta z w^*)}^j_{ip-\frac{1}{2}} \right] \right\}$$

$$= -\frac{1}{\Delta t} \int_{\Omega^*} \left(\Delta z^{n+1} - \Delta z^n\right) \Psi_i^* d\Omega^* + \int_{\Omega^*} \Delta z \, \vec{U} \, . \, \overrightarrow{grad}_{2D}(\Psi_i^*) d\Omega^*$$

$$- \int_{\Gamma^*_{liq}} \Delta z \vec{U} . \vec{n} \Psi_i^* d\Gamma^* \tag{5.61}$$

$\vec{U} \, . \, \overrightarrow{grad}_{2D}$ means that only the first two components of the 3D velocity are taken into account. In this form, the new unknowns $\overline{(\Delta z w^*)}^j_{ip+\frac{1}{2}}$ can be found plane by plane starting from the bottom, or in the same way from the surface.

The systems matrix is always the same mass matrix and is easily invertible. On the other hand, there are still too many equations and it is likely that, if we start from the bottom, the boundary condition at the free surface will not be retrieved. This would not happen if the 2D continuity equation was resolved in a compatible way to calculate the free surface. In fact, the sum of all equations, plane after plane, eliminates all the unknowns $\overline{(\Delta z w^*)}^j_{ip+\frac{1}{2}}$ and restores the mass conservation expression of the Saint-Venant equations (for more details, see next paragraph). If the latter has been correctly resolved, the linear system of the last plane is deduced from the others, and hence it cannot give rise to contradictions.

Important note: Equation 5.61 can be simplified by mass lumping. In this case the left-hand side is:

$$\int_{\Omega_{2D}} \Psi_i^h d\Omega_{2D} \left[\overline{(\Delta z w^*)}^i_{ip+\frac{1}{2}} - \overline{(\Delta z w^*)}^i_{ip-\frac{1}{2}} \right]$$

and the system is solved by a mere division of the right-hand side by $\int_{\Omega_{2D}} \Psi_i^h d\Omega_{2D}$. The same mass lumping must also be done when computing the corresponding terms in the tracer equation (see Section 6.4.2 and the computation of the terms b_i).

Compatibility between 2D and 3D continuity equations

In Equation 5.61, the term:

$$\frac{1}{\Delta t}\int_{\Omega^*}\left(\Delta z^{n+1}-\Delta z^{n}\right)\Psi_i^* d\Omega^*$$

is in fact treated in the real domain as:

$$\frac{1}{\Delta t}\left[\int_{\Omega^{n+1}}\Psi_i d\Omega^{n+1}-\int_{\Omega^n}\Psi_i d\Omega^n\right]$$

where the real domain Ω is indexed with n and $n+1$ as it changes with time. When summed over the vertical, these terms naturally yield the left-hand side of the 2D continuity equation, i.e.:

$$\frac{1}{\Delta t}\int_{\Omega_{2D}}(h^{n+1}-h^n)\Psi_i^{2D} d\Omega_{2D}$$

where Ψ_i^{2D} is the 2D basis in the 2D domain Ω_{2D}. However, the 2D continuity equation is sometimes computed with mass lumping, with a left-hand side approximated in the form:

$$\frac{1}{\Delta t}\int_{\Omega_{2D}}(h^{n+1}-h^n)\Psi_i^{2D} d\Omega_{2D}\simeq\frac{(h_i^{n+1}-h_i^n)}{\Delta t}\int_{\Omega_{2D}}\Psi_i^{2D} d\Omega_{2D}$$

then the terms $\int_{\Omega^{n+1}}\Psi_i d\Omega^{n+1}$ and $\int_{\Omega^n}\Psi_i d\Omega^n$ must be modified accordingly everywhere they are used. As a matter of fact, we must write:

$$\int_{\Omega}\Psi_i d\Omega=\int_{\Omega^*}h\Psi_i^* d\Omega^*\simeq h_i\int_{\Omega^*}\Psi_i^* d\Omega^*$$

to be compatible with the mass lumping in 2D.

Another problem is the term $\int_{\Omega^*}\Delta z\,\overrightarrow{U}\,.\,\overrightarrow{grad}_{2D}(\Psi_i^*)d\Omega^*$ which after summation over the vertical must give $\int_{\Omega_{2D}}h\,\overrightarrow{u}\,.\,\overrightarrow{grad}_{2D}(\Psi_i^{2D})d\Omega_{2D}$. If a generalized sigma transform is used, this is obtained only if we write:

$$\int_{\Omega^*}\Delta z\,\overrightarrow{U}\,.\,\overrightarrow{grad}_{2D}(\Psi_i^*)d\Omega^*=\int_{\Omega^*}h\,\frac{\Delta z\,\overrightarrow{U}}{h}\,.\,\overrightarrow{grad}_{2D}(\Psi_i^*)d\Omega^*$$

and consider h on the one hand and $\frac{\Delta z\overrightarrow{U}}{h}$ on the other hand as linear functions. By summing over the vertical we then see that we get the correct 2D term if $\overrightarrow{u}$ is a depth-averaged value of $\overrightarrow{U}$ obtained with the trapezoidal rule.

Nodal values of W^*?

One would now expect that we deduce W^* from ΔzW^* by dividing by Δz. This is not a good idea. As a matter of fact we need nodal values of W^*, e.g. for the advection in the transformed mesh with the method of characteristics, but if we write W in the transformed mesh, we get:

$$W=\frac{dz}{dt}=\frac{\partial z}{\partial t}+U\frac{\partial z}{\partial x}+V\frac{\partial z}{\partial y}+W^*\frac{\partial z}{\partial z^*}\tag{5.62}$$

and so we have:

$$\Delta z W^* = -\frac{\partial z}{\partial t} - U\frac{\partial z}{\partial x} - V\frac{\partial z}{\partial y} + W \tag{5.63}$$

We can deduce from this latter equation that if W is a linear function, then $\Delta z W^*$ is also a linear function. To get this linear function, we need to have nodal values of $\Delta z W^*$, which can be obtained by the following formula using the layer-averaged values of the previous paragraph:

$$\overline{(\Delta z w^*)}^j_{ip} = \frac{\overline{(\Delta z w^*)}^j_{ip-\frac{1}{2}} + \overline{(\Delta z w^*)}^j_{ip+\frac{1}{2}}}{2} \tag{5.64}$$

At the boundaries, i.e. at the bottom and at the free surface, we know that the impermeability condition imposes that $\Delta z W^* = 0$. We now have a linear function $\Delta z W^*$. When computing the characteristics in the transformed mesh, the vertical velocity in a layer will be this linear function $\Delta z W^*$ divided by Δz which is locally a function of x and y.

Passing from $\Delta z W^*$ to W

As the advection is done in the transformed mesh, W is actually never used, but users generally expect to see it in the results, and so it must be built.

This is done with Equation 5.63 which is discretized in time in the form:

$$W^{n+1} = (\Delta z W^*)^{n+1} + \frac{z^{n+1} - z^n}{\Delta t} + U\frac{\partial z^n}{\partial x} + V\frac{\partial z^n}{\partial y} \tag{5.65}$$

This equation is integrated layer by layer over the vertical in the transformed mesh, and then multiplied by a test function and integrated on the 2D mesh. It gives:

$$\int_{\Omega_{2D}} \left(\int_{z^*_{ip}}^{z^*_{ip+1}} W^{n+1} dz^* \right) \Psi_i^{2D} \, d\Omega_{2D} = \int_{\Omega_{2D}} \left(\int_{z^*_{ip}}^{z^*_{ip+1}} (\Delta z W^*)^{n+1} \, dz^* \right) \Psi_i^{2D} \, d\Omega_{2D}$$

$$+ \int_{\Omega_{2D}} \left(\int_{z^*_{ip}}^{z^*_{ip+1}} \left(\frac{z^{n+1} - z^n}{\Delta t} + U\frac{\partial z^n}{\partial x} + V\frac{\partial z^n}{\partial y} \right) dz^* \right) \Psi_i^{2D} \, d\Omega_{2D} \tag{5.66}$$

where ip is the number of the lower plane of the layer, and i is the global number of a node in the 2D mesh. As W is linear, we have simply:

$$\int_{z^*_{ip}}^{z^*_{ip+1}} W^{n+1} dz^* = (W_{ip}^{n+1} + W_{ip+1}^{n+1}) \frac{\Delta z^*_{ip+1/2}}{2}$$

By using similar formulae for $\frac{z^{n+1} - z^n}{\Delta t}$, we get:

$$\int_{\Omega_{2D}} (W_{ip}^{n+1} + W_{ip+1}^{n+1}) \Psi_i^{2D} \, d\Omega_{2D} = 2 \int_{\Omega_{2D}} (\Delta z W^*)^{n+1}_{ip+1/2} \Psi_i^{2D} \, d\Omega_{2D}$$

$$+\frac{1}{\Delta t}\int_{\Omega_{2D}}(z_{ip}^{n+1}+z_{ip+1}^{n+1}-z_{ip}^{n}-z_{ip+1}^{n})\Psi_i^{2D}\,d\Omega_{2D}$$

$$+\frac{2}{\Delta z^*_{ip+1/2}}\int_{\Omega_{2D}}\left(\int_{z^*_{ip}}^{z^*_{ip+1}}\left(U\frac{\partial z^n}{\partial x}+V\frac{\partial z^n}{\partial y}\right)dz^*\right)\Psi_i^{2D}\,d\Omega_{2D} \tag{5.67}$$

We recognize here a series of linear systems on the 2D mesh, the matrix being the mass matrix. For every layer in the mesh, we shall get $W_{ip}^{n+1}+W_{ip+1}^{n+1}$. If we start from $ip=1$, the first value W_1^{n+1} is given by the boundary condition at the bottom (see Equation 2.23), and then the boundary condition at the free surface should be naturally ensured. Another possibility is to start from the free surface, with Equation 2.21 and going down to the bottom. Lest the truncation errors or a lack of compatibility in the discretization should give a wrong boundary condition after going up or down, we build both solutions, creating a series of W_{ip}^{up} and W_{ip}^{down}, where W_1^{up} is obtained by the boundary condition at the bottom and $W_{np}^{down}$is obtained from the boundary condition at the free surface. The final value of the vertical velocity is then obtained by the formula:

$$W_{ip}^{n+1}=\frac{(ip-1)W_{ip}^{down}+(np-ip)W_{ip}^{up}}{np-1} \tag{5.68}$$

where np is still the number of superimposed meshes of triangles.

5.3 Non-hydrostatic Navier–Stokes equations

The hypothesis on hydrostatic pressure is put aside and we now solve:

Continuity:

$$div(\overrightarrow{U})=0 \tag{5.69}$$

Momentum equation:

$$\frac{\partial U}{\partial t}=-\frac{1}{\rho}\frac{\partial p}{\partial x}+\text{other terms} \tag{5.70}$$

$$\frac{\partial V}{\partial t}=-\frac{1}{\rho}\frac{\partial p}{\partial y}+\text{other terms} \tag{5.71}$$

$$\frac{\partial W}{\partial t}=-\frac{1}{\rho}\frac{\partial p}{\partial z}+g+\text{other terms} \tag{5.72}$$

The other terms mentioned in each equation are advection, turbulent diffusion and source terms, but not buoyancy terms. These terms have not been detailed as they are treated in the same manner as with the hydrostatic hypothesis. We now have an equation for the vertical component of velocity W, similar to the U and V equations, with a gravity term as well. The general idea of the resolution is again to use the fractional steps method, by simply adding a step after the resolution of the hydrostatic equations. Only the "continuity" steps are different. So, we break down pressure p into a hydrostatic term already denoted as p_h and a term called "dynamic pressure" denoted as p_d. We write:

$$p = p_h + p_d = \rho g(Z_s - z) + \rho_0 g \int_z^{Z_s} \frac{\Delta\rho}{\rho_0} dz' + p_d \tag{5.73}$$

After the advection and diffusion steps, the equations are solved in two other steps:
1) a "hydrostatic" step:

$$\frac{\vec{\widetilde{U}} - \vec{U}^n}{\Delta t} = -\frac{1}{\rho}\overrightarrow{grad}(p_h) + \text{other terms} \tag{5.74}$$

2) a continuity step:

$$div(\vec{U}^{n+1}) = 0 \tag{5.75}$$

$$\frac{\vec{U}^{n+1} - \vec{\widetilde{U}}}{\Delta t} = -\frac{1}{\rho}\overrightarrow{grad}(p_d) \tag{5.76}$$

The two steps are given in detail below.

5.3.1 Hydrostatic step

With the Boussinesq approximation:

$$p_h = \rho_o g\,(Z_s - z) + \rho_o g \int_z^{Z_s} \frac{\Delta\rho}{\rho_o} dz' \tag{5.77}$$

and the truncated series expansion of $1/\rho$, the equations of $\widetilde{U}$ and $\widetilde{V}$ become:

$$\frac{\widetilde{U} - U^n}{\Delta t} = -g\frac{\partial Z_s}{\partial x} + g\frac{\Delta\rho}{\rho_o}\frac{\partial Z_s}{\partial x} - g\frac{\partial}{\partial x}\left(\int_z^{Z_s} \frac{\Delta\rho}{\rho_o} dz'\right) + \text{other terms} \tag{5.78}$$

$$\frac{\widetilde{V} - V^n}{\Delta t} = -g\frac{\partial Z_s}{\partial y} + g\frac{\Delta\rho}{\rho_o}\frac{\partial Z_s}{\partial y} - g\frac{\partial}{\partial y}\left(\int_z^{Z_s} \frac{\Delta\rho}{\rho_o} dz'\right) + \text{other terms} \tag{5.79}$$

These equations are similar on all points to the "hydrostatic" version. The buoyancy terms included in the diffusion step are present. Again, for treating the free surface gradients the Saint-Venant equations are used, as in the hydrostatic option. A new free surface is computed.

The equation of W contains only diffusion and advection terms. In fact:

$$-\frac{1}{\rho}\frac{\partial p_h}{\partial z} + g = 0$$

by definition of hydrostatic pressure.

5.3.2 Continuity step

This step has been known as the "projection step" for the Navier–Stokes equations since its publication in 1968 [28], the gradient of dynamic pressure acting as a correction ensuring the divergence-free velocity. What is left is:

$$\frac{\vec{U}^{n+1} - \vec{\tilde{U}}}{\Delta t} = -\frac{1}{\rho}\overrightarrow{grad}(p_d) \tag{5.80}$$

The condition $div(\vec{U}^{n+1}) = 0$ is thus expressed as:

$$div\left(\frac{1}{\rho}\overrightarrow{grad}(p_d)\right) = \frac{1}{\Delta t}div(\vec{\tilde{U}}) \tag{5.81}$$

The variational formulation is written as:

$$\int_\Omega div\left(\frac{1}{\rho}\overrightarrow{grad}(p_d)\right)\Psi_i\, d\Omega = \frac{1}{\Delta t}\int_\Omega div(\vec{\tilde{U}})\Psi_i\, d\Omega \tag{5.82}$$

The left-hand side of this equation is integrated by parts:

$$\int_\Gamma \frac{\Psi_i}{\rho}\overrightarrow{grad}(p_d).\vec{n}\, d\Gamma - \int_\Omega \frac{1}{\rho}\overrightarrow{grad}(p_d).\overrightarrow{grad}(\Psi_i)\, d\Omega = \frac{1}{\Delta t}\int_\Omega div(\vec{\tilde{U}})\Psi_i\, d\Omega \tag{5.83}$$

and we arrive at the system:

$$\int_\Omega \frac{\Delta t}{\rho}\overrightarrow{grad}(p_d).\overrightarrow{grad}(\Psi_i)\, d\Omega = \int_\Gamma \frac{\Delta t \Psi_i}{\rho}\overrightarrow{grad}(p_d).\vec{n}\, d\Gamma - \int_\Omega div(\vec{\tilde{U}})\Psi_i\, d\Omega \tag{5.84}$$

If we assume that the lateral boundary conditions are already set on $\vec{\tilde{U}}$, then we must have $\overrightarrow{grad}(p_d).\vec{n} = 0$ on these boundaries. The boundary terms are thus non-zero only at the bottom and the free surface. They are deduced from the dynamic boundary conditions (Equations 2.22 and 2.24):

At the free surface:

$$\int_\Gamma \frac{\Delta t \Psi_i}{\rho}\overrightarrow{grad}(p_d).\vec{n}\ \ d\Gamma = \int_\Gamma \Psi_i \frac{\vec{\tilde{U}}_s.\vec{n}_s - \frac{\partial Z_s}{\partial t}}{\|\vec{n}_s\|}d\Gamma \tag{5.85}$$

At the bottom:

$$\int_\Gamma \frac{\Delta t \Psi_i}{\rho}\overrightarrow{grad}(p_d).\vec{n}\ \ d\Omega = \int_\Gamma \Psi_i \frac{\vec{\tilde{U}}_f.\vec{n}_f - \frac{\partial Z_f}{\partial t}}{\|\vec{n}_f\|}d\Gamma \tag{5.86}$$

The integrals here are computed on the 3D boundaries, but if they are computed on the 2D (flat) mesh, they become simpler:

At the free surface:

$$\int_\Gamma \frac{\Delta t \Psi_i}{\rho}\overrightarrow{grad}(p_d).\vec{n}\ \ d\Gamma = \int_{\Omega 2D} \Psi_i\left(\vec{\tilde{U}}_s.\vec{n}_s - \frac{\partial Z_s}{\partial t}\right)d\Omega \tag{5.87}$$

At the bottom:

$$\int_{\Gamma} \frac{\Delta t \Psi_i}{\rho} \overrightarrow{grad}(p_d).\vec{n} \; d\Omega = \int_{\Omega 2D} \Psi_i \left(\vec{\widetilde{U}}_f.\vec{n}_f + \frac{\partial Z_f}{\partial t} \right) d\Omega \tag{5.88}$$

The matrix of the linear system obtained is a diffusion-type matrix built with $1/\rho$ in the place of viscosity. The boundary condition on the free surface is $p_d = 0$. Setting such a Dirichlet boundary condition at the free surface will in fact cancel the boundary terms of Equation 5.85.

Once p_d is known, $\vec{U}^{n+1}$ is retrieved by the formula:

$$\frac{\vec{U}^{n+1} - \vec{\widetilde{U}}}{\Delta t} = -\frac{1}{\rho} \overrightarrow{grad}(p_d)$$

which can also be subjected to a variational formulation.

Li and Fleming [126] use such a projection step in finite differences, but the free surface gradient is also treated. The important point is that the step dealing with the free surface gradient should do so considering the constraint of mass conservation, which is the case when we use the Saint-Venant equations in the "hydrostatic" step.

5.3.3 Calculation of free surface

Once the final velocities $\vec{U}^{n+1}$ are known, the free surface which has already been updated by the hydrostatic step could be computed again, but is not, in order to maintain a compatibility between the depth and the depth-averaged continuity equation. This compatibility is necessary for tracer conservation. Moreover, tests have shown that computing a second time the free surface at this level is less stable, and adds no real accuracy to the results. Solving dynamic boundary conditions for the free surface would also spoil the depth-averaged continuity equation.

5.4 Pseudo wave equation, a variant of the hydrostatic step

5.4.1 With explicit diffusion

This is a very interesting variant which groups the diffusion step and the continuity equation of Section 5.2 in a single step, and replaces the hydrostatic step of the previous section. As has been done in 2D (see Section 4.12), here again the velocity is eliminated from the continuity equation averaged over the vertical. Let us consider resolution with a hydrostatic hypothesis. By assuming that an advection step (explicit or method of characteristics) has already given a velocity field $\vec{U}^C$, now the following equations have to be solved simultaneously:

$$\frac{h^{n+1} - h^n}{\Delta t} + div(h \left[\theta_u \vec{u}^{n+1} + (1 - \theta_u) \vec{u}^n\right]) = Sce$$

and:

$$\frac{\overrightarrow{U}^{n+1} - \overrightarrow{U}^{C}}{\Delta t} = -s1u\ \overrightarrow{U}^{n+1} - g\,\overrightarrow{grad}(Z_s) + \overrightarrow{F} + div(\nu_t grad(\overrightarrow{U}))$$

where $\overrightarrow{F}$ specifically contains the buoyancy terms. That is:

$$\overrightarrow{F} = g\frac{\Delta\rho}{\rho_o}\overrightarrow{grad}(Z_s) - g\,\overrightarrow{grad}\left(\int_z^{Z_s}\frac{\Delta\rho}{\rho_o}dz'\right)$$

The term $-s1u\ \overrightarrow{U}^{n+1}$ represents the manner in which any possible implicit source terms are treated. $s1u$ is a function discretized as the velocity.

Initially, we write the variational formulation of the depth-averaged continuity equation, with integration by parts of the divergence term:

$$\int_{\Omega_{2D}} \Psi_i^{2D}\left(\frac{h^{n+1}-h^n}{\Delta t}\right) d(\Omega_{2D}) + flux2d(i)$$

$$-\int_{\Omega_{2D}} h\left[\theta_u\overrightarrow{u}^{n+1} + (1-\theta_u)\overrightarrow{u}^{n}\right].\overrightarrow{grad}(\Psi_i^{2D})d(\Omega_{2D}) = s2d(i) \tag{5.89}$$

with:

$$flux2d(i) = \int_{\Gamma_{2D}} \Psi_i^{2D} h\left[\theta_u\overrightarrow{u}^{n+1} + (1-\theta_u)\overrightarrow{u}^{n}\right].\overrightarrow{n}_{2D}\, d(\Gamma_{2D})$$

and:

$$s2d(i) = \int_{\Omega_{2D}} \Psi_i^{2D}\ Sce\ d(\Omega_{2D})$$

This expression for $s2d(i)$ is actually slightly incorrect but is left as such here for simplicity. We shall see in Section 5.7.1 that this term has to be a sum over the vertical of 3D source terms, for reasons of compatibility between 2D and 3D continuity equations.

$flux2d(i)$ can be treated as in 2D, e.g. on solid walls by simply cancelling the fluxes, but in a 3D form:

$$flux2d(i) = \int_{\Gamma_{3D}} \Psi_i^{2D}\left[\theta_u\overrightarrow{U}^{n+1} + (1-\theta_u)\overrightarrow{U}^{n}\right].\overrightarrow{n}_{3D}\, d(\Gamma_{3D}) \tag{5.90}$$

the base Ψ_i^{2D} being a function of only x and y in the integrals. In fact, $flux2d(i)$ can be obtained by a summation over the vertical of i of the terms:

$$flux3d(j) = \int_{\Gamma_{3D}} \Psi_j^{3D}\left[\theta_u\overrightarrow{U}^{n+1} + (1-\theta_u)\overrightarrow{U}^{n}\right].\overrightarrow{n}_{3D}\, d(\Gamma_{3D}) \tag{5.91}$$

Then the velocities averaged over the vertical are replaced by their definition:

$$\int_{\Omega_{2D}} \Psi_i^{2D}\left(\frac{h^{n+1}-h^n}{\Delta t}\right) d(\Omega_{2D}) + flux2d(i)$$

$$-\int_{\Omega_{2D}} \left[\theta_u \int_{Z_f}^{Z_s} \overrightarrow{U}^{n+1} dz + (1-\theta_u) \int_{Z_f}^{Z_s} \overrightarrow{U}^n dz \right] .\overrightarrow{grad}(\Psi_i^{2D})\, d(\Omega_{2D}) = s2d(i) \tag{5.92}$$

For reasons of compatibility between 2D and 3D continuity equations, fluxes like:

$$\int_{\Omega_{2D}} \int_{Z_f}^{Z_s} \overrightarrow{U}.\overrightarrow{grad}(\Psi_i^{2D})\ d(\Omega_{2D})$$

will be treated as:

$$\sum_{j \text{ on the vertical of } i} \int_{\Omega_{3D}^*} \Delta z\ \overrightarrow{U}.\overrightarrow{grad}^*(\Psi_j^{3D})\ d(\Omega_{3D}^*)$$

We thus get:

$$\int_{\Omega_{2D}} \Psi_i^{2D} \left(\frac{h^{n+1} - h^n}{\Delta t} \right)\ d(\Omega_{2D}) + flux2d(i)$$

$$-\sum_{j \text{ on the vertical of } i} \int_{\Omega_{3D}^*} \Delta z\ \left(\theta_u \overrightarrow{U}^{n+1} + (1-\theta_u)\overrightarrow{U}^n \right) .\overrightarrow{grad}^*(\Psi_j^{3D})\ d(\Omega_{3D}^*) = s2d(i) \tag{5.93}$$

In the momentum equation, increments of the water depth are made to appear, by a semi-implicitation of the free surface as for the Saint-Venant equations:

$$\frac{\overrightarrow{U}^{n+1} - \overrightarrow{U}^C}{\Delta t} = -s1u\ \overrightarrow{U}^{n+1} - g\theta_h\, \overrightarrow{grad}(h^{n+1} - h^n)$$

$$-g\, \overrightarrow{grad}(Z_s^n) + \overrightarrow{F} + div(\nu_t \overrightarrow{grad}(\overrightarrow{U})) \tag{5.94}$$

Ideally we would like to obtain a variational formulation of this equation which provides an expression of $\overrightarrow{U}^{n+1}$ which can be introduced into the continuity equation, and, once h^{n+1} is known, helps to retrieve in a compatible manner the value of $\overrightarrow{U}^{n+1}$ itself.

The diffusion term will thus be written as:

$$\frac{\int_{\Omega_{3D}} div(\nu_t \overrightarrow{grad}(\overrightarrow{U}))\ \Psi_i^{3D}\ d(\Omega_{3D})}{\int_{\Omega_{3D}} \Psi_i^{3D}\ d(\Omega_{3D})}$$

a form in which the discretization in time is still not specified since we wish to retain the friction terms in an implicit form and the others in an explicit form.

The gradient term $-g\theta_h\, \overrightarrow{grad}(h^{n+1} - h^n)$, simplified in $-g\theta_h\, \overrightarrow{grad}(\delta h)$, will be kept in the continuous form for the term introduced into the continuity equation, and calculated just like $-g\, \overrightarrow{grad}(Z_s^n)$, as a nodal value for the final calculation of $\overrightarrow{U}^{n+1}$. That is:

$$\frac{\int_{\Omega_{2D}}[-g\theta_h\,\overrightarrow{grad}(\delta h) - g\,\overrightarrow{grad}(Z_s^n)]\ \Psi_i^{2D}\,d(\Omega_{2D})}{\int_{\Omega_{2D}}\Psi_i^{2D}\ d(\Omega_{2D})}$$

Let us now examine the diffusion term which should be integrated by parts:

$$\int_{\Omega_{3D}} div(\nu_t\overrightarrow{grad}(\vec{U}))\ \Psi_i^{3D}\,d(\Omega_{3D})$$

$$= \int_{\Gamma_{3D}} \Psi_i^{3D}\,\nu_t\,\overrightarrow{grad}(\vec{U}).\vec{n}\ \ d(\Gamma_{3D}) - \int_{\Omega_{3D}} \nu_t\,\overrightarrow{grad}(\vec{U})\ .\ \overrightarrow{grad}(\Psi_i^{3D}\,)\,d(\Omega_{3D}) \quad (5.95)$$

By specifying the discretization, we write, as mentioned:

$$\int_{\Omega_{3D}} div(\nu_t\overrightarrow{grad}(\vec{U}))\ \Psi_i^{3D}\,d(\Omega_{3D}) \simeq \int_{\Gamma_{3D}} \Psi_i^{3D}\,\nu_t\,\overrightarrow{grad}(\vec{U}^{n+1}).\vec{n}\ \ d(\Gamma_{3D})$$

$$- \int_{\Omega_{3D}} \nu_t\,\overrightarrow{grad}(\vec{U}^{n})\ .\ \overrightarrow{grad}(\Psi_i^{3D}\,)\,d(\Omega_{3D}) \quad (5.96)$$

Now, to deal with $\nu_t\,\overrightarrow{grad}(\vec{U}^{n+1}).\vec{n}$ we write that:

$$\nu_t\,\overrightarrow{grad}(U^{n+1}).\vec{n} = aubor\ U^{n+1} + bubor \quad (5.97)$$

and that:

$$\nu_t\,\overrightarrow{grad}(V^{n+1}).\vec{n} = avbor\ V^{n+1} + bvbor \quad (5.98)$$

where $avbor = aubor$ represents the stress due to friction and *bubor* and *bvbor* an explicit stress due for example to wind at the free surface or any other stress at the bottom or the lateral boundaries.

The term $\int_{\Gamma_{3D}} \Psi_i^{3D}\,\nu_t\,\overrightarrow{grad}(\vec{U}^{n+1}).\vec{n}\ \ d(\Gamma_{3D})$ finally takes the form of a diagonal matrix multiplied by the velocity vector plus an explicit stress denoted $\overrightarrow{bubor}$, with components *bubor* and *bvbor*. We note that:

$$\frac{\int_{\Gamma_{3D}}\Psi_i^{3D}\,aubor\ \vec{U}^{n+1}\ \ d(\Gamma_{3D})}{\int_{\Omega_{3D}}\Psi_i^{3D}\ d(\Omega_{3D})} = -\,fric3d\ \ \vec{U}^{n+1}$$

In the "mass-lumped" form, $fric3d$ is such that:

$$fric3d \simeq -\frac{\int_{\Gamma_{3D}}\Psi_i^{3D}\ \ d(\Gamma_{3D})}{\int_{\Omega_{3D}}\Psi_i^{3D}\ d(\Omega_{3D})}\,aubor = -\frac{aubor}{\cos(\alpha)}\,\frac{\int_{\Omega_{2D}}\Psi_i^{2D}\ d(\Omega_{2D})}{\int_{\Omega_{3D}}\Psi_i^{3D}\ d(\Omega_{3D})}$$

where α is the slope angle, such that:

$$\frac{1}{\cos(\alpha)} = \sqrt{1 + \left(\frac{\partial Z_f}{\partial x}\right)^2 + \left(\frac{\partial Z_f}{\partial y}\right)^2}$$

As a matter of fact we have:

$$d(\Gamma_{3D}) = \sqrt{1 + \left(\frac{\partial Z_f}{\partial x}\right)^2 + \left(\frac{\partial Z_f}{\partial y}\right)^2} d(\Omega_{2D})$$

on the bottom, because the 2D domain is flat, whereas there is a slope of the 2D boundary of the 3D domain.

We also note that:

$$\overrightarrow{diff} = \frac{\int_{\Omega 3D} \nu_t \overrightarrow{grad}(\overrightarrow{U}^n) . \overrightarrow{grad}(\Psi_i^{3D}) d(\Omega_{3D})}{\int_{\Omega 3D} \Psi_i^{3D} d(\Omega_{3D})} - \overrightarrow{bubor} \frac{\int_{\Gamma_{3D}} \Psi_i^{3D} \quad d(\Gamma_{3D})}{\int_{\Omega_{3D}} \Psi_i^{3D} d(\Omega_{3D})}$$

and finally, we get:

$$\frac{\int_{\Omega 3D} div(\nu_t \overrightarrow{grad}(\overrightarrow{U})) \Psi_i^{3D} d(\Omega_{3D})}{\int_{\Omega 3D} \Psi_i^{3D} d(\Omega_{3D})} = - fric3d \overrightarrow{U}^{n+1} - \overrightarrow{diff}$$

The expression $\overrightarrow{U}^{n+1}$ carried over in the mass conservation equation will be:

$$\overrightarrow{U}^{n+1} = D^{-1} \left[\overrightarrow{U}^C + \Delta t[-g\theta_h \overrightarrow{grad}(\delta h) - g \overrightarrow{grad}(Z_s^n) + \overrightarrow{F} - \overrightarrow{diff}]\right]$$

with:

$$D = 1 + \Delta t (fric3d + s1u)$$

We have:

$$D^{-1} = \frac{1}{1 + \Delta t (fric3d + s1u)}$$

but it is intentionally kept in the form D^{-1} because D will become a matrix in the next section.

The momentum equation is written as:

$$\overrightarrow{U}^{n+1} = D^{-1} \left[\Delta t \left(\frac{\int_{\Omega_{2D}} [-g\theta_h \overrightarrow{grad}(\delta h) - g \overrightarrow{grad}(Z_s^n)] \Psi_i^{2D} d(\Omega_{2D})}{\int_{\Omega_{2D}} \Psi_i^{2D} d(\Omega_{2D})}\right)\right]$$

$$+ D^{-1} \left(\overrightarrow{U}^C + \overrightarrow{F} - \overrightarrow{diff}\right) \tag{5.99}$$

Observing that, at the continuum level:

$$\theta_u \overrightarrow{U}^{n+1} + (1-\theta_u)\overrightarrow{U}^n = \theta_u D^{-1} \left[\overrightarrow{U}^C + \Delta t(\overrightarrow{F} - \overrightarrow{diff} - g \overrightarrow{grad}(Z_s^n))\right] + (1-\theta_u)\overrightarrow{U}^n$$

$$- \Delta t D^{-1} g\theta_h \theta_u \overrightarrow{grad}(\delta h) \tag{5.100}$$

and noting that:

$$\overrightarrow{U}^{aux} = D^{-1} \left[\overrightarrow{U}^C + \Delta t(\overrightarrow{F} - \overrightarrow{diff} - g \overrightarrow{grad}(Z_s^n))\right] \tag{5.101}$$

the continuity equation becomes:

$$\int_{\Omega_{2D}} \Psi_i^{2D} \left(\frac{h^{n+1} - h^n}{\Delta t} \right) d(\Omega_{2D})$$

$$+ \int_{\Omega_{3D}} \left[\Delta t D^{-1} g \theta_h \, \theta_u \overrightarrow{grad}(h^{n+1} - h^n) \right] .\overrightarrow{grad}(\Psi_i^{2D}) \, d(\Omega_{3D})$$

$$= \int_{\Omega_{3D}} \left[\theta_u \overrightarrow{U}^{aux} + (1 - \theta_u) \overrightarrow{U}^n \right] .\overrightarrow{grad}(\Psi_i^{2D}) d(\Omega_{3D}) + s2d - flux2d \qquad (5.102)$$

or even:

$$\int_{\Omega_{2D}} \Psi_i^{2D} \left(\frac{h^{n+1} - h^n}{\Delta t} \right) d(\Omega_{2D})$$

$$+ \int_{\Omega_{2D}} \left[\int_{Z_f}^{Z_s} \Delta t D^{-1} \, dz \right] \left[g \theta_h \, \theta_u \overrightarrow{grad}(h^{n+1} - h^n) \right] .\overrightarrow{grad}(\Psi_i^{2D}) \, d(\Omega_{2D})$$

$$= \int_{\Omega_{3D}} \left[\theta_u \overrightarrow{U}^{aux} + (1 - \theta_u) \overrightarrow{U}^n \right] .\overrightarrow{grad}(\Psi_i^{2D}) d(\Omega_{3D}) + s2d - flux2d \qquad (5.103)$$

In the second line of this equation there is a diffusion matrix built with a "viscosity" ν_{wave}:

$$\nu_{wave} = g \Delta t \, \theta_h \, \theta_u \int_{Z_f}^{Z_s} D^{-1} \, dz$$

Remark: the term:

$$\int_{\Omega_{3D}} \left[\theta_u \overrightarrow{U}^{aux} + (1 - \theta_u) \overrightarrow{U}^n \right] .\overrightarrow{grad}(\Psi_i^{2D}) d(\Omega_{3D})$$

is treated as:

$$\int_{\Omega_{2D}} \int_{Z_f}^{Z_s} \left[\theta_u \overrightarrow{U}^{aux} + (1 - \theta_u) \overrightarrow{U}^n \right] dz .\overrightarrow{grad}(\Psi_i^{2D}) d(\Omega_{2D})$$

Finally, the continuity equation is:

$$\int_{\Omega_{2D}} \Psi_i^{2D} \left(\frac{h^{n+1} - h^n}{\Delta t} \right) d(\Omega_{2D})$$

$$+ \int_{\Omega_{2D}} \nu_{wave} \, \overrightarrow{grad}(h^{n+1} - h^n) \, .\overrightarrow{grad}(\Psi_i^{2D}) \, d(\Omega_{2D})$$

$$= \int_{\Omega_{3D}} \left[\theta_u \overrightarrow{U}^{aux} + (1 - \theta_u) \overrightarrow{U}^n \right] .\overrightarrow{grad}(\Psi_i^{2D}) d(\Omega_{3D}) + s2d - flux2d \qquad (5.104)$$

And the momentum equation is:

$$\vec{U}^{n+1} = \vec{U}^{aux} + \Delta t\, D^{-1}\left[\frac{\int_{\Omega_{2D}}[-g\theta_h\, \overrightarrow{grad}(h^{n+1}-h^n)]\ \Psi_i^{2D}\, d(\Omega_{2D})}{\int_{\Omega_{2D}} \Psi_i^{2D}\ d(\Omega_{2D})}\right] \tag{5.105}$$

5.4.2 With implicit diffusion?

The fact that the diffusion is treated in an explicit way is a drawback of the method which can impose limitations on the time step. Let us now assume that the diffusion is implicit, and that we thus have $\overrightarrow{diff} = M\,\vec{U}^{n+1}$, where M is a matrix. All that has been written in the previous section remains valid if we erase the term $\overrightarrow{diff}$ and change D to $D + \Delta t\, M$. D would then be a matrix but, even if we can invert it by means of an iterative system for computing $\vec{U}^{aux}$ with Equation 5.101, terms like $D^{-1}\overrightarrow{grad}(h^{n+1}-h^n)$ in Equation 5.103 seem a nightmare. However, if we only have a vertical diffusion linking together the velocities over the verticals in the 3D mesh, D is composed of a series of tri-diagonal matrices, one for every vertical, and can be easily inverted by solving a series of independent tri-diagonal systems. $D^{-1}\overrightarrow{grad}(h^{n+1}-h^n)$ can be computed in an implicit way, i.e. keeping the unknown h^{n+1} because $\overrightarrow{grad}(h^{n+1}-h^n)$ is a 2D term which is constant on every vertical. Thus we have:

$$D^{-1}\overrightarrow{grad}(h^{n+1}-h^n) = \overrightarrow{grad}(h^{n+1}-h^n)D^{-1}(vector\ unity)$$

where *vector unity* is a vector with all components equal to 1.

It is thus possible to treat part of the diffusion terms in an implicit way; it suffices to remove from $\overrightarrow{diff}$ the terms:

$$\vec{U}_j^n \frac{\int_{\Omega_{3D}} \nu_t\, \overrightarrow{grad}(\Psi_i^{3D})\ .\ \overrightarrow{grad}(\Psi_j^{3D})\, d(\Omega_{3D})}{\int_{\Omega_{3D}} \Psi_i^{3D}\ d(\Omega_{3D})}$$

and replace them by:

$$\vec{U}_j^{n+1} \frac{\int_{\Omega_{3D}} \nu_t\, \overrightarrow{grad}(\Psi_i^{3D})\ .\ \overrightarrow{grad}(\Psi_j^{3D})\, d(\Omega_{3D})}{\int_{\Omega_{3D}} \Psi_i^{3D}\ d(\Omega_{3D})}$$

if we choose diagonal terms or extra-diagonal terms linking velocities of the same vertical. If there is vertical diffusion only, it can then, amazingly, be treated in a fully implicit way. If there is a horizontal diffusion, some extra-diagonal terms will have to remain explicit, and this different treatment would only be valid in steady state conditions. It is also worth noticing that when there is a vertical diffusion only, the depth integration of the diffusion terms should result only in the stresses at the bottom and at the free surface, so that the term denoted here by $\overrightarrow{diff}$ should have no influence on the 2D continuity equation. This is, however, not so obvious to show at the discrete level.

5.5 Tracer equation in 3D

The classical tracer equation:

$$\frac{\partial T}{\partial t} + \left(\vec{U}\,.\,\overrightarrow{grad}\right) T - div\left(\nu_T \;\; \overrightarrow{grad}(T)\right) + Q = 0$$

which is in fact in a moving grid:

$$\frac{\partial T}{\partial t} + \left(\left[\vec{U} - \vec{W}_{mesh}\right].\,\overrightarrow{grad}\right) T - div\left(\nu_T \;\; \overrightarrow{grad}(T)\right) + Q = 0$$

is resolved in two steps: advection, then diffusion and source terms. Advection takes place in the fixed transformed mesh with the methods presented in Chapter 6, or is coupled with diffusion (the case of the SUPG scheme). The advection equation is written as:

$$\int_{\Omega^*} \Delta z\,(T^{n+1} - T^n)\,\Psi_i^* d\Omega^* + \Delta t \int_{\Omega^*} \Delta z\,\vec{U}^*.\overrightarrow{grad}(T)\,\Psi_i^* d\Omega^* = 0 \tag{5.106}$$

We recall that the velocity of the mesh is also taken into account by this formula, as explained in Section 2.2.5.

In 1995, Janin (see [105]) found a remarkable solution for building a monotonic and at the same time conservative scheme:

- The use of a distributive scheme called the MURD scheme (see Section 6.4.2).
- The choice of advection in a mesh fixed in time t^{n+1}, with T written in an explicit way in the term $\overrightarrow{grad}(T)$ (see the demonstration of conservation of the tracer in Section 5.6.2).
- A compatible resolution of the mass conservation equation, with a variable $\Delta z W^*$ averaged per layer as seen in Section 5.2.7.

Only a good comprehension of these three points allows one to appreciate the interest and originality of this approach.

Diffusion has fewer constraints and is possible in the real mesh. In 2005 mass conservation of the tracer was extended to the SUPG advection scheme, using the same ideas.

5.6 The conservation of mass in 3D

5.6.1 Conservation of mass of water

In Section 5.3 we have seen that conservation of mass on a domain Ω is written as:

$$\int_{\Omega} \frac{\partial \rho}{\partial t}\, d\Omega + \int_{\Gamma} \rho\,\vec{U}.\vec{n}\, d\Gamma = 0 \tag{5.107}$$

Given the cylindrical form of Ω, and the definition of h, after discretization in time this equation is equivalent to:

$$\int_{\Omega_{2D}} h^{n+1} d\Omega_{2D} - \int_{\Omega_{2D}} h^n d\Omega_{2D} = -\Delta t \int_{\Gamma_{3D}} \vec{U}.\vec{n}\, d\Gamma_{3D} \qquad (5.108)$$

where Ω_{2D} is the mesh of triangles and Γ_{3D} the boundary of the 3D domain.

If we ensure a compatible calculation of the flux at the lateral boundaries:

$$\int_{\Gamma_{3D}} \vec{U}.\vec{n}\, d\Gamma_{3D} = \int_{\Gamma_{2D}} h\vec{u}.\vec{n}\, d\Gamma_{2D} \qquad (5.109)$$

we recognize the conservation of mass written for the Saint-Venant equations. Thus this condition will be ensured in the hydrostatic case by resorting to the Saint-Venant equations for calculating the free surface.

If this equality is not verified, the mass of water is not conserved. The difference between the left-hand side and the right-hand side constitutes the gain in mass between the instants t^n and t^{n+1}.

This expression is also equivalent to:

$$\int_{\Omega} \left(\frac{\partial U}{\partial x} + \frac{\partial V}{\partial y} + \frac{\partial W}{\partial z} \right) d\Omega = 0 \qquad (5.110)$$

but overall conservation of mass is possible without a local resolution of continuity, hence without a divergence-free 3D velocity field. A more localized conservation will be imposed for the conservation of the tracer.

The conservation of mass is thus naturally verified by the resolution of the Saint-Venant equations in the case of equations with a hydrostatic hypothesis. This is applicable also for non-hydrostatic equations if the free surface is calculated with the continuity equation. On the other hand, this conservation would become problematic with a dynamic boundary condition applied to the free surface.

5.6.2 Conservation of tracer

The conservation of the tracer is written as:

$$\int_{\Omega} (T^{n+1} - T^n)\, d\Omega + \Delta t \int_{\Gamma} T\vec{U}.\vec{n}\, d\Gamma - \Delta t \int_{\Gamma} \nu_t \frac{\partial T}{\partial n} d\Gamma - \Delta t \int_{\Omega} Q\, d\Omega = 0 \qquad (5.111)$$

where Q is a source term. If Dirichlet-type boundary conditions are imposed on the boundaries, this spoils the conservation and a correction is needed (see below in Section 5.7.3).

The non-conservative equation resolved for each degree of freedom is as follows:

$$\int_{\Omega} \left(\frac{\partial T}{\partial t} + \vec{U}.\overrightarrow{grad}(T) - \nabla\left(\nu_t \nabla T\right) - Q \right) \Psi_i\, d\Omega = 0 \qquad (5.112)$$

Hereafter, we shall drop the diffusion term and the creation term Q which are not really a problem for the conservation of mass. By summing over all the test functions, and after discretization in time, the advection equation of the tracer becomes:

$$\int_{\Omega} (T^{n+1} - T^n)\, d\Omega + \Delta t \int_{\Omega} \vec{U}.\overrightarrow{grad}(T)\, d\Omega = 0 \qquad (5.113)$$

To arrive at:

$$\int_{\Omega} (T^{n+1} - T^n)\, d\Omega + \Delta t \int_{\Gamma} T\,\vec{U}\,.\vec{n}\, d\Gamma = 0$$

it would be necessary to add to the right-hand side the term:

$$\Delta t \int_{\Omega} T\, div(\vec{U})\, d\Omega$$

which, in principle, is zero. This is where a compatibility condition with the continuity equation appears.

This formulation in the real domain masks the problem of the evolution of Ω and the fact that the advection step of the tracer should be done on a constant mesh. By using the sigma transform, the conservation of the tracer is:

$$\int_{\Omega^*} (\Delta z^{n+1} T^{n+1} - \Delta z^n T^n)\, d\Omega^* + \Delta t \int_{\Gamma^*} \Delta z T\,\vec{U}^*.\vec{n}\, d\Gamma^* = 0 \qquad (5.114)$$

Now, the sum of all the advection equations of the tracer with a fixed mesh at an instant taken between t^n and t^{n+1} gives:

$$\int_{\Omega^*} \Delta z\, (T^{n+1} - T^n)\, d\Omega^* + \Delta t \int_{\Omega^*} \Delta z\, \vec{U}^*.\overrightarrow{grad}(T)\, d\Omega^* = 0 \qquad (5.115)$$

In this equation, if the fixed mesh is chosen at instant $t^n + \theta_h \Delta t$, the following should be taken:

$$\Delta z = \theta_h \Delta z^{n+1} + (1 - \theta_h)\Delta z^n$$

We then need to choose an implicit expression of T for calculating its gradient. If we denote:

$$T = \theta_T T^{n+1} + (1 - \theta_T) T^n$$

the integral equation above, by adding the following null expression to it:

$$\Delta t \int_{\Gamma^*} \Delta z T\,\vec{U}^*.\vec{n}\, d\Gamma^* - \Delta t \int_{\Gamma^*} \Delta z T\,\vec{U}^*.\vec{n}\, d\Gamma^*$$

can then be written as:

$$\int_{\Omega^*} (\Delta z^{n+1} T^{n+1} - \Delta z^n T^n)\, d\Omega^* + \Delta t \int_{\Gamma^*} \Delta z T\,\vec{U}^*.\vec{n}\, d\Gamma^*$$

$$- \int_{\Omega^*} \theta_h T^n (\Delta z^{n+1} - \Delta z^n)\, d\Omega^* + (1 - \theta_T)\Delta t \int_{\Omega^*} \Delta z\, \vec{U}^*.\overrightarrow{grad}(T^n)\, d\Omega^*$$

$$-(1-\theta_T)\Delta t \int_{\Gamma^*} \Delta z T^n\,\vec{U}^*.\vec{n}\,d\Gamma^* - \int_{\Omega^*} (1-\theta_h)T^{n+1}(\Delta z^{n+1} - \Delta z^n)\,d\Omega^*$$

$$-\left(-\theta_T\Delta t \int_{\Omega^*} \Delta z\,\vec{U}^*.\overrightarrow{grad}(T^{n+1})\,d\Omega^* + \int_{\Omega^*} \theta_T\Delta t \int_{\Gamma^*} \Delta z T^{n+1}\,\vec{U}^*.\vec{n}\,d\Gamma^*\right) = 0 \tag{5.116}$$

It has to be proved that the last two terms are zero to guarantee the conservation of the tracer. By breaking down T^n and T^{n+1} on the basis Ψ_i^* of the transformed mesh, two conditions are sufficient:

$$\theta_h + \theta_T = 1 \tag{5.117}$$

and:

$$\int_{\Omega^*} (\Delta z^{n+1} - \Delta z^n)\Psi_i^* d\Omega^*$$
$$-\Delta t \int_{\Omega^*} \Delta z\,\vec{U}^*.\overrightarrow{grad}(\Psi_i^*)\,d\Omega^* + \Delta t \int_{\Gamma^*} \Psi_i^*\Delta z\,\vec{U}^*.\vec{n}\,d\Gamma^* = 0 \tag{5.118}$$

At this point the variational formulation of the mass conservation equation appears:

$$\frac{\partial h}{\partial t} + div(h\vec{U}) = 0$$

with an integration by parts of the divergence term identical to the one used for resolving the Saint-Venant equations.

We arrive at **two fundamental conclusions**:

- The conservation of the mass of the tracer requires a compatible treatment of the continuity equation.
- The condition $\theta_h + \theta_T = 1$ which signifies, for instance, that if an explicit scheme is used to calculate the value of $\int_{\Omega^*} \Delta z\,\vec{U}.\overrightarrow{grad}(T)\,d\Omega^*$, the advection step should be done (paradoxically) in the mesh fixed at time t^{n+1}. A more detailed demonstration of tracer conservation will be given in Section 5.7.2 below, with an example of the distributive scheme.

5.7 Sources and sinks in the Navier–Stokes equations

In order to deal with water or tracer inputs into a domain, e.g. drainage pipes whose dimensions are less than the size of finite elements, we use the concept of sources and sinks. The fluid flow, with or without tracer, is considered as entering the calculation domain at a single point. This concept, already treated for the Saint-Venant equations, is artificial in certain aspects and for the Navier–Stokes equations requires a thorough examination.

5.7.1 Sources of fluid

In 3D, it is now the continuity equation $div(\overrightarrow{U}) = 0$ which is no longer verified locally. In the transformed mesh, starting again from Equation 2.99, we now have to solve:

$$\int_{\Omega^*} \left[\left(\frac{\partial \Delta z}{\partial t} \right)_{x,y} + \left(\frac{\partial \Delta z U}{\partial x} \right)_{y,z^*,t} + \left(\frac{\partial \Delta z V}{\partial y} \right)_{x,z^*,t} + \left(\frac{\partial \Delta z W^*}{\partial z^*} \right)_{x,y,t} \right] \Psi_i^* \, d\Omega^*$$

$$= \int_{\Omega^*} \Delta z \, div(\overrightarrow{U}) \Psi_i^* \, d\Omega^* \tag{5.119}$$

by considering that on the right-hand side $div(\overrightarrow{U})$ will not be zero in certain points.

Now $\int_{\Omega^*} \Delta z \, div(\overrightarrow{U}) \Psi_i^* \, d\Omega^* = \int_{\Omega} div(\overrightarrow{U}) \Psi_i \, d\Omega$. The integral $div(\overrightarrow{U})$ around a domain surrounding a source point will be the discharge of the source Q_{sce}. As in 2D but taking into account the vertical, one solution consists of adding to the right-hand side of the continuity equations the term:

$$\sum_{i3d \text{ on the vertical of } i2d} \frac{\int_{\Omega_{3d}} Q_{sce} \Psi_{isce}^{3d} \Psi_{i3d}^{3d} \, d\Omega_{3d}}{\int_{\Omega_{3d}} \Psi_{isce}^{3d} \, d\Omega_{3d}}$$

where *isce* is the point with the source. We have distinguished here, for clarity if not conciseness, the 2D and 3D domains and basis functions by indexing them with $2d$ and $3d$. It is worth noticing that:

$$\sum_{i3d \text{ on the vertical of } i2d} \frac{\int_{\Omega_{3d}} Q_{sce} \Psi_{isce}^{3d} \Psi_{i3d}^{3d} \, d\Omega_{3d}}{\int_{\Omega_{3d}} \Psi_{isce}^{3d} \, d\Omega_{3d}} \neq \frac{\int_{\Omega_{2d}} Q_{sce} \Psi_{isce}^{2d} \Psi_{i2d} \, d\Omega_{2d}}{\int_{\Omega_{2d}} \Psi_{isce}^{2d} \, d\Omega_{2d}} \tag{5.120}$$

The domain Ω is considered at an instant in consistency with the other flux terms. The series of linear systems giving the vertical velocities averaged over each plane is then written for every point i in the 2D mesh as:

$$\sum_{j=1}^{npoin2} \left\{ \int_{\Omega_{2D}} \Psi_i^h \Psi_j^h d\Omega_{2D} \left[\overline{(\Delta z w^*)}_{ip+\frac{1}{2}}^{j} - \overline{(\Delta z w^*)}_{ip-\frac{1}{2}}^{j} \right] \right\}$$

$$= -\frac{1}{\Delta t} \int_{\Omega^*} \left(\Delta z^{n+1} - \Delta z^n \right) \Psi_i^* d\Omega^* + \int_{\Omega^*} \Delta z \, \overrightarrow{U} \, . \, \overrightarrow{grad}_{2D}(\Psi_i^*) \, d\Omega^*$$

$$- \int_{\Gamma_{liq}^*} \Delta z \overrightarrow{U} . \vec{n} \Psi_i^* d\Gamma^* + \sum_{i3d \text{ on the vertical of } i2d} \frac{\int_{\Omega 3d} Q_{sce} \Psi_{isce}^{3d} \Psi_{i3d}^{3d} \, d\Omega 3d}{\int_{\Omega 3d} \Psi_{isce}^{3d} \, d\Omega 3d} \tag{5.121}$$

5.7.2 Sources of tracer with a distributive scheme

Having seen both the conservation of tracer mass and the sources of water, we are now in a position to understand how to deal with sources of tracers. As a matter of fact it is necessary to rely on the proof of conservation of the tracer mass to find out which source terms apply to it. The treatment of the advection terms has its importance and we take here the example of the distributive scheme which will be presented in Section 6.4. It is an explicit scheme and it is about the only thing to recall in what follows. Let us denote:

$$FLUINT(i) = \int_{\Omega^*} \Delta z\, \vec{U}\,.\,\overrightarrow{grad}_{2D}(\Psi_i^*)\, d\Omega^*$$

$$FLUVER(i) = \int_{\Omega^*} \Delta z W^* \frac{\partial \Psi_i^*}{\partial z^*} d\Omega^*$$

$$SOURCE(i) = \frac{\int_\Omega Q_{sce} \Psi_{isce} \Psi_i\, d\Omega}{\int_\Omega \Psi_{isce}\, d\Omega} \alpha(i)$$

$$FLUEXT(i) = \int_{\Gamma^*_{liq}} \Delta z \vec{U} . \vec{n} \Psi_i^* d\Gamma^*$$

The continuity equation is then written as:

$$\frac{1}{\Delta t} \int_{\Omega^*} \left(\Delta z^{n+1} - \Delta z^n\right) \Psi_i^* d\Omega^*$$

$$= FLUINT(i) + FLUVER(i) - FLUEXT(i) + SOURCE(i) \tag{5.122}$$

The equation for the tracer T is as follows:

$$\frac{1}{\Delta t} \int_{\Omega^*} \Delta z^{n+1} \left(T^{n+1} - T^n\right) \Psi_i^* d\Omega^* =$$

$$- \int_{\Omega^*} \Delta z \vec{U} . \overrightarrow{grad}(T^n) \Psi_i^* d\Omega^* + SOURCE\_TRAC(i) \tag{5.123}$$

where the term $\int_{\Omega^*} \Delta z \vec{U} . \overrightarrow{grad}(T^n) \Psi_i^* d\Omega^*$ is explicit and calculated by the distributive scheme.

$SOURCE\_TRAC(i)$ is the source term we look for. The overall conservation of the tracer mass is shown by first multiplying all the continuity equations by T_i^n, then by adding them, and eventually by adding to them all the tracer equations, which gives:

$$\frac{1}{\Delta t} \int_{\Omega^*} \left(\Delta z^{n+1} - \Delta z^n\right) T^n d\Omega^* + \frac{1}{\Delta t} \int_{\Omega^*} \Delta z^{n+1} \left(T^{n+1} - T^n\right) d\Omega^*$$

$$= \sum_i FLUINT(i) T_i^n + \sum_i FLUVER(i) T_i^n - \sum_i FLUEXT(i) T_i^n$$

$$+ \sum_i SOURCE(i) T_i^n - \int_{\Omega^*} \Delta z \vec{U} . \overrightarrow{grad}(T^n) d\Omega^* + \sum_i SOURCE\_TRAC(i) \tag{5.124}$$

Now, the distributive scheme ensures that:

$$\sum_i FLUINT(i)T_i^n + \sum_i FLUVER(i)T_i^n - \int_{\Omega^*} \Delta z \vec{U}.\overrightarrow{grad}(T^n)d\Omega^* = 0$$

because it simply performs a redistribution of the first two terms without changing its sum. We finally arrive at:

$$\frac{1}{\Delta t}\int_{\Omega^*}\left(\Delta z^{n+1}T^{n+1} - \Delta z^n T^n\right)d\Omega^*$$

$$= -\sum_i FLUEXT(i)T_i^n + \sum_i SOURCE(i)T_i^n + \sum_i SOURCE\_TRAC(i) \quad (5.125)$$

In the case of a source with a tracer value T_{sce}, the conservation of tracer mass imposes that:

$$\sum_i SOURCE(i)T_i^n + \sum_i SOURCE\_TRAC(i) = Q_{sce}T_{sce}$$

One solution for a source is then:

$$SOURCE\_TRAC(i) = (T_{sce} - T_i^n)\,SOURCE(i)$$

If there are several sources, there should be a distinction between the $SOURCE(i)$ vectors for each source.

If the source is actually a sink, then $SOURCE(i)$ is negative and a different treatment is necessary. As a matter of fact, in a sink, the value of the tracer now exiting the domain is not T_{sce}, but rather the local value T_i, so that $SOURCE\_TRAC(i)$ must in fact be cancelled. Everywhere T_{sce} appears, it must be replaced by T_i (here T_i^n because we have an explicit scheme, but it could as well be T_i^{n+1} with an implicit scheme). To take into account sinks as well as sources, we must thus consider that:

$$SOURCE\_TRAC(i) = (T_{sce} - T_i^n)\,MAX(SOURCE(i), 0)$$

5.7.3 Dirichlet boundary conditions with a distributive scheme

As the previous subsections give a full proof of the mass conservation of the tracer, we are now in a position to deal with the problem of imposed boundary conditions. The distributive schemes pay no attention to prescribed values at the boundaries; they just take the tracer at time t^n and ensure mass conservation and monotonicity. A user may want to prescribe given values at an entrance to the domain, or a given tracer flux. If nothing special is done the flux of the tracer through the boundaries (here positive if entering the domain) will be:

$$observed\ flux = -\int_{\Gamma^*_{liq}} \Delta z T^n \vec{U}.\vec{n}\,d\Gamma^*$$

$$= -\sum_i T_i^n \int_{\Gamma^*_{liq}} \Delta z \vec{U}.\vec{n}\Psi_i^* d\Gamma^* = -\sum_i T_i^n FLUEXT(i) \quad (5.126)$$

and the imposed values will not be ensured. The summation on i must be understood as i being on the boundary with prescribed values. The user would in fact like to have:

$$wanted\ flux = -\int_{\Gamma^*_{liq}} \Delta z T^{imposed} \vec{U}.\vec{n}\, d\Gamma^* \quad \text{and} \quad T_i^{n+1} = T_i^{imposed} \ \text{on boundaries}$$

Both conditions are not compatible. Brutally setting $T_i^{n+1} = T_i^{imposed}$ or $T_i^n = T_i^{imposed}$ on boundaries will add or remove tracer mass in the domain, because the basis functions on the boundaries go inside the domain, thus creating an artificial flux. We propose here a relaxation of the Dirichlet conditions which consists of replacing T_i^n by:

$$T_i^n + \theta(T_i^{imposed} - T_i^n) = T_i^n + \Delta T^n \tag{5.127}$$

on liquid boundaries with imposed values, with $0 < \theta < 1$. θ is computed so that the sum of the *a priori* added mass and the observed flux gives the wanted flux. The added mass is:

$$added\ mass = \theta(T_i^{imposed} - T_i^n) \int_{\Omega^*} \Delta z^n \Psi_i^* d\Omega^* = \theta(T_i^{imposed} - T_i^n) \int_{\Omega^n} \Psi_i\, d\Omega^n$$

the new observed flux is:

$$observed\ flux = -\left[T_i^n + \theta(T_i^{imposed} - T_i^n)\right] FLUEXT(i)$$

As we want:

$$added\ mass\ +\ \Delta t * \ observed\ flux\ =\ \Delta t * \ wanted\ flux$$

we have to choose:

$$\theta = \frac{-FLUEXT(i)}{\left[-FLUEXT(i) + \frac{1}{\Delta t}\int_{\Omega^n} \Psi_i d\Omega^n\right]}$$

This local value of θ is considered only for entrances, i.e. when $FLUEXT(i)$ is negative, then θ is always between 0 and 1.

After applying the distributive scheme the value of the tracer initially set at the boundary, $T_i^n + \theta(T_i^{imposed} - T_i^n)$, will have been slightly modified by the algorithm but still ensures monotonicity, and the correct prescribed flux will be observed if a mass balance is done.

5.7.4 Sources of tracer with the SUPG advection scheme

We have taken the example of a distributive scheme for our explanations in the last two subsections. The theory with a SUPG advection scheme (see Section 6.2) would be similar but with a few modifications detailed hereafter, due to the fact that this scheme is implicit, the tracer considered being $\theta_T T^{n+1} + (1-\theta_T)T^n$. The mesh used for the advection equation must be taken at time $(1-\theta_T)\,t^{n+1}+\theta_T\,t^n$. The derivative in time of the tracer is thus in the transformed mesh:

$$\frac{1}{\Delta t}\int_{\Omega^*}\left[(1-\theta_T)\Delta z^{n+1}+\theta_T\Delta z^n\right]\left(T^{n+1}-T^n\right)\Psi_i^* d\Omega^*$$

Consequently, in the proof of tracer mass conservation, Equation 5.122 is no longer multiplied by T_i^n but by $\theta_T T_i^{n+1}+(1-\theta_T)T_i^n$ so that the combination and sum of all equations form on the left-hand side the term:

$$\frac{1}{\Delta t}\int_{\Omega^*}\left(\Delta z^{n+1}T^{n+1}-\Delta z^n T\right)d\Omega^*$$

As will be seen later, the SUPG theory introduces an extra term into the tracer equation, in the form:

$$-\int_{\Omega^*}\Delta z\overrightarrow{U}.\overrightarrow{grad}(T^n)\,k\frac{\overrightarrow{u}}{|\overrightarrow{u}|}\,.\,\overrightarrow{grad}(\Psi_i^*)d\Omega^*$$

on the right-hand side, but this term is cancelled when summed over all bases Ψ_i^* because their sum equals 1.

As in our proof $SOURCE(i)$ in the continuity equation is now multiplied by $\theta_T T_i^{n+1}+(1-\theta_T)T_i^n$, the source of the tracer is accordingly modified in the tracer equation:

$$SOURCE\_TRAC(i)=\left[T_{sce}-\theta_T T_i^{n+1}+(1-\theta_T)T_i^n\right]\,SOURCE(i)$$

5.7.5 Dirichlet boundary conditions with a SUPG advection scheme

In the case of an implicit advection scheme, advection and diffusion are solved together. The difficulty of ensuring a correct flux corresponding to prescribed boundary conditions is that the flux cannot be *a priori* known, as it was with an explicit scheme. An *a posteriori* correction is thus necessary. However, we choose to start with the *a priori* correction of the distributive scheme, thus adding the increment ΔT^n to T^n as in Equation 5.127. Then the advection–diffusion algorithm is applied, starting from $T_i^n+\Delta T^n$, and yields a result denoted T^D. Eventually an increment ΔT^{n+1} is added to T^D, such that:

$$a\ posteriori\ added\ mass$$

$$=\Delta T^{n+1}\int_{\Omega^n}\Psi_i d\Omega^{n+1}=-\Delta t\,FLUEXT(i)\,\theta\,(T_i^n+\Delta T^n-T_i^D)$$

As the observed flux is:

$$observed\ flux=-\Delta t\left[T_i^n+\Delta T^n+\theta(T_i^D-T_i^n-\Delta T^n)\right]FLUEXT(i)$$

and as:

$$a\ priori\ added\ mass = \Delta T_i^n \int_{\Omega^n} \Psi_i d\Omega^n = -\Delta t\, FLUEXT(i) \left[T_i^{imposed} - T_i^n - \Delta T^n \right]$$

we can check that we recognize the required property:

$$a\ priori\ \ added\ \ mass + a\ \ posteriori\ \ added\ \ mass + observed\, flux$$

$$= -T_i^{imposed}\, \Delta t\, FLUEXT(i)$$

which is the wanted flux.

5.8 Treatment of uncovered beds in 3D

5.8.1 Option 1: correction of free surface gradients

This option used in 2D has been considered impossible for a long time in 3D, on account of severely distorted elements which appear when, in the absence of water, the free surface and the bottom become identical. It appeared later that the work with these severely distorted elements was manageable on condition that some divisions by 0 were avoided. For example, to calculate the vertical average of the horizontal velocity, calculation with the trapezoidal rule should be replaced on the exposed beds by an arithmetic average. Apart from a few numerical problems, for which a remedy exists, the correction of the free surface gradients is formally the most elegant because, when a free surface gradient is treated with a resolution of the Saint-Venant equations, nothing else is left in the actual 3D part of the resolution for both the hydrostatic hypothesis and the non-hydrostatic equations. Plate 9, shown earlier, shows that exposure on a beach can be treated in 3D, in this case with 10 planes along the vertical, with zero water depth on the right side of the calculation domain. In the non-hydrostatic case, we notice a trough in the free surface. This trough becomes more prominent when the flow at the exit of the pool which is emptied, becomes supercritical. The water trapped in the puddle seems to take longer to return to rest. This is because the dynamic pressure gradient is not subject to any specific treatment on the exposed banks.

5.8.2 Option 2: masking of the uncovered elements

As in the case of 2D, the columns of dry or partially dry elements have to be extracted from the calculation domain. This is done with the help of an array equal to 1 or 0, called a mask. The detailed description of this technique is mainly a catalogue of dissuasive difficulties.

Untimely masking–unmasking

In situations at the limit between covering and exposing, with fluxes of water changing signs from one time step to another, the elements balance endlessly between a normal situation and a situation where they are removed from the mesh. In the presence of a bottom slope, the flow is guided by the slope and those elements whose bottom elevation is the highest will dry first. On the other hand, on a flat bottom, the flow is much more erratic and the problem is more acute. It is therefore necessary to mask the entire flat zone when one of the elements of this zone has the criterion for masking. This choice also partially helps in solving the following difficulty.

Occurrence of singular points

The removal of an element does not always give a topologically correct mesh. A point can form an isthmus with zero width where two different coasts touch but without a chance of flux. This type of point should be eliminated by extending the masked zone around the point at one of the two angular sectors not yet masked. This process should be iterated till a correct topology is obtained. A prior analysis of the topography helps to accelerate the process. In exposed zones, the free surface is considered locally quasi-horizontal, and by knowing its elevation at a given instant, all the dry or partially dry elements can be deduced from it. In fact, if an element is masked because the water depth is below a given criterion, all the neighbouring elements with a greater or equal bottom elevation should also be masked and so on from one to the next. Unfortunately, this hypothesis is not valid in zones with high variations in the free surface, but it helps to eliminate the risk of a singular point, if from the start certain regularity is assured in the bathymetry. In practice, it has to be ensured that, by locally changing a bottom elevation if necessary, if we turn around an internal point in the domain by going over the elements to which it belongs, we pass over a single minimum and a single maximum of the bottom.

Criterion for masking

The third difficulty encountered concerns determining the criterion for masking. There is no criterion for unmasking, since, in the algorithm, we start by unmasking all the elements. The criterion for masking needs the computation of the water depth per element which cannot be less than a given threshold value. This calculation of the height only requires an estimation of the free surface elevation for each element, which is compared with one bottom elevation per element. The first idea would be to retain the elevation of the lowest free surface of the three vertices of each triangle in order to be certain of masking all the elements that require masking. However, if the vertex in question is surrounded by masked elements, there is no chance that its elevation will change and the masked elements will remain the same whatever the evolution of the situation around them. To avoid this, or to limit this risk as far as possible, the elevation of the free surface per element is taken as the average of the elevations of the three vertices. Thereupon, almost all the elements become unmasked when the level of the free surface rises, except in some specific places with which we have not yet dealt, such as basins with a single node of mesh at the centre. If some isolated elements have not been correctly unmasked at the time of rise in level of the free surface, we should refine the mesh locally (a very cumbersome solution for a study that is already under way) or resort to a simplification of the bathymetry.

Treatment of masked nodes

Formally, as the technique of masking presented here consists of excluding from the calculation domain those elements that fit the criterion mentioned above, there may be nodes surrounded by masked elements, called masked nodes, for which the equations are no longer resolved. The variables in these nodes (u, v, T, k, ε) are then arbitrarily modified. When unmasking these nodes, these values are once again taken into account in the calculation of the solution. To remedy this shortcoming, we should retain, during the entire duration of the masking of a node, the values of the variables calculated just before this masking. So, it is the increments of the variables and not the variables themselves which are arbitrarily forced to zero.

Chapter 6

Solving transport equations

The treatment of transport equations is one of the strategic points in the numerical resolution of fluid mechanics equations. The hyperbolic nature of these equations implies that the space derivatives need an upwind treatment. A number of numerical methods that have been developed on structured meshes, e.g. Quickest, Ultimate, Nirvana by Leonard, cannot be applied to the finite elements, because they are dependent on mesh regularity. We shall describe below three methods that can be applied to the finite elements:

- The method of characteristics.
- The SUPG (Streamline Upwind Petrov–Galerkin) method that we shall also apply to the continuity equation.
- Distributive schemes, which are a link with techniques that are used mainly for finite volumes.

6.1 Method of characteristics

The main advantages of the method of characteristics are unconditional stability and monotonicity of the solutions. In its "weak" form mass conservation should also be added. The basic principle is to temporarily adopt a Lagrangian approach and to write that the value of a function at time t^{n+1} on a node M of the mesh is equal to the value that the same function had at time t^n and at point Q, identified by going backwards in time along the trajectory from point M and for a length of time Δt. Thus Riemann's invariant is used in association with the characteristic equation:

$$\frac{d\overrightarrow{x}}{dt} = \overrightarrow{U}$$

We call the point Q "the foot of the characteristic".

For each point M, the method of characteristics consists of two steps:

1) Finding the foot of the characteristic (point Q) by creating the trajectory that passes by M at the moment t^{n+1}.

2) Interpolation within the arrival element.

Hydrodynamics of Free Surface Flows: Modelling with the Finite Element Method J.-M. Hervouet

Calculating the trajectory poses the problem of time order. The velocity field, which should be used to calculate the trajectory, is:

$$\vec{U}(t) = \vec{U}^n + \frac{t - t^n}{\Delta t}\vec{U}^{n+1} \tag{6.1}$$

Unfortunately the advection field is unknown at time t^{n+1} before the end of the time step. There are two options: either one chooses the advection field at time t^n (hypothesis of a fixed field), or one chooses to approach the advection field at time t^{n+1} and the same time step is repeated several times. The latter option is time consuming because the number of iterations required multiplies the execution time. The hypothesis of a fixed field is thus retained for general cases. Moreover, within the framework of the method of characteristics, the notion of time order is complex because it is also dependent on the calculation of the characteristics.

The characteristics (trajectories) are calculated by a Runge–Kutta order 1 method and interpolation at the foot of the characteristic conforms to the type of finite element chosen. The second-order Runge–Kutta method has already been tested but does not add any significant improvement; it is also expensive. A higher order interpolation would certainly bring about a greater improvement than a higher order Runge–Kutta method.

6.1.1 Trajectory calculation

Here the principle is explained in two dimensions. Each time step Δt is itself divided into sub time steps $\frac{\Delta t}{nsub}$ during which one considers that the velocity remains constant ($nsub$ is calculated locally, so that it takes for example three substeps to pass through each element). If X_{start} and Y_{start} are the coordinates of a point D on the trajectory being calculated, the next point A, with coordinates $X_{arrival}$, $Y_{arrival}$, is found by means of the following formulae (the minus sign is used because one is moving backwards from time t^{n+1} to time t^n):

$$X_{arrival} = X_{start} - u_{start} * \frac{\Delta t}{nsub} \tag{6.2}$$

$$Y_{arrival} = Y_{start} - v_{start} * \frac{\Delta t}{nsub} \tag{6.3}$$

where u_{start} and v_{start} are the velocity components of point D. As this point is not always a domain node, u_{start} and v_{start} are obtained by interpolation within an element. The main problem in finite elements is to pinpoint the arrival point A; that is to say, to locate which element it belongs to. It is essential to know this element in order to find the velocity at this point by interpolation. Localization at the foot of the characteristic Q is also necessary, in order to interpolate the quantities under advection.

While calculating the trajectory, two events can occur before the arrival at time t^n:

- One reaches a liquid boundary: in this case the trajectory theoretically continues outside the domain. As the information to continue the calculation of this trajectory is lacking, the point Q is taken at the point of intersection of the boundary and the trajectory. The interpolation should be done as precisely as possible, at the moment of leaving the domain which is intermediary between t^n and t^{n+1}.

- One reaches a rigid boundary: this theoretically impossible case can occur due to approximations in the Runge–Kutta method. The point of intersection of the trajectory with the boundary then replaces point A, which would fall outside the domain. The velocity field is then projected onto the boundary in order to avoid another exit of the domain at the next sub time step.

Navigation and localization in the mesh during the construction of the characteristics is the cornerstone of the whole method. One employs a data structure that allows the numbers of elements located behind each side of a given element to be identified (see the IFABOR table in Section 7.2). The algorithmic geometry required and the management of truncation errors make the method of characteristics a difficult technique to use.

6.1.2 Interpolation

As indicated above, interpolation is compatible with the type of element chosen in the second stage of the calculation. If these interpolations are found to be too diffusive for certain cases of application, it would then be necessary to consider higher order elements. For more information about the method of characteristics, one can refer to the following works: [45] and [69]. Advection programmes following the method of characteristics have to be carefully optimized and can be partially vectorized.

Important note: this presentation is consistent with the notion of fractional steps, but gives no indication about the treatment of other terms which normally figure in the right-hand side of the equations. These can in fact be included in the characteristics treatment. In this case the variation of a function f along its characteristic curve during the time interval taken between t^n and t^{n+1} is no longer zero, but equal to $\int_{t^n}^{t^{n+1}} (other\ terms)_C\, dt$ where the notation $()_C$ signifies that these other terms apply to the particle transported by the current and thus refers to the characteristic curve. One also sees more clearly the need for a semi-implicitation of all the terms that do not belong to the Lagrangian derivative. If the expression of these terms at time t^{n+1} poses no problem, one notices, on the other hand, that the explicit terms (taken at time t^n) should be taken at the foot of the characteristic.

6.1.3 Method of characteristics in weak formulation

The two main shortcomings of the method of characteristics are the diffusive nature and the lack of exact mass conservation. These two problems are corrected by a "weak" approach, where the method is applied within a framework of finite elements extended to time. This method was first presented in 1982 by Benque *et al.* [10]. A more detailed description can be found in reference [70]. Here we shall resolve a conservative transport equation:

$$\frac{\partial f}{\partial t} + div(f\overrightarrow{u}) = 0 \tag{6.4}$$

which, after variational formulation extended to time, becomes:

$$\int_{t^n}^{t^{n+1}} dt \left[\int_\Omega \left(\frac{\partial f}{\partial t} + div(f\vec{u}) \right) \varphi_i \, d\Omega \right] = 0 \tag{6.5}$$

Upon integration by parts this equation leads formally to:

$$\int_\Omega f^{n+1} \varphi_i^{n+1} d\Omega - \int_\Omega f^n \varphi_i^n d\Omega - \int_{t^n}^{t^{n+1}} dt \int_\Omega f \left(\frac{\partial \varphi_i}{\partial t} + \vec{u}\,.\overrightarrow{grad}(\varphi_i) \right) d\Omega$$

$$= \int_{t^n}^{t^{n+1}} dt \int_\Gamma \varphi_i \, f \, \vec{u} \,.\, \vec{n} \, d\Gamma \tag{6.6}$$

In this equation, one has intentionally distinguished the test function φ_i at time t^n and at time t^{n+1}. In fact, the originality of this method lies in its mobile mesh, which follows the water particles in their movement during the time step Δt. Under these conditions, one can say that there is an advection of the test functions, and thus that:

$$\frac{\partial \varphi_i}{\partial t} + \vec{u}\,.\overrightarrow{grad}(\varphi_i) = 0$$

Here one is dealing with a non-conservative form of the advection equation for the test functions, though we started from a conservative form. Equation 6.6 can then be simplified to:

$$\int_\Omega f^{n+1} \varphi_i^{n+1} d\Omega = \int_\Omega f^n \varphi_i^n d\Omega + \int_{t^n}^{t^{n+1}} dt \int_\Gamma \varphi_i f \vec{u} \,.\, \vec{n} \, d\Gamma \tag{6.7}$$

The evolution of the test functions is done from time t^n to time t^{n+1} and one considers that the mesh at time t^{n+1} coincides with the classic mesh. The test functions at time t^n are thus developed by going up the characteristic curves and by calculating the position of their foot at time t^n. This is done by the classic method of characteristics ("strong" form). The left-hand side of Equation 6.7 is a well-known term: a mass matrix multiplied by the vector of unknowns f_i^{n+1}. On the right-hand side, there is a term denoting a flux through the boundary, which is given by the boundary conditions, and the term $\int_\Omega f^n \varphi_i^n d\Omega$ which is problematic because it is equal to $\sum_{j=1}^{npoin} \int_\Omega f_j^n \Psi_j^{n+1} \varphi_i^n d\Omega$. The decomposition of f^n is effectively known on the basis Ψ_j of the classic mesh. Therefore, one has to know how to calculate a matrix whose general term is $\int_\Omega \varphi_i^n \, \Psi_j^{n+1} d\Omega$ when the bases are staggered between the time t^n and the time t^{n+1}. The calculation domain is thus a union of intersections of two triangles, arbitrarily laid on the map. Reference [68] proposes a means of calculating these terms precisely, by referring to the distribution theory. However, this method is expensive and sensitive to computer-generated truncation errors. In 1983, Hasbani *et al.* [66] came up with an original but approximate solution that consisted of calculating the integral with a Gauss quadrature. This technique was adopted in the USA where the method of characteristics in its weak form is used under the name ELLAM (Eulerian Lagrangian Localized Adjoint Method).

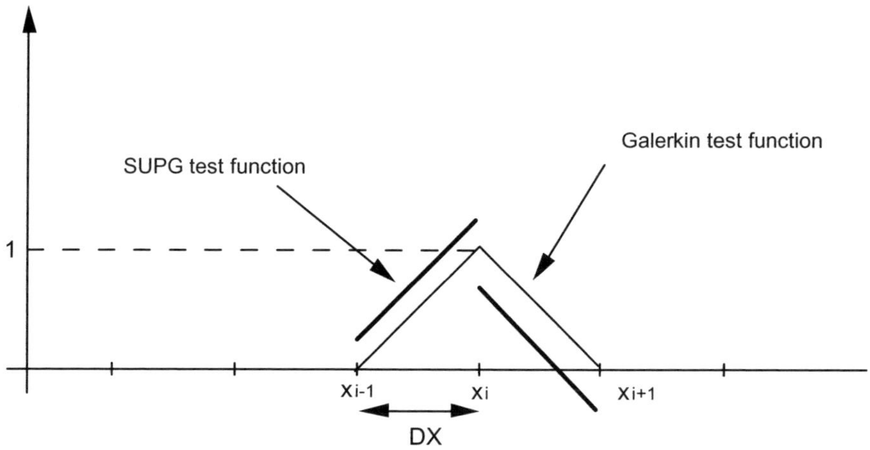

Figure 6.1: a SUPG test function

6.2 Streamline Upwind Petrov–Galerkin method

The principle of the SUPG method developed by Brooks and Hughes [18] is to use test functions deformed by the direction of the current. With the Petrov–Galerkin method, these test functions are obtained with quadratic functions: the originality of the SUPG method is to employ linear but discontinuous functions, thus the test function φ_i^h is replaced by:

$$\varphi_i^h + k\frac{\overrightarrow{u}_i}{\|\overrightarrow{u}_i\|}.\overrightarrow{grad}(\varphi_i^h) \tag{6.8}$$

where $\overrightarrow{u}_i$ is the velocity taken at node i of the domain and k a coefficient, which is yet to be defined. On a one-dimensional mesh, the shape of these new test functions is as given in Figure 6.1.

In theory these test functions should be applied to all the terms of an equation for the sake of conformity with the finite element principle. We have been compelled to modify three points of the method, because of the specificities of our equations:

1) The SUPG test function cannot be applied to all the terms of the continuity equation. It seems that its application to propagation term $div(h\overrightarrow{u})$ produces aberrant terms. There is a problem yet to be resolved, probably due to the fact that the propagation terms should not be subject to an upwind treatment.

2) The vector $\overrightarrow{u}_i$ (vector equal to the nodal value of the velocity vector at point i) had to be replaced, either by a vector $\overrightarrow{u}_{el}$ constant by element (this can, for example, be the average velocity at each node), or by the real vector velocity $\overrightarrow{u}$ taken at time t^n, which is itself a variable function in space. This minor modification does not change the stability analysis, but is essential in order to obtain an exact conservation of the mass. The two options $\overrightarrow{u}$ and $\overrightarrow{u}_{el}$ have been programmed and the more elegant, and paradoxically cheaper, choice of $\overrightarrow{u}$ was finally retained. In the SUPG formulation, $\overrightarrow{u}_i$ will thus henceforth be replaced by $\overrightarrow{u}$.

3) The choice of k: the one Brooks and Hughes recommended results from the heat equation analysis which does not apply to our case. Our choice of k will now be explained.

It is well known that one can obtain an upwind advection term of second-order accuracy, by adding an artificial diffusion $\frac{uDX}{2}$ to a centred term. Let us now look at the effect of SUPG bases on finite elements, when they are applied only to the advection term (here we take the example of advection of a function f). Thus $\int_\Omega \vec{u}.\overrightarrow{grad}(f)\,\varphi_i^h\,d\Omega$ becomes:

$$\int_\Omega \vec{u}.\overrightarrow{grad}(f)\left(\varphi_i^h + k\frac{\vec{u}}{\|\vec{u}\|}.\overrightarrow{grad}(\varphi_i^h)\right)d\Omega$$

Compared with the centred case we thus have to add the term:

$$\int_\Omega k\,\vec{u}.\frac{\vec{u}}{\|\vec{u}\|}\;\overrightarrow{grad}(f).\;\overrightarrow{grad}(\varphi_i^h)\,d\Omega$$

which then modifies the matrices and the right-hand sides, because of the semi-implicitation of f. This term, in one dimension, is like a diffusion coefficient ku. Thus we see that the coefficient k which gives the diffusion $\frac{uDX}{2}$ (which adds an upwind treatment to a centred advection term) is $k = \frac{DX}{2}$. We shall henceforth write k in the form $k = \frac{\delta\,DX}{2}$, where δ is the upwind coefficient, which choice needs to be clarified.

The supplementary term resulting from SUPG can be interpreted as an artificial diffusion that stabilizes a centred scheme. In two dimensions, because of the effect of dot products, this diffusion applies only in the direction of the flow. Through Fourier analysis one can show that in order to ensure the stability of the scheme (see [93]), it is sufficient for the coefficient δ to be equal to the Courant number $\frac{\|\vec{u}\|\Delta t}{\Delta x}$; k is then equal to $\frac{\|\vec{u}\|\Delta t}{2}$.

Having made this choice, it is possible to show that the SUPG method, thus modified, does not alter the mass conservation of our mass conservation equation. The sum of modified bases remains equal to 1 because:

$$\sum_{i=1}^{nel}\varphi_i^h + \sum_{i=1}^{nel}\frac{\delta\,DX}{2}\frac{\vec{u}}{\|\vec{u}\|}.\overrightarrow{grad}(\varphi_i^h) = 1 + \frac{\delta\,DX}{2}\frac{\vec{u}}{\|\vec{u}\|}.\overrightarrow{grad}(\sum_{i=1}^{nel}\varphi_i^h) = 1$$

We then note that one can treat each term of a conservative equation with test functions of different nature, without altering the mass conservation, as long as the sum of these test functions is equal to 1 on the mesh. For the SUPG test functions, the latter quality is ensured by the choice of a coefficient δ which is constant per element, and by the choice of $\vec{u}$ instead of $\vec{u}_i$.

6.3 Conservative scheme for the continuity equation

This scheme is based on a conservative continuity equation, in the form:

$$\frac{\partial h}{\partial t} + div(h\vec{u}) = \frac{\partial h}{\partial t} + div\left[h_{prop}(\theta_u\,\vec{u}^{n+1} + (1-\theta_u)\,\vec{u}^{n})\right] = 0$$

If one uses the SUPG method, the test functions of this method can be applied only to advection terms, so that one actually resolves:

$$\int_\Omega \frac{\partial h}{\partial t}\varphi_i^h \, d\Omega + \theta_u \int_\Omega div(h_{prop}\overrightarrow{u}^{n+1})\,\varphi_i^h\, d\Omega$$

$$+\theta_u \int_\Omega \overrightarrow{u}^{n+1}.\overrightarrow{grad}(h_{prop})\, k\, \frac{\overrightarrow{u}}{\|\overrightarrow{u}\|}.\overrightarrow{grad}(\varphi_i^h)\, d\Omega$$

$$= -(1-\theta_u)\int_\Omega div(h_{prop}\overrightarrow{u}^{n})\,\varphi_i^h\, d\Omega$$

$$-(1-\theta_u)\int_\Omega \overrightarrow{u}^{n}.\overrightarrow{grad}(h_{prop})\, k\, \frac{\overrightarrow{u}}{\|\overrightarrow{u}\|}.\overrightarrow{grad}(\varphi_i^h)\, d\Omega \qquad (6.9)$$

This is a conservative scheme without sub-iteration.

6.4 Distributive schemes

Distributive schemes were first published by Roe and Sidilkover [167] under the name of "fluctuation splitting schemes". While they were first limited to regular meshes, they were later extended to triangular meshes (see [37] and [151]).

In all that follows, we shall pay particular attention to resolving the equation:

$$\frac{\partial f}{\partial t} + \overrightarrow{u}.\overrightarrow{grad}(f) = 0$$

by first limiting ourselves to triangular meshes. We start by examining the classic centred Galerkin method, to show the drawbacks we could remedy by using distributive schemes.

Galerkin's method

For each test function of the domain denoted φ_i, we resolve:

$$\int_\Omega \varphi_i \left[\frac{\partial f}{\partial t} + \overrightarrow{u}.\overrightarrow{grad}(f)\right] d\Omega = 0 \qquad (6.10)$$

The derivative with respect to time will thus classically produce a term:

$$\frac{M}{\Delta t}(F^{n+1} - F^n)$$

where M is the mass matrix, Δt the time step, and F^{n+1} and F^n represent the vectors whose components are the f_i taken respectively at the instants t^{n+1} and t^n. Since the term $\overrightarrow{u}.\overrightarrow{grad}(f)$ is constant per element, if we consider that the velocity is constant on a triangle (which will be done with the distributive schemes), this gives a contribution on each triangle of:

$$w_i = \overrightarrow{u}.\overrightarrow{grad}(f)\int_\Omega \varphi_i \, d\Omega$$

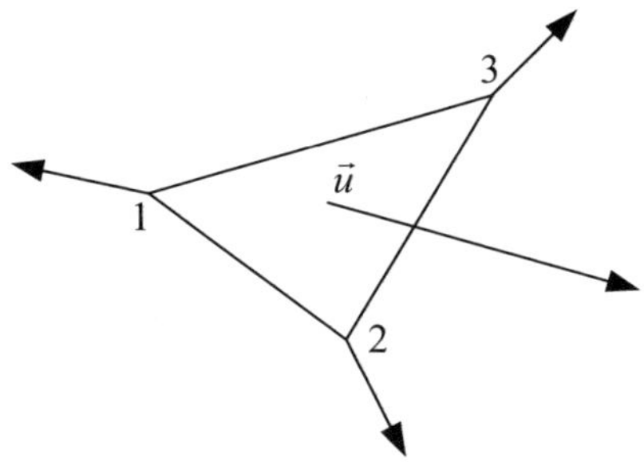

Figure 6.2: case with two downstream points

Now, on a triangle, $\int_\Omega \varphi_i \, d\Omega$ does not depend on the considered point and is equal to $S_T/3$, where S_T is the surface of the triangle. The contributions of each point are thus equal, independently of their upstream or downstream nature, and give rise to a drawback. These contributions then need to be assembled in order to obtain the expression of $\vec{u}.\overrightarrow{grad}(f)$ as projected onto a test function φ_i.

6.4.1 N & PSI schemes in two dimensions

Notations and definitions

For each triangle one defines three normal vectors (not unit vectors) at the vertices:

$$\vec{n}_1 = \begin{pmatrix} y_2 - y_3 \\ x_3 - x_2 \end{pmatrix} \qquad \vec{n}_2 = \begin{pmatrix} y_3 - y_1 \\ x_1 - x_3 \end{pmatrix} \qquad \vec{n}_3 = \begin{pmatrix} y_1 - y_2 \\ x_2 - x_1 \end{pmatrix}$$

One notes that: $\sum_{i=1}^{3} \vec{n}_i = 0$.

For reasons of simplicity we shall consider that the velocity $\vec{u}$ is constant on each triangle. This hypothesis is made systematically in the literature. Actually, this is not necessary, as will be shown later.

The surface of the triangle is denoted S_T. It is:

$$S_T = \frac{x_1y_2 - x_2y_1 + x_2y_3 - x_3y_2 + x_3y_1 - x_1y_3}{2} \tag{6.11}$$

or in form of a determinant:

$$S_T = \frac{\begin{vmatrix} 1 & x_1 & y_1 \\ 1 & x_2 & y_2 \\ 1 & x_3 & y_3 \end{vmatrix}}{2} \tag{6.12}$$

Figure 6.3: definition of f_{in}

For i=1, 2 and 3 we define the quantities $k_i = \frac{1}{2}\overrightarrow{u}.\overrightarrow{n}_i$. If k_i is positive, the velocity $\overrightarrow{u}$ is exiting at point i; the segment opposite point i is an entry. The total $k_1 + k_2 + k_3$ is zero and there are thus two possible cases:

- One of the k_i is positive and the two others are negative (the case with one entry).
- One of the k_i is negative and the two others are positive (the case with two entries).

In the one-entry case, there are two upstream points and one downstream point.

In the two-entry case, there are one upstream point and two downstream points (Figure 6.2).

We shall now define two new quantities: f_{in}, the value of f at entry, and f_{out} the value of f at exit.

If there is only one upstream point, f_{in} will denote the value of f at this point. If there are two upstream points, f_{in} will be defined as the value of f at the point of the upstream side where the velocity $\overrightarrow{u}$ turns towards the downstream point (both cases are illustrated in Figure 6.3).

Simple geometrical calculations in this case, at a downstream point (Figure 6.3 on the right), give the following value for f_{in}:

$$f_{in} = \frac{(\overrightarrow{u}.\overrightarrow{n}_1)\,f_1 + (\overrightarrow{u}.\overrightarrow{n}_2)\,f_2}{(\overrightarrow{u}.\overrightarrow{n}_1) + (\overrightarrow{u}.\overrightarrow{n}_2)} \tag{6.13}$$

All cases can in fact be covered by the following formula:

$$f_{in} = \frac{\sum\limits_{i=1}^{3} \min(0, k_i) f_i}{\sum\limits_{i=1}^{3} \min(0, k_i)} \tag{6.14}$$

In the same way one finds:

$$f_{out} = \frac{\sum\limits_{i=1}^{3} \max(0, k_i) f_i}{\sum\limits_{i=1}^{3} \max(0, k_i)} \tag{6.15}$$

Let us now try to evaluate the gradient of f. It can be written as:

$$\overrightarrow{grad}(f) = \frac{1}{2S_T} \sum_{i=1}^{3} f_i \overrightarrow{n}_i \tag{6.16}$$

If the function $f(x, y)$ is written in the linear form $a + bx + cy$, for each vertex of the triangle one should have $f(x_i, y_i) = a + bx_i + cy_i$. The components of the gradient of f are b and c, which are solutions of a Kramer system:

$$b = \frac{\begin{vmatrix} 1 & f_1 & y_1 \\ 1 & f_2 & y_2 \\ 1 & f_3 & y_3 \end{vmatrix}}{\begin{vmatrix} 1 & x_1 & y_1 \\ 1 & x_2 & y_2 \\ 1 & x_3 & y_3 \end{vmatrix}} \text{ and } c = \frac{\begin{vmatrix} 1 & x_1 & f_1 \\ 1 & x_2 & f_2 \\ 1 & x_3 & f_3 \end{vmatrix}}{\begin{vmatrix} 1 & x_1 & y_1 \\ 1 & x_2 & y_2 \\ 1 & x_3 & y_3 \end{vmatrix}}$$

The denominator is $2S_T$ and the two numerators are effectively the components of $\sum_{i=1}^{3} f_i \overrightarrow{n}_i$. This gradient is constant on the triangle. The quantity $\overrightarrow{u}.\overrightarrow{grad}(f)$ can now be expressed as:

$$\overrightarrow{u}.\overrightarrow{grad}(f) = \frac{1}{2S_T} \sum_{i=1}^{3} f_i \overrightarrow{u}.\overrightarrow{n}_i \tag{6.17}$$

or, with our definition of k_i:

$$\overrightarrow{u}.\overrightarrow{grad}(f) = \frac{1}{S_T} \sum_{i=1}^{3} f_i k_i \tag{6.18}$$

This can be reorganized as:

$$\overrightarrow{u}.\overrightarrow{grad}(f) = \frac{1}{S_T} \left[k_1(f_1 - f_3) + k_2(f_2 - f_3) \right] \tag{6.19}$$

which only shows up differences of the function f. The quantity $\sum_{i=1}^{3} f_i k_i$ is denoted Φ_T. By separating this sum into two parts according to the sign of k_i, we obtain:

$$\Phi_T = \sum_{i=1}^{3} \max(0, k_i) f_i + \sum_{i=1}^{3} \min(0, k_i) f_i \tag{6.20}$$

This, by using the expressions for f_{in} and f_{out}, and the fact that:

$$\sum_{i=1}^{3} \max(0, k_i) = -\sum_{i=1}^{3} \min(0, k_i) \tag{6.21}$$

finally produce:

$$\Phi_T = (f_{out} - f_{in}) \sum_{i=1}^{3} \max(0, k_i) \tag{6.22}$$

This expression shows that the contribution Φ_T will be zero if f is in the same state at the entry and exit. (The name "Streamwise Invariant" of the PSI scheme that we shall look at below comes from this property.)

Upwind multidimensional schemes

With the Galerkin method, we saw that each contribution for the three vertices of a triangle in the assembly of the term:

$$\int_\Omega \varphi_i(\overrightarrow{u}.\overrightarrow{grad}(f))\, d\Omega$$

is $\Phi_T/3$. This quantity ignores the fact that some triangle vertices are entries and others exits. We shall denote w_1, w_2 and w_3 as these contributions. Mass conservation is ensured as long as:

$$w_1 + w_2 + w_3 = \Phi_T$$

One can thus modify the w_i if the upstream and downstream points in the triangle are taken into consideration, by writing $w_i = \beta_i\, \Phi_T$, as long as $\sum_{i=1}^{3} \beta_i = 1$. The coefficients β_i are called "distribution coefficients".

If there is only one downstream point, and if one considers that the upstream points in a triangle should not depend on what happens in this triangle, one can cancel their contributions and the total contribution Φ_T will be affected to the downstream point.

In the case of two downstream points, we cancel the upstream point contribution, but one still has to decide how to share Φ_T between the two downstream points. This choice will determine different variants of the schemes and is done based on considerations regarding positivity, the behaviour in the presence of linear functions, or mass conservation. We shall now study the different properties of distributive schemes.

Positivity: in the case of PSI schemes, the problem of positivity is considered in the hypothesis of a "lumped" mass matrix; that is to say, equal to a diagonal matrix whose general term is:

$$S_i = \int_\Omega \Psi_i\, d\Omega$$

Moreover, two ideas must be distinguished: local and global positivity.

Local positivity demands that the function f remains positive at a point i, while only considering the contribution of a single triangle. With mass lumping, and considering only one triangle, our advection equation will be written as:

$$\frac{S_i(f_i^{n+1} - f_i^n)}{\Delta t} = -\beta_i^T\, \Phi_T \tag{6.23}$$

or even:

$$\frac{S_i(f_i^{n+1} - f_i^n)}{\Delta t} = -\beta_i^T \sum_{j=1}^{3} f_j k_j \tag{6.24}$$

which gives for f_i^{n+1}:

$$f_i^{n+1} = f_i^n \left(1 - \frac{\Delta t\, k_i \beta_i^T}{S_i}\right) - \frac{\Delta t}{S_i} \beta_i^T \sum_{j \neq i} f_j k_j \tag{6.25}$$

This scheme takes the form $f_i^{n+1} = \sum_{j=1}^{3} c_j f_j$. In order to have local positivity, all the c_j have to be positive. If i is the only downstream point (k_i positive and both the other k_j negative), one chooses $\beta_i^T = 1$. All the c_j are then positive if $\Delta t \leq \frac{S_i}{k_i}$; the time step is thus limited. Local positivity in the case of two downstream points is less easily verified.

In order to distinguish the property of global positivity, one looks at the contributions of all the triangles to which the point belongs. We find the same positivity constraint that applies to coefficient c_j, but for reasons of consistency one also has to have $\sum_{j=1}^{3} c_j = 1$.

Local positivity is a stronger condition than global positivity.

Preservation of linearity: this property demands that if f is linear, the problem gives an exact solution in steady flow. This property is guaranteed if the distribution coefficients β_i are bounded.

Continuity: this property requires that the distribution coefficients vary continuously if the direction of the vector $\overrightarrow{u}$ is changed. In the opposite case, the permanent case could converge towards cyclical solutions.

N scheme

Let us go back to the expression $\Phi_T = \sum_{i=1}^{3} f_i k_i$. By using the fact that $\sum_{i=1}^{3} k_i = 0$, we can propose different variants of this expression. We shall put Φ_T into a form where only the differences of f values appear, attributed with positive coefficients. That is to say:

$$\Phi_T = \sum_{i=1}^{3} \sum_{j=1}^{3} \lambda_{ij}(f_i - f_j) \tag{6.26}$$

where the coefficients λ_{ii} are zero. It is always possible to write Φ_T this way. For example, if k_1 and k_2 are positive and k_3 negative, one has:

$$\Phi_T = k_1(f_1 - f_3) + k_2(f_2 - f_3)$$

which gives λ_{13} and λ_{23} positive and the others zero. If k_1 and k_2 were negative, one would take λ_{31} and λ_{32} positive, and the others zero. These two solutions actually cover all possible situations.

The formula $\lambda_{ij} = \max(\min(k_i, -k_j), 0)$ guarantees the desired result (we can check it with two examples).

The N scheme (N for narrow) chooses, for the contribution of point i, the inner sum of the expression of Φ_T: that is to say, $\sum_{j=1}^{3} \lambda_{ij}(f_i - f_j)$. This expression respects the monotonicity, as all the f_j coefficients are negative (they will thus appear with a positive coefficient in the formula that gives f_i). According to the notations adopted above, the distribution coefficients are thus:

$$\beta_i = \frac{\sum\limits_{j=1}^{3} \lambda_{ij}(f_i - f_j)}{\sum\limits_{l=1}^{3} \sum\limits_{j=1}^{3} \lambda_{lj}(f_l - f_j)} \tag{6.27}$$

It is, however, preferable to preserve the form $\beta_i \Phi_T = \sum_{j=1}^{3} \lambda_{ij}(f_i - f_j)$ which carries no risk of division by zero.

This scheme does not respect linearity, as the distribution coefficients are not bounded. On the other hand, it is truly monotonous. As far as mass lumping is concerned, the scheme yields, at time t^{n+1}:

$$f_i^{n+1} = f_i^n \left(1 - \sum_{\text{neighbouring triangles}} \lambda_{ij} \frac{\Delta t}{S_i} \right) + \sum_{\text{neighbouring triangles}} \frac{\Delta t}{S_i} \lambda_{ij} f_j^n \tag{6.28}$$

The sum of all the coefficients is equal to 1 and positivity is guaranteed if the condition:

$$\Delta t \leq \frac{S_i}{\sum\limits_{\text{neighbouring triangles}} \lambda_{ij}} \tag{6.29}$$

is checked. This latter condition is thus the criterion for N scheme stability.

For the case of non-constant velocity on the element, in order to construct the N scheme, we have only used the fact that:

$$\Phi_T = \sum_{i=1}^{3} f_i k_i \text{ and } \sum_{i=1}^{3} k_i = 0$$

Now $\Phi_T = \int_{1\ triangle} \vec{u}.\overrightarrow{grad}(f)\, d\Omega$ and the two preceding properties are thus always checked, even if $\vec{u}$ is not constant on each element. However, for the triangles the change to a linear velocity does not bring any improvement, as the formulae obtained only show the average velocity on the element.

Perthame gave a proof of convergence of the N scheme in L_2 [154].

PSI scheme

The Positive Streamwise Invariant (PSI) scheme is obtained merely by limiting the distribution coefficients in the N scheme, so that they fall between 0 and 1. To do this, one applies the function called the "MinMod limiter" which is nothing more than:

$$MinMod(x) = \max(\min(x, 1), 0) \tag{6.30}$$

This limitation can seem arbitrary, but in fact it does not change the sum of coefficients and the scheme remains conservative (at least one of the three distribution coefficients in scheme N is zero, and the corrections applied to the other two compensate each other).

We then have:

$$\beta_i^{PSI} = \max(\min(\beta_i^N, 1), 0) \tag{6.31}$$

We have seen, nonetheless, that it is preferable for the N scheme to form the contributions $\beta_i^N \Phi_T$ directly, in order to avoid division. We do the same in this case.

If Φ_T is positive, we obtain:

$$\beta_i^{PSI} \Phi_T = \max(\min(\beta_i^N \Phi_T, \Phi_T), 0) \tag{6.32}$$

If Φ_T is negative, we obtain:

$$\beta_i^{PSI} \Phi_T = -\max(\min(-\beta_i^N \Phi_T, -\Phi_T), 0) \tag{6.33}$$

The two formulae can be expressed as:

$$\beta_i^{PSI} \Phi_T = sign(\Phi_T) \max(\min(sign(\Phi_T) \beta_i^N \Phi_T, sign(\Phi_T) \Phi_T), 0) \tag{6.34}$$

6.4.2 MURD scheme in three dimensions

The MURD scheme of Janin [105], which signifies "Multidimensional Upwind Residual Distribution", is an adaptation to prisms of the N and PSI schemes that we have just described for the triangular meshes. We shall look successively at the two variants: N and PSI.

N-type MURD scheme

We work with the transformed fixed mesh and we want to guarantee the global conservation of the function f by showing that one has:

$$\int_{\Omega^*} \Delta z^{n+1} \frac{(f^{n+1} - f^n)}{\Delta t} d\Omega^* + \int_{\Omega^*} \Delta z \overrightarrow{U}^* . \overrightarrow{grad}(f^n) \, d\Omega^* = 0 \tag{6.35}$$

based on our local equations. We recall that $\overrightarrow{U}^*$ is made up of the components (U, V, W^*).

The first term is initially written as:

$$\int_{\Omega^*} \Delta z^{n+1} \frac{(f^{n+1} - f^n)}{\Delta t} d\Omega^* = \sum_{ip=1}^{np} \frac{(f_{ip}^{n+1} - f_{ip}^n)}{\Delta t} \int_{\Omega^*} \Psi_{ip}^* \, \Delta z^{n+1} d\Omega^* \tag{6.36}$$

We later denote:

$$\int_{\Omega^*} \Psi_{ip}^* \, \Delta z^{n+1} d\Omega^* = S_{ip}$$

to obtain $\sum_{ip=1}^{np} \frac{(f_{ip}^{n+1} - f_{ip}^n) S_{ip}}{\Delta t}$, the first term of the equation of a point ip.

For the second term, $\int_{\Omega^*} \Delta z \overrightarrow{U} \,.\, \overrightarrow{grad}(f^n) \, d\Omega^*$, we shall calculate it element by element, and for each element share the contributions over the points it is made up of. The sum of all our equations will again give us the global equation desired.

By limiting ourselves to a prism P^* of the transformed mesh, we shall thus try to calculate:

$$\Phi_P = \int_{P^*} \Delta z \overrightarrow{U} . \overrightarrow{grad}(f) \; dP^*$$

by trying to put it into the form:

$$\Phi_P = \sum_{i=1}^{6} \sum_{j=1}^{6} \lambda_{ij} (f_i^n - f_j^n) \tag{6.37}$$

with the coefficients λ_{ij} positive or zero, as we did for the N scheme. By decomposing f, we already know that:

$$\Phi_P = \sum_{i=1}^{6} f_i^n \int_{P^*} \Delta z \overrightarrow{U} . \overrightarrow{grad}(\Psi_i) \; dP^* \tag{6.38}$$

Now the terms:

$$\int_{P^*} \Delta z \overrightarrow{U} . \overrightarrow{grad}(\Psi_i) \; dP^*$$

can be decomposed in horizontal or vertical gradients in the form $a_i + b_i$ with:

$$a_i = \int_{P^*} \Delta z U \frac{\partial \Psi_i}{\partial x} \, dP^* + \int_{P^*} \Delta z V \frac{\partial \Psi_i}{\partial y} \, dP^*$$

and:

$$b_i = \int_{P^*} \Delta z W^* \frac{\partial \Psi_i}{\partial z^*} \, dP^*$$

As we did in Section 5.2.7, we now decompose the bases $\Psi_i(x, y, z^*)$ of the prism into the product $\Psi_i^h(x, y) \Psi_i^v(z^*)$, which gives:

$$a_i = \frac{\partial \Psi_i^h}{\partial x} \int_{P^*} \Delta z U \Psi_i^v \, dP^* + \frac{\partial \Psi_i^h}{\partial y} \int_{P^*} \Delta z V \Psi_i^v \, dP^* \tag{6.39}$$

and:

$$b_i = \frac{\partial \Psi_i^v}{\partial z^*} \int_{P^*} \Delta z W^* \Psi_i^h \, dP^* \tag{6.40}$$

The properties of the bases $\Psi_i^h(x, y)$ and $\Psi_i^v(z^*)$ then give us a group of properties for coefficients a_i and b_i:

- $a_1 + a_2 + a_3 = 0$ inferred by $\Psi_1^h + \Psi_2^h + \Psi_3^h = 1$
- $a_4 + a_5 + a_6 = 0$ inferred by $\Psi_4^h + \Psi_5^h + \Psi_5^h = 1$
- $b_1 + b_4 = 0$ inferred by $\Psi_1^v + \Psi_4^v = 1$
- $b_2 + b_5 = 0$ inferred by $\Psi_2^v + \Psi_5^v = 1$
- $b_3 + b_6 = 0$ inferred by $\Psi_3^v + \Psi_6^v = 1$

Here, we have again used the local numbering of the points in a prism.

Based on the N scheme, which is built on triangles, a first set of coefficients λ_{ij} could be the following:

- $\lambda_{ij} = \max(\min(a_i, -a_j), 0)$ for the points i and j of the lower triangle of the prism, or for those of the upper triangle.
- $\lambda_{i\,i+3} = \max(b_i, 0)$ for the points i of the lower triangle.
- $\lambda_{i+3\,i} = \max(b_{i+3}, 0)$ for the points i of the upper triangle.
- $\lambda_{ij} = 0$ for the other terms (which are "crossed" terms, linking the points which are neither at the same level, nor on the same vertical line).

We already possess all the good properties to construct an N-type scheme (all the coefficients are positive or zero). However, one could obtain a less diffusive scheme by transforming the sums of horizontal and vertical fluxes into crossed fluxes, which can be done by noticing that for a set of three coefficients λ_{ij}, λ_{jk} and λ_{ik}, one does not change the value of Φ_P by carrying out the following series of operations:

- $\lambda_{ik} = \lambda_{ik} + \min(\lambda_{ij}, \lambda_{jk})$
- $\lambda_{ij} = \lambda_{ij} - \min(\lambda_{ij}, \lambda_{jk})$
- $\lambda_{jk} = \lambda_{jk} - \min(\lambda_{ij}, \lambda_{jk})$

One in fact observes that the coefficients of f_i^n, f_j^n and f_k^n remain unchanged. We now have a supplementary property, which is that if a coefficient λ_{ij} is strictly positive, then all the coefficients λ_{jk} are zero.

We shall choose, as in two dimensions, the distribution coefficients β_i as:

$$\beta_i \Phi_P = \sum_{j=1}^{6} \lambda_{ij}(f_i^n - f_j^n) \tag{6.41}$$

in order to obtain an N-type scheme. Here again, we find a stability criterion in order to ensure the monotonicity of the scheme:

$$\Delta t \leq \frac{S_i}{\sum\limits_{\text{neighbouring prisms}} \lambda_{ij}} \tag{6.42}$$

PSI-type MURD scheme

Based on our newly found N-type scheme, we need to construct the PSI-type scheme that will also guarantee that the distribution coefficients fall between 0 and 1. To do this, in the calculation of Φ_P we need to separate the positive and negative contributions:

$$\Phi_P = \Phi_P^+ - \Phi_P^- = \sum_{j=1}^{6} \lambda_{ij} \max(f_i^n - f_j^n, 0) - \sum_{j=1}^{6} \lambda_{ij} \max(f_j^n - f_i^n, 0) \tag{6.43}$$

By comparing Φ_P^+ and Φ_P^-, one can modify the scheme N coefficients:
If $\Phi_P^+ \geq \Phi_P^-$:
If $f_j^n \geq f_i^n$, then λ_{ij} is cancelled,
otherwise λ_{ij} is replaced by $\frac{\Phi_P}{\Phi_P^+}\lambda_{ij}$.
If $\Phi_P^+ \leq \Phi_P^-$:
If $f_j^n \geq f_i^n$, then λ_{ij} is replaced by $-\frac{\Phi_P}{\Phi_P^-}\lambda_{ij}$,
otherwise λ_{ij} is cancelled.

The following is an explanation of these manipulations. In the case where the positive contributions are the most important, we multiply them by $\frac{\Phi_P}{\Phi_P^+}$ and we cancel the negative contributions. What was Φ_P^- is thus cancelled, what was Φ_P^+ becomes $\Phi_P^+ \frac{\Phi_P}{\Phi_P^+} = \Phi_P$. The final expression for Φ_P remains unchanged. The case $\Phi_P^+ \leq \Phi_P^-$ can easily be deduced.

Following this modification of the coefficients λ_{ij}, all the terms $\lambda_{ij}(f_i^n - f_j^n)$ are positive and all the terms β_i fall between 0 and 1. Actually:

$$\beta_i = \frac{\sum_{j=1}^{6} \lambda_{ij}(f_i^n - f_j^n)}{\Phi_P} = \frac{\sum_{j=1}^{6} \lambda_{ij}(f_i^n - f_j^n)}{\sum_{l=1}^{6}\sum_{j=1}^{6} \lambda_{ij}(f_l^n - f_j^n)} \tag{6.44}$$

All the terms of the summations are positive and the numerator is smaller than the denominator.

Remarks on mass conservation with the MURD scheme

The MURD scheme is perfectly compatible with the proof of tracer mass conservation set out in Section 5.6.2 and with the continuity step described in Section 5.2.7. We have in fact said in this section that:

$$\Delta z W^* = \sum_{j=1}^{npoin2} \left[\Delta z W^*(z^*)\right]_j \Psi_i^h(x, y)$$

and $\Delta z W^*$ is necessary here to calculate the coefficients:

$$b_i = \frac{\partial \Psi_i^v}{\partial z^*} \int_{P^*} \Delta z W^* \Psi_i^h \, dP^*$$

We thus find that:

$$b_i = \sum_{j=1}^{npoin2} \int_{\Omega_{2D}} \Psi_i^h \Psi_j^h \, d\Omega_{2D} \left\{ \frac{\partial \Psi_i^v}{\partial z^*} \int_{z_{ip}^*}^{z_{ip+1}^*} [\Delta z W^*]_j \, dz^* \right\} \tag{6.45}$$

We see that the term in brackets is (ignoring the sign):

$$\frac{1}{(z_{ip}^* - z_{ip+1}^*)} \int_{z_{ip}^*}^{z_{ip+1}^*} [\Delta z W^*(z^*)]_j \, dz^*$$

which we have denoted $\overline{(\Delta z w^*)}_{ip+\frac{1}{2}}^{j}$. These terms are produced to resolve the continuity equation.

If mass lumping has been used when computing $\overline{(\Delta z w^*)}_{ip+\frac{1}{2}}^{j}$ (see the note at the end of Section 5.2.7), then the coefficients b_i must be computed in a compatible way, which is:

$$b_i = \int_{\Omega_{2D}} \Psi_i^h \, d\Omega_{2D} \left\{ \frac{\partial \Psi_i^v}{\partial z^*} \int_{z_{ip}^*}^{z_{ip+1}^*} [\Delta z W^*]_i \, dz^* \right\} \tag{6.46}$$

Chapter 7

Modern techniques in finite elements

We shall describe in this chapter the finite elements used in Telemac, the mesh and the storage of the finite element matrices.

7.1 Description of finite elements

7.1.1 Triangle with linear interpolation

The reference triangle (see Figure 7.1) is made up of the points whose coordinates are (0,0), (1,0) and (0,1), and whose numbering is 1, 2 and 3 respectively. On this reference element, the basis functions are:

$$P_1(\alpha, \beta) = 1 - \alpha - \beta \tag{7.1}$$

$$P_2(\alpha, \beta) = \alpha \tag{7.2}$$

$$P_3(\alpha, \beta) = \beta \tag{7.3}$$

7.1.2 Quasi-bubble triangle

The quasi-bubble element (see Figure 7.2) was published in 1993 (see [142]). It is derived from a triangle, by adding to the three vertices P1, P2, P3 an extra point P4. The centre of gravity of the triangle is a natural choice for this fourth point. The initial triangle is thus divided into three sub-triangles, T1, T2 and T3.

By choosing a linear discretization, the basis functions of the quasi-bubble triangle (in the finite element theory) are the four piecewise linear functions Ψ_i (for each point P1 to P4), specified on the triangle and verifying:

$$\Psi_i(P_j) = \delta_{ij} \tag{7.4}$$

Hydrodynamics of Free Surface Flows: Modelling with the Finite Element Method J.-M. Hervouet

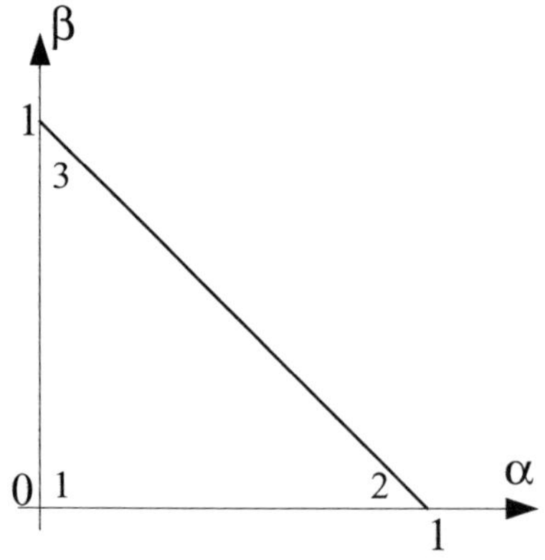

Figure 7.1: reference triangle

7.1.3 Tetrahedron

The reference element is defined by the four points whose coordinates are (0,0,0), (1,0,0), (0,1,0) and (0,0,1) (see Figure 7.3). The basis functions are:

$$P_1(\alpha, \beta, \gamma) = 1 - \alpha - \beta - \gamma \tag{7.5}$$

$$P_2(\alpha, \beta, \gamma) = \alpha \tag{7.6}$$

$$P_3(\alpha, \beta, \gamma) = \beta \tag{7.7}$$

$$P_4(\alpha, \beta, \gamma) = \gamma \tag{7.8}$$

7.1.4 Prism

The prism is made up of three vertical rectangular sides and two triangular sides, one upper, one lower, not necessarily horizontal (see Figure 7.4). The digits in the figure refer to the local numbering of the nodes.

Every function f is approximated by:

$$f = \sum_{i=1}^{6} f_i \Psi_i \tag{7.9}$$

where f_i and Ψ_i are, respectively, the values of f and the bilinear basis functions, defined on the nodes i of the mesh of domain Ω (Ψ_i equals 1 on the node i and 0 on any other node).

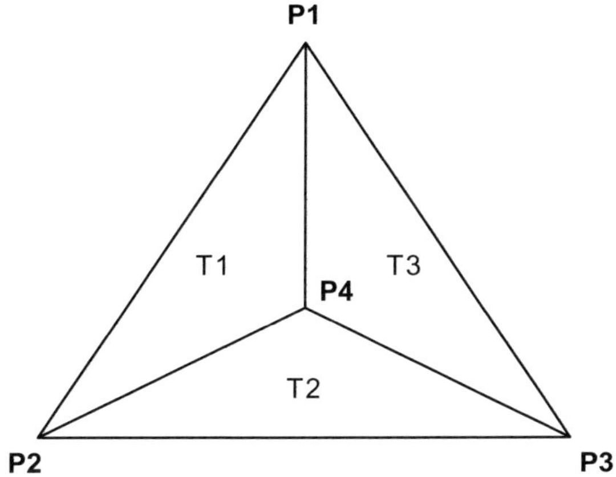

Figure 7.2: quasi-bubble element

In order to simplify the calculations, and for evaluating the basis functions Ψ_i on each element of the triangularization, one first operates on a simpler prism that constitutes the reference element (see Figure 7.5).

The basis function P_i associated with the nodes of the reference element are:

$$P_1 = (1 - \alpha - \beta)\left(\frac{1-\gamma}{2}\right)$$

$$P_4 = (1 - \alpha - \beta)\left(\frac{1+\gamma}{2}\right)$$

$$P_2 = \alpha\left(\frac{1-\gamma}{2}\right)$$

$$P_5 = \alpha\left(\frac{1+\gamma}{2}\right)$$

$$P_3 = \beta\left(\frac{1-\gamma}{2}\right)$$

$$P_6 = \beta\left(\frac{1+\gamma}{2}\right)$$

The restrictions of the basis functions Ψ_i on any prism of the mesh of domain Ω are derived from the composition of the functions P_i with the isoparametric transformation F, which transforms the reference prism into the prism under consideration.

For any prism of mesh Ω, whose vertex coordinates are (x_i, y_i, z_i), F associates a node M_0 with coordinates (α, β, γ) of the reference element with a node M with coordinates (x, y, z) of the prism under consideration, on account of the relations:

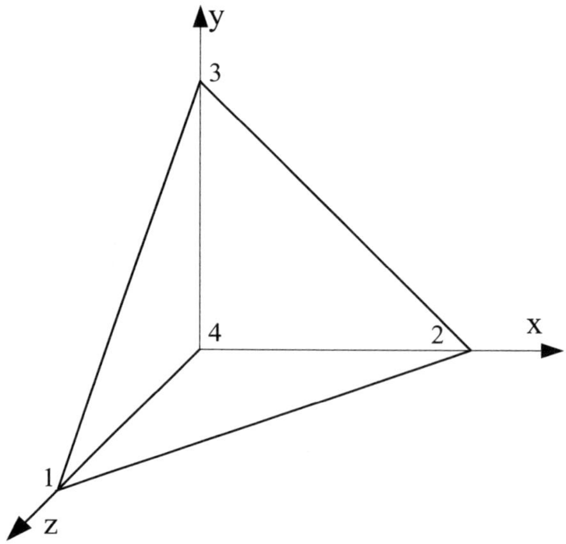

Figure 7.3: tetrahedron element

$$\left\{\begin{array}{l} x = \sum_{i=1}^{6} x_i P_i(\alpha, \beta, \gamma) \\ y = \sum_{i=1}^{6} y_i P_i(\alpha, \beta, \gamma) \\ z = \sum_{i=1}^{6} z_i P_i(\alpha, \beta, \gamma) \end{array}\right.$$

The functions P_i appearing in the definition of F are the basis functions specified on the reference element, as this chosen reference element is isoparametric (the nodes of interpolation are the geometric nodes).

The definition of F can be simplified because:

$$\left\{\begin{array}{ll} x_1 = x_4 & y_1 = y_4 \\ x_2 = x_5 & y_2 = y_5 \\ x_3 = x_6 & y_3 = y_6 \end{array}\right.$$

We thus get the following relations for F:

$$x = (1 - \alpha - \beta)x_1 + \alpha x_2 + \beta x_3$$

$$y = (1 - \alpha - \beta)y_1 + \alpha y_2 + \beta y_3$$

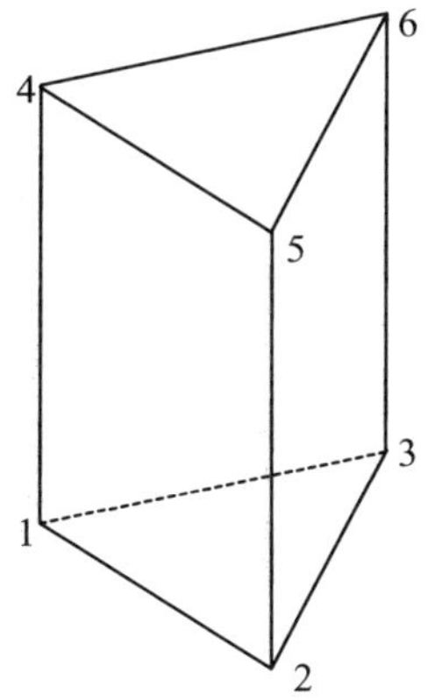

Figure 7.4: prismatic element

$$z = [(1-\alpha-\beta)z_1 + \alpha z_2 + \beta z_3]\left(\frac{1-\gamma}{2}\right)$$
$$+ [(1-\alpha-\beta)z_4 + \alpha z_5 + \beta z_6]\left(\frac{1+\gamma}{2}\right) \quad (7.10)$$

The functions Ψ_i are defined by $\Psi_i(x, y, z) = P_i(F^{-1}(x, y, z))$. The bilinearity of the transformation implies that, generally, the basis functions in the actual domain have a non-polynomial definition. Note that a plane with level z=constant in the actual mesh may not be a plane in the reference space. That may have far-reaching consequences for the treatment of stratifications.

7.2 Mesh description

7.2.1 Two-dimensional mesh

In this subsection we shall define the notations and data structure necessary for navigating inside the mesh. The variables in capital letters are the ones to be used in the FORTRAN program examples given later.

A mesh is made up of NELEM elements and NPOIN vertices, defined by their coordinates X,Y,Z. Any kind of element (triangle with linear interpolation, quasi-bubble triangle, prism with constant value, etc.) can be referred to by a code IELM (e.g. 11 for linear triangles), and has NDP nodes (NDP=3 for a linear triangle). The nodes of each element are associated to a local number, ranging from 1 to NDP. The links between this numbering specific to one element (local numbering), and the numbering specific to the whole mesh nodes ranging from 1 to NPOIN (global numbering), are specified by the connectivity table IKLE. The global number of a node whose local number is IDP in the element IELEM is IKLE(IELEM,IDP). For

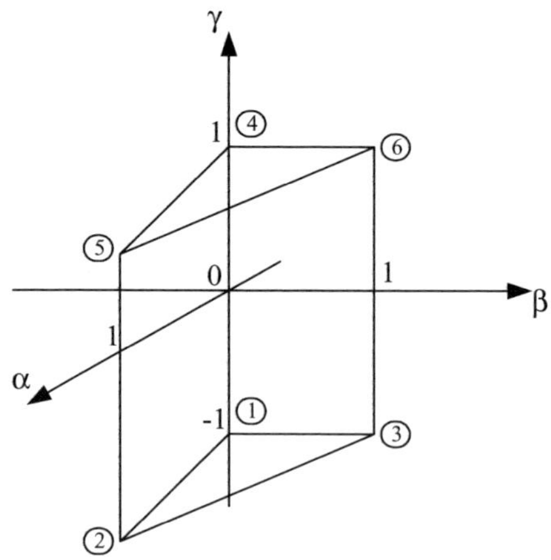

Figure 7.5: reference prism

each triangle, the list of local points is provided anticlockwise, so as to compute the triangle area without mistaking the sign. Knowledge of IKLE provides also an implicit local numbering of the sides of each triangle: side number 1 is made up of local points 1 and 2, and so on.

The tables X,Y,Z and IKLE are sufficient for a full description of the mesh. However, it is useful to define other tables which will help in the elaboration of algorithms, mainly when dealing with boundary conditions.

The points on the boundary of the mesh are numbered from 1 to NPTFR. The link with the global numbering is made by the table NBOR. NBOR(IPTFR) is the global number of the point IPTFR of the boundary. By choosing this data structure, any point on the boundary can be quickly located. In two dimensions the list of these points provides more information: when following the increasing numbers, one moves along the boundary with the domain at one's left (anticlockwise displacement except for islands whose numbering is clockwise). The point IPTFR+1 of the edge follows the point IPTFR except when the numbering has to shift from the external edge to an island or from one island to another. Whatever the case, the table KP1BOR will be used to determine the next point on the boundary edge. Usually, KP1BOR(IPTFR)=IPTFR+1, but, in the case of a mesh without an island, KP1BOR(NPTFR)=1. In two dimensions, on the boundary, the numbering of points, may be used for numbering the segments. Segment I is limited by points I and KP1BOR(I). Sometimes, it may be useful to add a second dimension to KP1BOR, to describe the boundary in the opposite direction.

KP1BOR(IPTFR,1) is the node just after IPTFR, while KP1BOR(IPTFR,2) is the node just before it.

To describe the boundary, other tables are available: IKLBOR is a connectivity table for the boundary sides (it gives the global numbers of the points constituting the side), NELBOR associates the numbers of the segments on the boundary to the numbers of the elements to which they belong, and NULONE is a table which connects the local number of the boundary points within the sides of which they are a part to the local number of the same points within the element to which they belong. Figure 7.6 shows how to use these tables for a triangular element.

Let us consider the triangle P1, whose number is IELEM, built on the three nodes whose global numbers are NG1, NG2 and NG3. The side limited by the points NG2 and NG3 pertains to the boundary and has the number IFAFR, equal to IPTFR1. The nodes NG2 and NG3 are on the boundary. Their numbers as boundary points are IPTFR1 and IPTFR2. The nodes NG1, NG2 and NG3 have local numbers 1, 2 and 3 within the triangle.

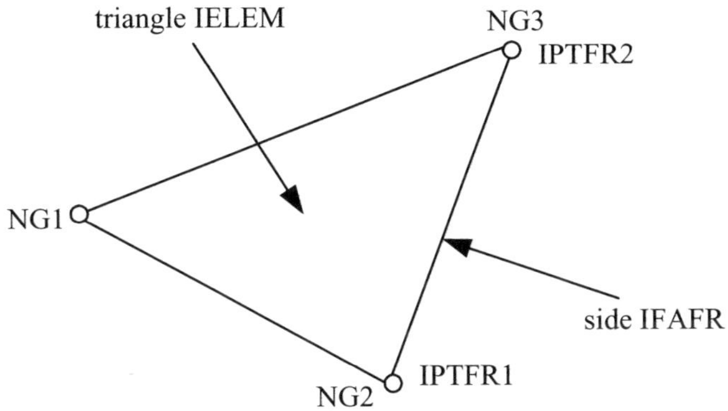

Figure 7.6: global numbering and boundary point numbering

One gets:
IKLE(IELEM,1) = NG1
IKLE(IELEM,2) = NG2
IKLE(IELEM,3) = NG3
NBOR(IPTFR1) = NG2
NBOR(IPTFR2) = NG3
NELBOR(IFAFR) = IELEM
IKLBOR(IFAFR,1) = NG2
IKLBOR(IFAFR,2) = NG3
NULONE(IFAFR,1) = 2
NULONE(IFAFR,2) = 3
KP1BOR(IPTFR1,1)=IPTFR2
KP1BOR(IPTFR1,2)=IPTFR1

7.2.2 Three-dimensional mesh

Most of the concepts established in two dimensions will be useful since the three-dimensional mesh is a pile of meshes made of triangles.

The numbering of the nodes and elements in the 3D mesh is deduced from the numbering of the 2D mesh. The first plane nodes and the first layer of prisms in the 3D mesh have the same numbers as the nodes and triangles in the 2D mesh. For the planes and layers above, the numbering is simply shifted for the nodes by the number of nodes in the 2D mesh, and for the prisms by the number of elements in the 2D mesh.

The boundary sides of the prisms are of two types. They may be triangles or quadrilaterals. A dimension must be added to the tables NELBOR, IKLBOR and NULONE in order to identify the kind of side under consideration (triangle, quadrilateral, etc.), which may be necessary for calculating the boundary matrix. TYPELB is then used for this purpose. TYPELB(IELM,1) provides the code with the first type of side of the IELM-type element. Prisms with linear interpolation, referred to by the code 41, have two types of sides, quadrilaterals and triangles. Consequently, TYPELB(41,1)=21 (code IELM for the linear quadrilateral) and TYPELB(41,2)=11 (code IELM for the linear triangle).

7.3 Matrix storage

Within the usual techniques for storing matrices in finite elements, we can point out the diagonal storage with optimization of the matrix bandwidth and the compact storage that saves memory space but makes the symmetric matrix calculations more difficult. Though we shall not use these two types of storage, we shall describe them, in order to show their imperfections. These methods are, slowly but surely, dropped for the benefit of more efficient ones. These are endowed with a data structure whose regularity is very similar to the regularity of structured meshes. We shall then detail successively EBE (Element-By-Element) storage, which uses vector computer architecture performances in the best possible manner, and edge-based storage, which seems like the optimum and final solution, and is very close to the data structure used in finite volume methods.

7.3.1 Diagonal storage

This method assumes, and it is one of its main disadvantages, that the matrix has a low number of non-zero diagonals (diagonals are here understood as lines parallel to the main diagonal). That requires, except in the simple cases of regular meshes, a previous numbering of the nodes in order to get a reduced bandwidth for the matrix. If n is the order of the matrix and k the number of such non-zero diagonals, the matrix is made up of:

- a table W(n,k) containing the elements of the non-zero diagonals,
- a table IY(n,k) specifying the column index of the elements of W,
- a table pointing to the location of the non-zero diagonal.

A great number of zero values may be stored by this method, whose efficiency depends largely on the quality of optimization of the bandwidth. Furthermore, working out the matrices is very difficult.

7.3.2 Compressed Sparse Row

This method, also known by its acronym CSR, allows one to limit the storage to the non-zero elements of the matrix. It needs:

- a table W containing the non-zero elements, stored row by row,
- a table IY specifying the column index of the elements of W,
- a table pointing to the location in the array W of the non-zero elements on each row.

The major disadvantage of the CSR method is the difficulty of directly accessing an element of the matrix, referred to by indices i and j. This is because, in the finite element processing, the number of neighbours for each point is variable from point to point, and hence the number of non-zero elements within a row is variable too. As a consequence, a symmetric matrix should be stored fully. This disadvantage vanishes with the edge-based storage, though it stores also the non-zero elements of the matrices only. A variant of CSR, called MSR (Modified Sparse Row), puts aside the diagonal of the matrices. This is what will be done in the following two kinds of storage.

7.3.3 EBE storage

Within finite element software with iterative methods of computation, the matrices are generally multiplied by vectors. Any other kind of operation with matrices is less frequent and, as will be seen in Section 7.4, these operations can be elaborated on the same architecture as the matrix–vector product. The EBE technique of storage has been conceptualized to make the matrix–vector product as efficient as possible. The fundamental idea states that, in finite element theory, it is not necessary to assemble a matrix in order to multiply it by a vector. Within a mesh with NELEM elements, a matrix M is expressed via the element matrices M_e on each element:

$$M = \sum_{e=1}^{nelem} P_e M_e {}^t P_e \tag{7.11}$$

where P_e is a matrix which makes the transition between the element and the overall mesh. P_e is derived from the connectivity table. As an example, for a triangular element whose number is e and whose vertices have the global numbers NG1, NG2 and NG3, M_e will be a matrix of dimension 3*3 and P_e a matrix of dimension NPOIN*3. The coefficient of P_e located at the intersection of row I and column J equals 1 if I=IKLE(e,J) and 0 otherwise:

$$P_e = \begin{pmatrix} 0 & 0 & 0 \\ 1 & 0 & 0 \\ \cdots & \cdots & \cdots \\ 0 & 1 & 0 \\ \cdots & \cdots & \cdots \\ 0 & 0 & 0 \\ 0 & 0 & 1 \end{pmatrix} \quad \begin{matrix} \\ \text{row IKLE(e,1)} \\ \\ \text{row IKLE(e,2)} \\ \\ \\ \text{row IKLE(e,3)} \end{matrix}$$

If X is a vector, the product $M\,X$ is expressed as:

$$M\,X = \sum_{e=1}^{nelem} \left(P_e M_e\, {}^t P_e\right) X$$

This is equivalent to multiplying M_e by the components of X associated with the nodes of element e (elementary products), then summing over the whole elements of the mesh (assembly). It is never necessary to build P_e itself, which is just a sparse form of the connectivity table IKLE.

A matrix M can be stored as NELEM matrices M_e. For a mesh made of linear triangles, that leads to 9*NELEM coefficients. This number can be reduced by keeping only the extra-diagonal entries for each element matrix and by assembling the diagonal entries, as described hereafter.

If the diagonal and extra-diagonal part of M_e are denoted D_e and X_e respectively ($M_e = D_e + X_e$), one can write:

$$MX = \sum_{e=1}^{nelem} \left(P_e D_e\, {}^t P_e\right) X + \sum_{e=1}^{nelem} \left(P_e X_e\, {}^t P_e\right) X = DX + \sum_{e=1}^{nelem} \left(P_e X_e\, {}^t P_e\right) X \quad (7.12)$$

where D is the diagonal of M, resulting from the assembly of diagonals D_e.

Any matrix can be stored within two tables, one D_{mat} containing the assembled diagonal, and the other X_{mat} containing the extra-diagonal entries of the element matrices. For a matrix associated with a mesh of triangular elements, 6*NELEM + NPOIN coefficients will have to be stored, which makes an economy of about 2.5*NELEM coefficients in the memory space compared with the storage of the whole element matrices. For a symmetric matrix, only half of the extra-diagonal terms will be stored, and the economy is of 3*NELEM more coefficients.

Furthermore, the use of element matrices allows one to vectorize the matrix–vector product on vector computer architectures. The loop on the elementary products can be vectorized and the assembling loop as well, as long as the numbering has been handled with care. The detailed computation of matrix–vector products and the process of assembling will be examined further in Section 7.4 which deals with matrix operations.

We shall use the following convention for the storage of extra-diagonal elements. Let e be an element made up of NDP nodes. Then an element matrix M_e has NDP*(NDP−1) extra-diagonal entries X_{ij}, located at the intersection of row i and column j of M_e.

We shall denote:

$$\begin{aligned}
&XMAT(e,1) = X_{12}\\
&XMAT(e,2) = X_{13}\\
&\ldots\\
&XMAT(e,NDP-1) = X_{1\ ndp}\\
&\ldots\\
&XMAT(e,NDP*(NDP-1)/2) = X_{ndp-1\ ndp}
\end{aligned}$$

for the entries of the upper triangular part of M_e. If M_e is symmetric, $XMAT$

Table 7.1: memory space depending on storage and element

Element type	EBE storage	CSR
Linear quadrilateral	NPOIN+12 NELEM≈13 NPOIN	19 NPOIN
Linear triangle	NPOIN+ 6 NELEM≈13 NPOIN	15 NPOIN
Quadratic triangle	NPOIN+30 NELEM≈16 NPOIN	24 NPOIN
Quadratic quadrilateral	NPOIN+72 NELEM≈19 NPOIN	33 NPOIN
Linear brick	NPOIN+56 NELEM≈57 NPOIN	55 NPOIN
Linear prismrism	NPOIN+30 NELEM≈61 NPOIN	43 NPOIN

is fully defined. Otherwise, it is necessary to store the lower triangular part of M_e as follows:

$$XMAT(e, NDP*(NDP-1)/2+1) = X_{21}$$
$$XMAT(e, NDP*(NDP-1)/2+2) = X_{31}$$
$$\dots$$
$$XMAT(e, NDP*(NDP-1)/2+NDP-1) = X_{ndp\ 1}$$
$$\dots$$
$$XMAT(e, NDP*(NDP-1)) = X_{ndp\ ndp-1}$$

In the matrix M_e hereafter, made up of a triangle with linear interpolation, the different parts of the table XMAT have been located as they should (the * in the table refers to the diagonal entries that are assembled and not stored at the elementary level):

$$M_e = \begin{pmatrix} \dots & XMAT(e,1) & XMAT(e,2) \\ XMAT(e,4) & \dots & XMAT(e,3) \\ XMAT(e,5) & XMAT(e,6) & \dots \end{pmatrix} \tag{7.13}$$

Table 7.1 summarizes, for some kinds of elements, the memory space necessary for the storage of a matrix (some values are purely statistical and depend on the organization of the elements and on the number of neighbours).

7.3.4 Edge-based storage

This is a new technique enabling an assembled matrix to be stored in a regular and optimal way. The concept is that, for a mesh made of triangles with linear interpolation, the products $\int_\Omega \Psi_i \Psi_j \, d\Omega$ of two basis functions Ψ_i and Ψ_j with distinct i and j will be non-zero only if i and j are linked by a segment of the mesh. Every segment is then the ideal "place" to store the extra-diagonal entries of the matrices. For a non-symmetric matrix, a segment represents two entries. This idea is very simple for all simplicial elements like triangles or tetrahedra. It can be extended to all other elements, by adding extra imaginary segments. For a quasi-bubble triangle, three segments linking the three vertices to the central point will be necessary.

The data structure necessary to handle and build edge-based stored matrices is as follows:

- An array, named GLOSEG, similar to IKLE for the triangles, gives the numbers of the two points defining the segment. Its dimension is (NSEG,2), where NSEG is the number of segments (in the case of triangles, NSEG= $\frac{3NELEM+NPTFR}{2}$). The second dimension refers to the two points of the segments.

- An array named ELTSEG, whose dimension is (NELEM,NS), where NS is the number of segments constituting an element (three for a triangle). For each element, ELTSEG provides the numbering of its associated segments.

- An array named ORISEG, whose dimension is (NELEM,NS). ORISEG is linked to ELTSEG, and indicates the orientation of the segment within the element. Its value is 1 if one moves anticlockwise around the element (following the increasing local numbering of nodes) and 2, otherwise.

An edge-based stored assembled matrix will be defined by a diagonal and two tables XA1 and XA2 (rank NSEG) for a non-symmetric matrix or only one table XA1 for a symmetric matrix.

XA1 includes the coefficient of point 2, in the point 1 related equation.

XA2 includes the coefficient of point 1, in the point 2 related equation.

The way of assembling a matrix for edge-based storage as well as the expression for the matrix–vector product will be described when required.

7.4 Operations on matrices

This chapter describes how to build the matrices and if necessary, as in the case of edge-based storage, how to assemble them. It also describes the main operations on matrices performed in finite element software, in the case of EBE storage. We shall discuss successively:

- The construction of a non-assembled matrix.

- The assembly process for edge-based storage.

- The product of a non-symmetric matrix and a vector.

- The product of a symmetric matrix and a vector.

- The product of a transposed matrix and a vector.

- The treatment of Dirichlet-type boundary conditions.

- The diagonal preconditioning.

- The most important algorithm, namely the matrix–vector product algorithm. Any other can be based on or reduced to this one.

7.4.1 Building the matrices

Unlike the techniques based on Gauss quadrature, the matrices here are calculated by exact analytical integration. The entries of a finite element matrix are usually polynomial integrals, which can be evaluated by such an analytical integration. Writing the integration is a long, cumbersome and risky task even though some tricks can be used. This is the reason why we have chosen to use formal computation software (REDUCE, MAPLE V) which can provide the exact result of an integral calculation.

An example of such a computation is given next.

Example of a matrix computation

Let us calculate the mass matrix for a mesh made of quadrilaterals with linear interpolation. We voluntarily choose a non-simplicial element whose Jacobian is not constant, in order to avoid too simple an isoparametric transformation that would hide the difficulties.

The example of computation will be made on an arbitrary quadrilateral Q determined by its vertices P1, P2, P3 and P4 (Figure 7.7), whose coordinates are (x1,y1), (x2,y2), (x3,y3), (x4,y4). The entry M_{ij} of the mass matrix will be written $M_{ij} = \int_\Omega \Psi_i \Psi_j \, d\Omega$, where Ψ_i is the basis function associated to the node i (i=1, 2, 3 or 4).

The computation of the integral can then be made on a reference element, thanks to an isoparametric transformation T. Any node of the reference element Q0, with coordinates (ξ, η), is associated with one node of the quadrilateral Q, whose coordinates are (x,y) such that:

$$x(\xi, \eta) = t_1 + t_2\xi + t_3\eta + t_4\xi\eta$$

$$y(\xi, \eta) = t_1^{'} + t_2^{'}\xi + t_3^{'}\eta + t_4^{'}\xi\eta$$

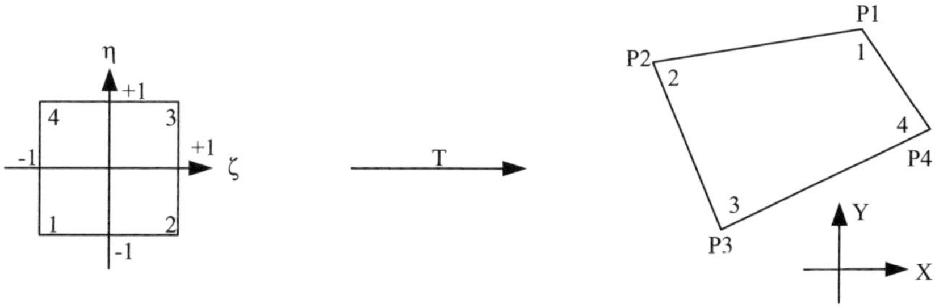

Figure 7.7: isoparametric transformation for a quadrilateral

The image per T of every vertex in the reference element is thus a vertex of the quadrilateral. By a process of identification, one can deduce the coefficients t_1, t_2, etc., of the transformation, with respect to the coordinates x_1, y_1, x_2, y_2, x_3, y_3, x_4, y_4 of the vertices:

$$t_1 = (+x_1 + x_2 + x_3 + x_4)/4$$

$$t_2 = (-x_1 + x_2 + x_3 - x_4)/4$$

$$t_3 = (-x_1 - x_2 + x_3 + x_4)/4$$

$$t_4 = (+x_1 - x_2 + x_3 - x_4)/4$$

$$t_1^{'} = (+y_1 + y_2 + y_3 + y_4)/4$$

$$t_2^{'} = (-y_1 + y_2 + y_3 - y_4)/4$$

$$t_3^{'} = (-y_1 - y_2 + y_3 + y_4)/4$$

$$t_4^{'} = (+y_1 - y_2 + y_3 - y_4)/4$$

A basis function Ψ in the actual element corresponds to a polynomial Pi such as:

$$\Psi(x, y) = Pi(T^{-1}(x, y)) \tag{7.14}$$

In the example above, there are four polynomials associated to the four bases of the actual element:

$$P1(\xi, \eta) = (1 - \xi - \eta + \xi\eta)/4$$

$$P2(\xi, \eta) = (1 + \xi - \eta - \xi\eta)/4$$

$$P3(\xi, \eta) = (1 + \xi + \eta + \xi\eta)/4$$

$$P4(\xi, \eta) = (1 - \xi + \eta - \xi\eta)/4$$

As for the functions Ψ, the value of each polynomial is 1 at one vertex and 0 for the others.

In the reference element, the integral is:

$$\int_Q \Psi_i \Psi_j \, dQ = \int_{-1}^{+1} \int_{-1}^{+1} P_i P_j \, |J| \, d\xi d\eta \tag{7.15}$$

where J is the Jacobian of the transformation T, equal to the determinant of the Jacobian matrix:

$$\begin{pmatrix} \frac{\partial x}{\partial \xi} & \frac{\partial x}{\partial \eta} \\ \frac{\partial y}{\partial \xi} & \frac{\partial y}{\partial \eta} \end{pmatrix}$$

So:

$$J = (t_2 + t_4\eta)(t_3^{'} + t_4^{'}\xi) - (t_2^{'} + t_4^{'}\eta)(t_3 + t_4\xi) \tag{7.16}$$

We shall assume a positive value for J, which is achieved with an anticlockwise local numbering of the nodes along the edge. Because of the polynomial nature of J, the computation is reduced to the calculation of a polynomial (whose highest order is expressed as $\xi^3\,\eta^3$).

These data are sufficient to achieve an analytical computation with formal calculation software. The result may be included in FORTRAN software. For example, the formal calculation of a mass matrix on a quadrilateral will give:

$$
\begin{aligned}
MAT(1,1) &= (X2*(2*Y4+Y3)+X3*(Y4-Y2)+X4*(-Y3-2*Y2))/36\\
MAT(1,2) &= (X2*(Y4+2*Y3)+X3*(Y4-2*Y2)-X4*(Y3+Y2))/72\\
MAT(1,3) &= (X2*Y3+X3*(Y4-Y2)-X4*Y3)/72\\
MAT(1,4) &= (X2*(Y4+Y3)+X3*(2*Y4-Y2)+X4*(-2*Y3-Y2))/72\\
MAT(2,1) &= (X2*(Y4+2*Y3)+X3*(Y4-2*Y2)-X4*(Y3+Y2))/72\\
MAT(2,2) &= (3*X2*Y3+X3*(Y4-3*Y2)-X4*Y3)/36\\
MAT(2,3) &= (X2*(-Y4+3*Y3)+X3*(2*Y4-3*Y2)+X4*(-2*Y3+Y2))/72\\
MAT(2,4) &= (X2*Y3+X3*(Y4-Y2)-X4*Y3)/72\\
MAT(3,1) &= (X2*Y3+X3*(Y4-Y2)-X4*Y3)/72\\
MAT(3,2) &= (X2*(-Y4+3*Y3)+X3*(2*Y4-3*Y2)+X4*(-2*Y3+Y2))/72\\
MAT(3,3) &= (X2*(-2*Y4+3*Y3)+3*X3*(Y4-Y2)+X4*(-3*Y3+2*Y2))/36\\
MAT(3,4) &= (X2*(-Y4+2*Y3)+X3*(3*Y4-2*Y2)+X4*(-3*Y3+Y2))/72\\
MAT(4,1) &= (X2*(Y4+Y3)+X3*(2*Y4-Y2)+X4*(-2*Y3-Y2))/72\\
MAT(4,2) &= (X2*Y3+X3*(Y4-Y2)-X4*Y3)/72\\
MAT(4,3) &= (X2*(-Y4+2*Y3)+X3*(3*Y4-2*Y2)+X4*(-3*Y3+Y2))/72\\
MAT(4,4) &= (X2*Y3+X3*(3*Y4-Y2)-3*X4*Y3)/36
\end{aligned}
$$

With a vector computer architecture, the previous expression included in a loop can be vectorized. No line depends on the result of a previous one: there is no backward dependence.

The demonstration just shown can be applied to any matrix that leads to the integral of a polynomial expression. This would be the case for mass matrices and divergence-type matrices. It would not always be appropriate for diffusion-type matrices, where a Jacobian appears as the denominator.

For non-simplicial elements, the analytical calculation will not be suitable for the diffusion matrix (and the Gauss quadrature will just give an approximate solution). Unfortunately, the prism is of this type and requires approximations.

Appendix B shows the file of MAPLE V instructions for the analytical calculation of a diffusion matrix in a tetrahedron with linear interpolation. In this case the Jacobian is constant on every element.

Properties of finite element matrices

Besides being "sparse", i.e. containing a lot of zero entries, the finite element matrices have interesting properties which can be used in order to optimize CPU time or the memory space, or which can explain a certain behaviour of the numerical schemes.

Symmetry: this is the case of the mass matrix:

$$\int_\Omega \Psi_i \Psi_j \, d\Omega \tag{7.17}$$

and of the diffusion matrix:

$$\int_\Omega \nu \, \overrightarrow{grad}(\Psi_i) \, . \, \overrightarrow{grad}(\Psi_j) \, d\Omega \tag{7.18}$$

Thanks to the kinds of storage used, only half the extra-diagonal terms are stored.

"Quasi-magic" square: a diffusion matrix is a quasi-magic square. On every row, as well as on every column, the sum of all the terms is zero. The comparison stops here because adding up the terms of the diagonal gives a non-zero total. These properties can be demonstrated by using the equation $\sum_{i=1}^{npoin} \Psi_i = 1$. They help to optimize the calculations while reducing truncation errors to the minimum. Thus, for the diffusion matrix of one element, with n degrees of freedom, $\frac{n(n-1)}{2}$ elementary terms should be calculated instead of $\frac{n(n+1)}{2}$. For a triangle with linear interpolation, the extra-diagonal terms of the diffusion matrix A are fully calculated from the values of the terms on the diagonal:

$$A_{13} = \frac{1}{2}(A_{22} - A_{11} - A_{33}) \qquad A_{12} = \frac{1}{2}(A_{33} - A_{11} - A_{22}) \qquad A_{23} = \frac{1}{2}(A_{11} - A_{22} - A_{33})$$

M-matrix: if a matrix A can be written as the sum of a diagonal DA whose entries are strictly positive and of extra-diagonal terms XA all strictly negative, A is known as an M-matrix. It can be easily proved that the system $AX = B$ then has a positive solution if all the entries of B are positive. A diffusion matrix should be an M-matrix, but some pathological cases can be found for which this property is not verified.

In such cases, the diffusion is no longer monotonous. For example, let us consider a triangle comprising three points M_1, M_2 and M_3, whose coordinates are (x_1, y_1), (x_2, y_2) and (x_3, y_3). The linear basis functions on this triangle are:

$$\Psi_1(x, y) = \frac{(y_2 - y_3)}{2S} x + \frac{(x_3 - x_2)}{2S} y + \frac{(x_2 y_3 - x_3 y_2)}{2S} \tag{7.19}$$

$$\Psi_2(x, y) = \frac{(y_3 - y_1)}{2S} x + \frac{(x_1 - x_3)}{2S} y + \frac{(x_3 y_1 - x_1 y_3)}{2S} \tag{7.20}$$

$$\Psi_3(x, y) = \frac{(y_1 - y_2)}{2S} x + \frac{(x_2 - x_1)}{2S} y + \frac{(x_1 y_2 - x_2 y_1)}{2S} \tag{7.21}$$

where S is the area of the triangle, equal to:

$$\frac{1}{2}(x_1 y_2 - x_2 y_1 + x_2 y_3 - x_3 y_2 + x_3 y_1 - x_1 y_3)$$

We get:

$$\int_{triangle} \left(\frac{\partial \Psi_1}{\partial x} \frac{\partial \Psi_2}{\partial x} + \frac{\partial \Psi_1}{\partial y} \frac{\partial \Psi_2}{\partial y} \right) d\Omega$$

$$= \frac{1}{4S}\left[(y_2 - y_3)(y_3 - y_1) + (x_3 - x_2)(x_1 - x_3)\right] \tag{7.22}$$

or:

$$\int_{triangle} \left(\frac{\partial \Psi_1}{\partial x}\frac{\partial \Psi_2}{\partial x} + \frac{\partial \Psi_1}{\partial y}\frac{\partial \Psi_2}{\partial y} \right) d\Omega = -\frac{\overrightarrow{M_1M_3} \, . \, \overrightarrow{M_2M_3}}{4S} \tag{7.23}$$

This quantity is negative only if the angle at M_3 is acute. A diffusion matrix can thus lose the characteristics of an M-matrix if one angle in the mesh is obtuse. However, we have considered only one element, but it would be necessary to examine the extra-diagonal terms after assembly.

Diffusion matrix with prismatic element

For diffusion matrices of the form:

$$M_{ij} = \int_\Omega \nu \, \overrightarrow{grad}(\Psi_i). \, \overrightarrow{grad}(\Psi_j) \, d\Omega$$

the Jacobian J of the transformation T (see Section 7.4.1 above) appears in the denominator, due to the two gradients within the integral. In that case, if J is not constant in space, the function to be integrated is not a polynomial and there is no hope of finding an exact integral. Therefore an approximation must be done, e.g. taking an average constant Jacobian on the prism. Another problem is the fact that this diffusion matrix is far from being an M-matrix. First the same problem arises as with triangles, but there is also the extra difficulty that some of the extra-diagonal terms stemming from vertical diffusion are obviously always positive. For example, the term:

$$\int_\Omega \nu_z \frac{\partial \Psi_1}{\partial z}\frac{\partial \Psi_2}{\partial z} \, d\Omega$$

with indices 1 and 2 referring to our local numbering of points in a prism, will be positive because both derivatives with respect to z are negative. To cope with this problem we remark that because our prisms are based on three vertical lines, it should be easy to compute the second derivative with respect to z of a function, to use points on a single vertical. In other words, when we compute terms such as:

$$\int_\Omega \nu_z \frac{\partial f}{\partial z}\frac{\partial \Psi_1}{\partial z} \, d\Omega$$

it is strange to use the value of f at point 2 whereas $\frac{\partial f}{\partial z}$ could be taken more simply as $\frac{f_4 - f_1}{\Delta z_1}$, where Δz_1 is the distance between points 4 and 1 (we define in the same way Δz_2 for points 2 and 5, and Δz_3 for points 3 and 6). Following this idea, and if we denote:

$$R_1 = \int_\Omega \nu_z \frac{\partial \Psi_1}{\partial z} \, d\Omega, \;\; R_2 = \int_\Omega \nu_z \frac{\partial \Psi_2}{\partial z} \, d\Omega \text{ and } R_3 = \int_\Omega \nu_z \frac{\partial \Psi_3}{\partial z} \, d\Omega$$

we get to a vertical part of the diffusion matrix in the prism in the form:

$$\begin{pmatrix} -\frac{R_1}{\Delta z_1} & 0 & 0 & \frac{R_1}{\Delta z_1} & 0 & 0 \\ 0 & -\frac{R_2}{\Delta z_2} & 0 & 0 & \frac{R_2}{\Delta z_2} & 0 \\ 0 & 0 & -\frac{R_3}{\Delta z_3} & 0 & 0 & \frac{R_3}{\Delta z_3} \\ \frac{R_1}{\Delta z_1} & 0 & 0 & -\frac{R_1}{\Delta z_1} & 0 & 0 \\ 0 & \frac{R_2}{\Delta z_2} & 0 & 0 & -\frac{R_2}{\Delta z_2} & 0 \\ 0 & 0 & \frac{R_3}{\Delta z_3} & 0 & 0 & -\frac{R_3}{\Delta z_3} \end{pmatrix}$$

We now have an M-matrix. It is also clear that the "quasi-magic" square properties are conserved, and only points which are on the same vertical line are linked. This latter property leads to the important fact that if there is only vertical diffusion, it can be computed by solving a series of tri-diagonal systems corresponding to as many vertical lines in the mesh of prisms.

Matrices with a quasi-bubble element

The matrices under consideration are defined as follows:

$$M_{ij} = \int_\Omega f(\varphi_i, \Psi_j, F, U, V, W)\, d\Omega \tag{7.24}$$

where F, U, V, W are different functions, scalar or vectorial, which the matrices depend on. For these matrices, the test functions φ and the basis functions Ψ may be of different types (linear or quasi-bubble) just like the discretization functions of variables F,U,V, etc.

Two cases may occur:

1) φ and Ψ are of a quasi-bubble type:

The element matrices have 16 entries. The calculation for these entries is easy thanks to the partition of triangle T into three sub-triangles. So, we get, if φ_i^{Tj} is the restriction of the basis function φ_i to the sub-triangle Tj:

$$M_{ij} = \int_\Omega f(\varphi_i^{T1}, \Psi_j^{T1}, F, U, V, W)\, d\Omega + \int_\Omega f(\varphi_i^{T2}, \Psi_j^{T2}, F, U, V, W)\, d\Omega$$

$$+ \int_\Omega f(\varphi_i^{T3}, \Psi_j^{T3}, F, U, V, W)\, d\Omega \tag{7.25}$$

which is a sum of three integrals denoted $I1$, $I2$ and $I3$.

Now, the basis functions of the quasi-bubble element restricted to the sub-triangles are the very basis functions P1 on the same cells. The calculation of the three integrals is then standard.

The next step is to add up these integrals. Furthermore, the domains of the quasi-bubble basis functions rarely intersect on the three sub-triangles but most often on only one (extra-diagonal terms). As a consequence, one or two of the integrals $I1$, $I2$ and $I3$ are zero.

One can observe that only the function Ψ_4 is defined on the whole triangle T.

Considering the term M_{11}, one gets:

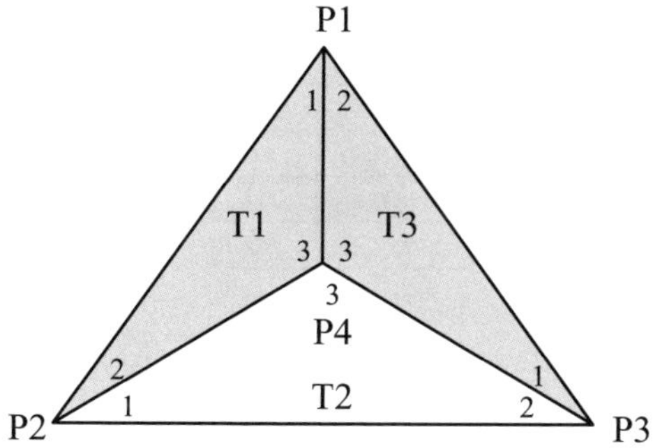

Figure 7.8: domain of definition of the quasi-bubble function $\Psi 1$ (shaded)

$$M_{11} = \int_{\Omega} f(\varphi_i^{T1}, \Psi_j^{T1}, F, U, V, W, \text{etc.})\, d\Omega + \int_{\Omega} f(\varphi_i^{T3}, \Psi_j^{T3}, F, U, V, W, \text{etc.})\, d\Omega$$

$$= I1 + I3 = m1_{11} + m3_{22}$$

where $m1$, $m2$, $m3$ are the linear element matrices computed on the sub-triangles $T1$, $T2$, $T3$. As a matter of fact, Ψ_1^{T1} is the basis function associated to the first point of sub-triangle $T1$ and Ψ_1^{T3} is the basis function associated to the second point of sub-triangle $T3$ (see Figure 7.8).

The quasi-bubble matrix M is then a result of the pre-assembling process of linear sub-matrices. The lists of operations is given below:

$$\begin{aligned}
M_{11} &= m1_{11} + m3_{22}\\
M_{12} &= m1_{12}\\
M_{13} &= m3_{21}\\
M_{14} &= m1_{13} + m3_{23}\\
M_{21} &= m1_{21}\\
M_{22} &= m1_{22} + m2_{11}\\
M_{23} &= m2_{12}\\
M_{24} &= m1_{23} + m2_{13}\\
M_{31} &= m3_{12}\\
M_{32} &= m2_{21}\\
M_{33} &= m2_{22} + m3_{11}\\
M_{34} &= m2_{23} + m3_{13}\\
M_{41} &= m1_{31} + m3_{32}\\
M_{42} &= m1_{32} + m2_{31}\\
M_{43} &= m2_{32} + m3_{31}\\
M_{44} &= m1_{33} + m2_{33} + m3_{33}
\end{aligned}$$

2) φ and Ψ are of different types:

Table 7.2: values of the linear basis functions on the sub-triangles

Function P	$P(1)$	$P(2)$	$P(3)$
$P1(\xi,\eta) = 1 - \xi - \frac{2}{3}\eta$	1	0	1/3
$P2(\xi,\eta) = \xi + \frac{1}{3}\eta$	0	1	1/3
$P3(\xi,\eta) = \frac{1}{3}\eta$	0	0	1/3

The element matrices are rectangular and contain 12 entries. We consider here the case where φ is linear and Ψ quasi-bubble; the symmetrical situation is automatically deduced from this case. The relation $M_{ij} = I1 + I2 + I3$ is always true but, unlike the previous case, the functions φ restricted to the sub-triangles no longer correspond to the basis functions P1 on these sub-triangles. In fact, $\varphi_i(P_j) = \delta_{ij}$ for $1 \leq j \leq 3$ and $\varphi_i(P_4) = \frac{1}{3}$ because P4 is the centre of gravity of the triangle T. Within the internal numbering of the sub-triangles Ti, P4 always corresponds to the point number P3, and so the basis functions in the reference triangle are those given in Table 7.2.

The isoparametric transformation saves the values of the functions on the nodes, and the expected result is finally obtained within the actual mesh.

The matrix coefficients related to the triangle T are obtained by an assembly on the sub-triangles:

$$
\begin{aligned}
M_{11} &= m1_{11} + m3_{22}\\
M_{12} &= m1_{12} + m2_{31}\\
M_{13} &= m3_{21} + m2_{32}\\
M_{14} &= m1_{13} + m3_{23} + m2_{33}\\
M_{21} &= m1_{21} + m3_{32}\\
M_{22} &= m1_{22} + m2_{11}\\
M_{23} &= m2_{12} + m3_{31}\\
M_{24} &= m1_{23} + m2_{13} + m3_{33}\\
M_{31} &= m3_{12} + m1_{31}\\
M_{32} &= m2_{21} + m1_{32}\\
M_{33} &= m2_{22} + m3_{11}\\
M_{34} &= m2_{23} + m3_{13} + m1_{33}
\end{aligned}
$$

We would get for the inverse case of a quasi-bubble $\times$ linear matrix:

$$
\begin{aligned}
M_{11} &= m1_{11} + m3_{22}\\
M_{12} &= m1_{12} + m3_{23}\\
M_{13} &= m3_{21} + m1_{13}\\
M_{21} &= m1_{21} + m2_{13}\\
M_{22} &= m1_{22} + m2_{11}\\
M_{23} &= m1_{23} + m2_{12}\\
M_{31} &= m2_{23} + m3_{12}\\
M_{32} &= m2_{21} + m3_{13}\\
M_{33} &= m2_{22} + m3_{11}\\
M_{41} &= m1_{31} + m2_{33} + m3_{32}\\
M_{42} &= m1_{32} + m2_{31} + m3_{33}
\end{aligned}
$$

$M_{43} = m1_{33} + m2_{32} + m3_{31}$

7.4.2 Assembling a matrix for edge-based storage

We describe belowr how to implement a storage by segment, from a matrix initially stored in a non-assembled EBE manner. As indicated in Section 7.3.4, the extra-diagonal terms of the matrix are stored in two tables, XA1 and XA2, themselves included in the 2-dimension table XMS (S as in segment), initially set to 0. The table XMT (T as in triangle) refers to the EBE storage. The value of the table ORISEG (with index 1, 2 or 3) indicates if it is necessary to add one element of the matrix to XA1 or to XA2. The value 3−ORISEG equals 2 and 1 respectively, when ORISEG equals 1 and 2. It is used to store the lower part of the matrix:

```
C
C INITIALIZATION
C
DO ISEG=1,NSEG
  XMS(ISEG,1) = 0.D0
  XMS(ISEG,2) = 0.D0
ENDDO
C
C ASSEMBLY.
C
DO IELEM=1,NELEM
!  term 1 2
   XMS(ELTSEG1(IELEM), ORISEG1(IELEM))
= XMS(ELTSEG1(IELEM), ORISEG1(IELEM)) + XMT(IELEM,1)
!  term 2 3
   XMS(ELTSEG2(IELEM), ORISEG2(IELEM))
= XMS(ELTSEG2(IELEM), ORISEG2(IELEM)) + XMT(IELEM,3)
!  term 3 1
   XMS(ELTSEG3(IELEM), ORISEG3(IELEM))
= XMS(ELTSEG3(IELEM), ORISEG3(IELEM)) + XMT(IELEM,5)
!  term 2 1
   XMS(ELTSEG1(IELEM),3-ORISEG1(IELEM))
= XMS(ELTSEG1(IELEM),3-ORISEG1(IELEM)) + XMT(IELEM,4)
!  term 3 2
   XMS(ELTSEG2(IELEM),3-ORISEG2(IELEM))
= XMS(ELTSEG2(IELEM),3-ORISEG2(IELEM)) + XMT(IELEM,6)
!  term 1 3
   XMS(ELTSEG3(IELEM),3-ORISEG3(IELEM))
= XMS(ELTSEG3(IELEM),3-ORISEG3(IELEM)) + XMT(IELEM,2)
ENDDO
```

This algorithm is easily simplified for a symmetric matrix. When the matrix is rectangular, e.g. if it has been formed with linear bases and quasi-bubble test functions, different numbers of segments have to be considered (see NSEG11 and NSEG12 hereafter), and the program must be completed with:

```
C
C QUASI-BUBBLE ASSEMBLY
C
DO IELEM=1,NELEM
    XMS(ELTSEG4(IELEM),2) = XMT(IELEM,07)        ! term 4 1
    XMS(ELTSEG5(IELEM),2) = XMT(IELEM,08)        ! term 4 2
    XMS(ELTSEG6(IELEM),2) = XMT(IELEM,09)        ! term 4 3
ENDDO
```

As the XMT terms are already assembled, this is in fact just a copy.

7.4.3 Assembly of an elementary vector

We now come back to the EBE storage. The product of an EBE stored matrix and a vector will result in a non-assembled vector, also stored element by element. Before detailing the matrix–vector product, we shall study the process of assembling an elementary vector. This process is also necessary to assemble the diagonal of any matrix.

With a vector being defined on each element by a vector W_e (dimension NDP, as, for example, the diagonal terms A11, A22 and A33 of a matrix on a triangle), the target is to construct the global vector R (dimension npoin). With the notations of Section 7.3.3:

$$R = \sum_{e=1}^{nelem} P_e W_e \tag{7.26}$$

The triangles with linear interpolation will again be used as an example: W1, W2 and W3 are the components of vector W, stored element by element; W1(IELEM) is the component of the vector at node 1 of element IELEM, etc.

In the FORTRAN programming language, the steps for constructing vector R are (R has been previously set to 0):

```
DO IELEM=1,NELEM
    R(IKLE(IELEM,1)) = R(IKLE(IELEM,1)) + W1(IELEM)
    R(IKLE(IELEM,2)) = R(IKLE(IELEM,2)) + W2(IELEM)
    R(IKLE(IELEM,3)) = R(IKLE(IELEM,3)) + W3(IELEM)
ENDDO
```

where IKLE is the connectivity table: IKLE(IELEM,I) is the global number of the Ith local point of element IELEM.

With a vector computer, this loop is not automatically vectorized, because of the risk of a simultaneous assembly at the same location in vector R. Vectorization will, however, be possible, under a few conditions as follows.

The assembly loop is first split into three consecutive loops:

```
C
DO IELEM=1,NELEM
    R(IKLE(IELEM,1))=R(IKLE(IELEM,1))+W1(IELEM)
ENDDO
C
DO IELEM=1,NELEM
```

```
        R(IKLE(IELEM,2))=R(IKLE(IELEM,2))+W2(IELEM)
    ENDDO
    C
    DO IELEM=1,NELEM
        R(IKLE(IELEM,3))=R(IKLE(IELEM,3))+W3(IELEM)
    ENDDO
    C
```

These three loops are still not automatically vectorized on a vector computer. In fact, vectorization works simultaneously on a certain number of indices in the loop. This number is known as the length of the computer vector register, and varies from one machine to another. It is 64 for a CRAY YMP and 1024 for a FUJITSU machine. So, on a CRAY machine, a loop ranging from 1 to NELEM will be performed by groups of 64 elements. If the first loop had been vectorized on a CRAY, the statement:

$$\mathrm{R(IKLE(IELEM,1))=R(IKLE(IELEM,1))+W1(IELEM)}$$

would have been executed at the same time for elements 1 to 64, then 65 to 128, etc. The result is correct only if, within a group of 64 elements, each component of vector R is used once and only once. Two elements IELEM1 and IELEM2 such that:

$$\mathrm{IKLE(IELEM1,1)=IKLE(IELEM2,1)}$$

should not appear in the same group of 64 elements.

When compiling, the vector computer which is unaware of the content of IKLE identifies this conflict of backward dependence, and does not vectorize the loop.

It is possible to force vectorization by an order to the compiler, but it must be guaranteed that no backward dependence occurs within the mesh.

The advantage of splitting the initial loop assembly into three loops becomes clear. Without this, respecting the backward dependence condition would have been more difficult to achieve. IKLE would have to verify:

$$\mathrm{IKLE(IELEM1,I)} \neq \mathrm{IKLE(IELEM2,J)}$$

for I, J=1, 2 and 3 and for all different elements IELEM1 and IELEM2, within a group of 64 elements.

When the loop is split, a risk of backward dependence remains if, for two different elements IELEM1 and IELEM2, IELEM1/64 = IELEM2/64 (/ denotes integer division) and if there is an index I=1, 2 or 3 such that:

$$\mathrm{IKLE(IELEM1,I)=IKLE(IELEM2,I)}$$

The idea behind the vectorization of the three loops is to find a numbering of elements that would eliminate any backward dependence occurrence. This is impossible for meshes made of too few elements, but it is very easy to perform when the number of elements exceeds a few hundreds (for a vector register length of 64).

We shall detail consecutively a “counter-example”, then a “heuristic” algorithm, efficient in the case of sufficient number of elements.

Counter-example:

Imagine a mesh of triangles, with fewer than 64 elements in which one point belongs to four different triangles. This point will have the same local number (1, 2 or 3) in, at least, two of these triangles. That will result in a backward dependence, which cannot be avoided.

Heuristic algorithm for numbering the elements:

The idea is to find out the dependence situations and eliminate them when they occur. We start from an existing numbering (the number of elements should be greater than 64) and progress in this numbering, examining all the groups of 64 elements. As soon as a critical element appears, its number is exchanged with a higher rank element (the conflicts are thus pushed forward as the progression goes on). This algorithm fails if the highest rank is reached and dependences remain. In this case, the exchanges are made with lower rank elements, which will invalidate the previous corrections. The whole algorithm is to be done again until it works!

Practically, this algorithm is very efficient and rarely needs more than two iterations to succeed. This is due to the fact that, as soon as the number of elements exceeds a few hundreds, the combinatorics provides many appropriate solutions. When the numbering is done, the compiler has to be informed that no more backward dependence remains in the loops (the order is CDIR$IVDEP on a CRAY machine). If the vector register length is high, the solution may not exist for medium-size meshes (some thousands of elements). However, vectorization can still be performed, by splitting the loops into sub-loops, containing groups of elements without backward dependence. The vectorization should be forced for these sub-loops.

The following loop, which can be vectorized on a CRAY machine (if there is no backward dependence in any group of 64 elements):

```
C
DO IELEM=1,NELEM
    R(IKLE(IELEM,1))=R(IKLE(IELEM,1))+W1(IELEM)
ENDDO
C
```

can be transposed, without any change in the mesh, to a FUJITSU machine whose vector register length is 1024, as follows:

```
C
M64 = NELEM/64
N64 = MOD(NELEM,64)
C
DO 1 K=1,M64
DO 2 IELEM=(K-1)*64+1,K*64
    R(IKLE(IELEM,1))=R(IKLE(IELEM,1))+W1(IELEM)
2 CONTINUE
1 CONTINUE
C
DO 3 IELEM=M64*64+1,M64*64+N64
    R(IKLE(IELEM,1))=R(IKLE(IELEM,1))+W1(IELEM)
3 CONTINUE
C
```

The vectorization of loops 2 and 3 may be forced.

Some tests on the CRAY YMP reveal that vectorization provides considerable economy in computation time for the vector assembly:

- ratio 19 for quadrilaterals with bilinear interpolation (four loops),

- ratio 12 for triangles with linear interpolation (three loops).

As the assembly of vectors is intensively used in various algorithms, the economy in time for the overall computation is quite noteworthy, and can reach a ratio of 11.

Note: the heuristic algorithm may look too challenging because the backward dependences are eliminated for all the groups of 64 elements: 1 to 64, 2 to 65, 3 to 66, etc. *A priori*, it would be enough to consider the groups 1 to 64, 65 to 128, etc. However, it is a protection against the behaviour of the compiler, which, for some internal architectural reasons, can decide the number of elements to be handled in each group, and thus introduce some arbitrary shifts. Furthermore, dealing with the groups 1 to 64, 65 to 128, etc., seems to make the convergence of the algorithm more difficult.

7.4.4 Product of a non-symmetric EBE matrix and a vector

The vector R to be computed is the product of matrix M and vector V. The following equality must be remembered:

$$R = MV = DV + \sum_{e=1}^{nelem} (P_e E_e P_e^t)V \tag{7.27}$$

The example will continue with triangles. The matrix M is stored in two tables, DM(NPOIN) and XM(NELEM,6).

The FORTRAN instructions are as follows:

Diagonal input: Product DV (vectorizable loop)

```
DO 1 I=1,NPOIN
    R(I) = DM(I) * V(I)
1  CONTINUE
```

Input of extra-diagonal terms, stored in workspace W1, W2 and W3. This loop can also be vectorized.

```
DO 2 IELEM=1,NELEM
    Global numbering of the nodes of element (given by table IKLE)
    I1 = IKLE(IELEM,1)
    I2 = IKLE(IELEM,2)
    I3 = IKLE(IELEM,3)
    At the element level, the results of nodes with different local numbers
    are stored separately (for number 1 : W1, etc.)
    W1(IELEM) = XM(IELEM, 1) * V(I2) + XM(IELEM, 2) * V(I3)
    W2(IELEM) = XM(IELEM, 4) * V(I1) + XM(IELEM, 3) * V(I3)
    W3(IELEM) = XM(IELEM, 5) * V(I1) + XM(IELEM, 6) * V(I2)
2  CONTINUE
```

The vector defined by W1, W2 and W3 is assembled (see Section 7.4.3) then added up to R, which already contains the diagonal input.

The meaning of this algorithm is very simple. Let us consider XM(IELEM,1) which is, by convention, equal to MAT(1,2) for the element IELEM. It is part of the coefficient of point 2, pertaining to the element IELEM, within the equation of point 1, pertaining to the same element. The product XM(IELEM,1)*V(I2) must then be added to the result R(I1). This is done in the second loop and during the assembly stage, via the workspace W1.

As a summary, one could say that the assembly process is no longer performed on the matrices but on the vectors themselves. With classical techniques working on a classical compact storage, the vectorization of the matrix–vector product is "broken up" by an internal loop on the neighbours of each point (except for the "frontal" product, which will be used also for the edge-based storage).

7.4.5 Product of a symmetric EBE matrix and a vector

When the matrix is symmetric, the terms:
XM(IELEM,NDP*(NDP-1)/2+1) to XM(IELEM,NDP*(NDP-1))
are not stored because they are respectively equal to:
XM(IELEM,1),...,XM(IELEM,NDP*(NDP-1)/2)
They are substituted in the previous loop, which leads to:
W1(IELEM) = + XM(IELEM, 1) * V(I2) + XM(IELEM, 2) * V(I3)
W2(IELEM) = + XM(IELEM, 1) * V(I1) + XM(IELEM, 3) * V(I3)
W3(IELEM) = + XM(IELEM, 2) * V(I1) + XM(IELEM, 3) * V(I2)

7.4.6 Product of a transposed EBE matrix and a vector

The elementary products XM(IELEM,I) will simply be exchanged with:

XM(IELEM,I+NDP*(NDP-1)/2)

if I≤NDP*(NDP-1)/2 and with:

XM(IELEM,I-NDP*(NDP-1)/2)

if I>NDP*(NDP-1)/2, which will lead, for matrices built on triangles, to:
W1(IELEM) = XM(IELEM, 4) * V(I2) + XM(IELEM, 5) * V(I3)
W2(IELEM) = XM(IELEM, 1) * V(I1) + XM(IELEM, 6) * V(I3)
W3(IELEM) = XM(IELEM, 2) * V(I1) + XM(IELEM, 3) * V(I2)

These last two examples prove that the problems of symmetry and transposition in EBE storage are very simply solved.

7.4.7 Processing of the Dirichlet boundary conditions

The classical finite element techniques eliminate the known and imposed values (points with Dirichlet conditions) from the linear systems, which results in a renumbering of the unknown variables.

In order to avoid this renumbering and its cost, the points with Dirichlet conditions are not removed from the matrices. They are kept and assigned with an equation of the type x = imposed value. In the other equations, the coefficient associated to a Dirichlet condition node will be set to 0, and the right-hand sides of the equations will be changed accordingly. By doing so, the symmetry of the matrix is not affected.

In other words, for a point with a Dirichlet condition, the values on the diagonal of the matrix are set to 1 whereas the extra-diagonal terms are set to 0, after having taken into account the known values on the right-hand sides. The problem is that, for a given point, no data structure exists that could help in finding quickly the elements to which this point belongs. It thus seems impossible to set to 0 the associated extra-diagonal terms, if they have been stored element by element. However, this can be done using the following loop of instructions, based on the concept of the matrix–vector product.

In what follows, every component of vector V equals 1 for a standard point and 0 for a Dirichlet condition point. By doing so, any element of the matrix, which would "touch" a Dirichlet point when computing a matrix–vector product, would be set to 0, the other elements remaining unchanged, which is the expected result. For a matrix built on linear triangles, the loop is as follows:

```
DO IELEM=1,NELEM
   I1 = IKLE(IELEM,1)
   I2 = IKLE(IELEM,2)
   I3 = IKLE(IELEM,3)
   XM(IELEM, 1) = XM(IELEM, 1) * V(I2) * V(I1)
   XM(IELEM, 2) = XM(IELEM, 2) * V(I3) * V(I1)
   XM(IELEM, 3) = XM(IELEM, 3) * V(I2) * V(I3)
   XM(IELEM, 4) = XM(IELEM, 4) * V(I1) * V(I2)
   XM(IELEM, 5) = XM(IELEM, 5) * V(I1) * V(I3)
   XM(IELEM, 6) = XM(IELEM, 6) * V(I2) * V(I3)
ENDDO
```

In the non-assembled matrices, the values which will form the line and the column related to a point with a Dirichlet condition will be set to 0, except for the diagonal terms.

As a specific table exists for the diagonal, it is very easy to set to 1 any term associated with a Dirichlet point. In the same way, the imposed values must be written afterwards on the right-hand side of the linear system.

The method described above will be detailed again and formalized in Section 9.3 of the chapter related to parameter estimation.

7.4.8 Products between a diagonal matrix and a matrix

These products appear during the diagonal or block diagonal preconditioning of a linear system, where the matrix M is changed into DMD, D being a diagonal matrix (see Section 7.7.1).

It is necessary to know how to compute the product DM on one hand, and the product MD on the other.

In the following FORTRAN program examples, D is declared as a real array of dimension NPOIN, and D(I) denotes the Ith element of the diagonal matrix. Since the computation of the diagonal of the resulting matrix is obvious, only the algorithms for the extra-diagonal terms are given below (in the case of linear triangles):

Product DM:

```
C
DO IELEM=1,NELEM
C
   I1 = IKLE(IELEM,1)
   I2 = IKLE(IELEM,2)
```

```
      I3 = IKLE(IELEM,3)
C
      XM(IELEM, 1) = XM(IELEM, 1) * D(I1)
      XM(IELEM, 2) = XM(IELEM, 2) * D(I1)
      XM(IELEM, 3) = XM(IELEM, 3) * D(I2)
      XM(IELEM, 4) = XM(IELEM, 4) * D(I2)
      XM(IELEM, 5) = XM(IELEM, 5) * D(I3)
      XM(IELEM, 6) = XM(IELEM, 6) * D(I3)
C
ENDDO
C
```

For assembled matrices, the formula would be:

$$(DM)_{mn} = D_m * M_{mn}$$

XM(IELEM,1) denotes, for instance, a part of term M(I1,I2), and it will be logically multiplied by D(I1). As for product MD, it will be multiplied by D(I2):

Product MD:

```
C
DO IELEM=1,NELEM
C
      I1 = IKLE(IELEM,1)
      I2 = IKLE(IELEM,2)
      I3 = IKLE(IELEM,3)
C
      XM(IELEM, 1) = XM(IELEM, 1) * D(I2)
      XM(IELEM, 2) = XM(IELEM, 2) * D(I3)
      XM(IELEM, 3) = XM(IELEM, 3) * D(I3)
      XM(IELEM, 4) = XM(IELEM, 4) * D(I1)
      XM(IELEM, 5) = XM(IELEM, 5) * D(I1)
      XM(IELEM, 6) = XM(IELEM, 6) * D(I2)
C
ENDDO
C
```

The result itself takes the form of a non-assembled matrix, and the two previous loops are vectorized.

7.4.9 Matrix–vector product with edge-based storage

As a matrix with edge-based storage is fully assembled, elaborating a matrix–vector product is no longer a problem. Multiplying A by Y to get a vector X first requires that we compute $X = DA\ Y$ (diagonal contribution), then execute the following loop:

```
DO ISEG =1,NSEG
      X(GLOSEG(ISEG,1))=
      X(GLOSEG(ISEG,1))+XA1(ISEG) * Y(GLOSEG(ISEG,2))
      X(GLOSEG(ISEG,2))=
      X(GLOSEG(ISEG,2))+XA2(ISEG) * Y(GLOSEG(ISEG,1))
ENDDO
```

This loop is a little more complex for rectangular matrices because some values of X are not initialized by the contribution of the diagonal, and need to be set to 0 first.

However, as the edge-based storage provides assembled matrices, it allows a more efficient matrix–vector product, the frontal product, detailed in the following subsection.

7.4.10 Matrix–vector frontal product

The idea is to write the matrix–vector product directly in its classical form:

$$X(i) = \sum_{j=1}^{npoin} Aij\, Y(j)$$

As the matrix A is sparse, we do not want really to write loops from 1 to NPOIN, but we want to use an optimal storage, like the edge-based storage.

For each point, a data structure will be defined containing the number of neighbours and their addresses, and the entry of the matrix related to it. In the classical versions of this kind of matrix–vector product, the internal loop is a loop on the number of neighbours for each point, which inhibits any attempt at vectorization. With the frontal product, the points are numbered according to the number of their neighbours, in ascending order. This results in long sequences of nodes, with the same number of neighbours, for which a specific loop is written. The maximum number of neighbours for one point, denoted MXPTVS, is thus restricted by what has been specified in the program.

A 2-dimension table, denoted LIMVOI, determines the arrangement of the nodes:

LIMVOI(K,1) is the global number of the first point having K neighbours.

LIMVOI(K,2) is the global number of the last point having K neighbours.

The address of the neighbours of points is denoted IY, with size (NPOIN,MXPTVS).

The address of the elements of the matrix, related to the neighbours, is denoted IW, which has the same dimension as IY.

The table X is, as usual, initialized by the contribution of the diagonal $DA\,Y$. The loop for those nodes having three neighbours is given below. In this case, the extra-diagonal terms are stored in the table W:

```
DO I=LIMVOI(3,1),LIMVOI(3,2)
 X(I)=X(I)+W(IW(I,1))*Y(IY(I,1))+W(IW(I,2))*Y(IY(I,2))
 +W(IW(I,3))*Y(IY(I,3))
ENDDO
```

Unfortunately, W changes when the matrix is symmetric, and IW must be written in two versions. Transposed products, like the computation of $X = {}^tA Y$, requires two other versions.

The whole data structure is thus stored in an array named IKLEM1 with four dimensions: NPOIN, MXPTVS, 4 and 2. The third dimension is coded as follows:

1: IW for a normal product
2: IY for a normal product
3: IW for a transposed product
4: IY for a transposed product

The fourth dimension, i.e. 2, is coded as follows:
1: non-symmetric matrix
2: symmetric matrix

7.5 Linear system resolution

We give here an overview of iterative techniques for solving linear systems, including preconditioning.

$AX = B$ is the linear system to be solved. A is a matrix resulting from the variational formulation in the finite element method. A is thus a sparse matrix, whose storage may be done in diverse forms: EBE or edge-based. Furthermore, considering the resolution of the Saint-Venant system of equations or any other resolution coupling many variables, A may be broken down into blocks of several matrices, e.g. four or nine.

The direct resolution methods, too costly as soon as the variables are too numerous, are not appropriate because they require knowledge of the assembled values of the matrix. On the contrary, the iterative methods, based on the minimization of a functional, need only matrix–vector and dot products, and are perfectly relevant to any kind of storage. On the other hand, these methods are very sensitive to matrix conditioning (see section below). This is why it will be necessary to perform a previous preconditioning, in order to reduce the number of iterations required to reach a given precision.

The principles of the methods of descent are given next, followed by some of their algorithms, then by preconditioning.

7.6 Methods of descent and preconditioning

From now on, except if specifically mentioned, A will be a symmetric, positive definite matrix. $AX - B$ is the vector gradient of the quadratic functional:

$$J(X) = \frac{1}{2}(AX, X) - (B, X) \tag{7.28}$$

where the brackets stand for the dot product. Equating this vector to 0 means finding an extremum of $J(X)$ that will be a minimum if J is positive (hence the interest in a positive definite matrix).

Let Z be the function $Z = J(X) = J(x_1, x_2, x_3, ..., x_n)$, n being the number of unknowns in the system (Figure 7.9).

The gradient of $J(X)$ is normal to the iso-J lines. If J is convex (which is the case for a quadratic, positive definite functional), J decreases in the direction opposite to its gradient. The idea behind the gradient methods is to approach the minimum of J, by progressing in the direction opposite to the gradient. That leads to an iterative method of the kind:

$$X^{m+1} = X^m - \rho(AX^m - B) \tag{7.29}$$

where the choice of ρ must be specified.

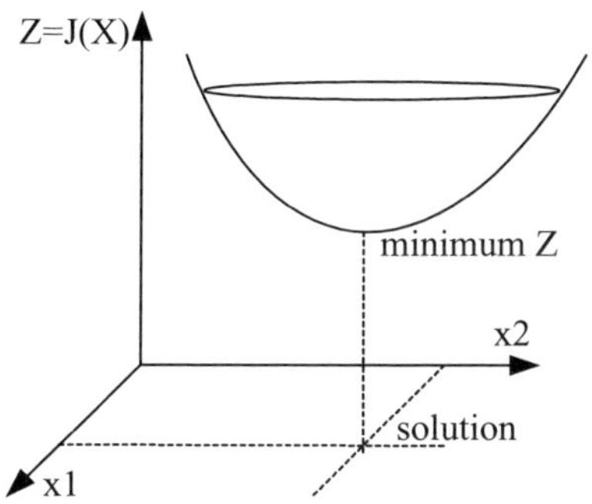

Figure 7.9: minimum of the quadratic functional

The convergence criterion of this method becomes clear when writing these iterations as follows:

$$\frac{1}{\rho}X^{m+1} = \left(\frac{I}{\rho} - A\right)X^m + B \tag{7.30}$$

In fact, an iterative method such as:

$$MX^{m+1} = NX^m + B$$

converges if the spectral radius of $M^{-1}N$, i.e. $I - \rho A$, is lower than 1. Furthermore, the norm of the residual $\|AX^m - B\|$ is multiplied at each iteration by the spectral radius of $I - \rho A$. If the eigenvalues of A are denoted by λ_i (they are real and positive for a symmetric positive definite matrix), the eigenvalues of $I - \rho A$ are equal to $1 - \rho\lambda_i$. The best method with constant ρ is to choose:

$$\rho = \frac{\lambda_{\max} - \lambda_{\min}}{\lambda_{\max} + \lambda_{\min}} \tag{7.31}$$

where $\lambda_{\max}$ and $\lambda_{\min}$ are respectively the highest and the lowest eigenvalue of A. That highlights a fundamental parameter for the resolution of linear systems, namely the "conditioning", also called "condition number":

$$\frac{\lambda_{\max}}{\lambda_{\min}}$$

Preconditioning aims at changing the system $AX = B$ to a simpler one $C^{-1}AX = C^{-1}B$, for which the conditioning of $C^{-1}A$ will be better (closer to 1) than for A. The computation of C^{-1} must be simple enough to keep the benefit obtained by accelerating the resolution.

Other gradient methods are, in fact, far more efficient than a constant optimal value of ρ. Optimizing ρ at each iteration and minimizing:

$$J(X^{m+1}) = J(X^m - \rho(AX^m - B)) \tag{7.32}$$

which is a quadratic function of ρ, leads to the optimal value of ρ:

$$\rho = \frac{(g^m, g^m)}{(g^m, Ag^m)}$$

where g^m stands for $AX^m - B$. The knowledge of the eigenvalues is no longer necessary.

Most of the following methods arise from this one.

The different algorithms detailed below, formally include preconditioning with a matrix C which is not defined (but will be in Section 7.7).

The stop test is the same whatever the method. Iterations go on until the precision (or tolerance) *epsilon* depending on the test:

$$\frac{\|AX^{m+1} - B\|}{\|B\|} \leq epsilon \quad \text{if } \|B\| \geq 1 \text{ (relative precision)} \tag{7.33}$$

or:

$$\|AX^{m+1} - B\| \leq epsilon \quad \text{if } \|B\| \leq 1 \text{ (absolute precision)} \tag{7.34}$$

We choose the Euclidean norm L_2:

$$\|X\| = \sqrt{\sum_{i=1}^{n} (x_i)^2}$$

The L_∞ norm:

$$\|X\| = \max(|x_i|)$$

would be safer, as it ensures a convergence at any point, but it shows no coherence with the norm used in the descent methods and consequently with the demonstrations showing that the residual is decreasing.

7.6.1 Conjugate gradient method

The weakness of the optimal gradient algorithm which minimizes:

$$J(X^{m+1}) = J(X^m - \rho(AX^m - B))$$

is that the minimization following the direction of descent $g^m = AX^m - B$ (which is also the direction of the gradient) is forgotten at the next iteration. The idea behind the conjugate gradient method (see [98]) is to distinguish the residual g^m and the direction of descent d^m and guarantee that every new iteration will lead to a minimum for any direction of descent previously tested. With recurrence formulae, this can be achieved without storing all these directions. Convergence is guaranteed if A is a symmetric, positive definite matrix. Orthogonality relations, such as $(g^i, g^j) = 0$ if $i \neq j$, are used to prove that the exact solution is reached in n iterations if n is the number of unknowns. Practically, for solving fluid mechanics systems, it takes a few iterations (between 5 and 100) to converge with a precision ranging between 10^{-4} and 10^{-8}.

Conjugate gradient algorithm
Initialization:

$$r^0 = AX^0 - B$$

$$\text{resolution of } Cg^0 = r^0$$

$$d^0 = g^0$$

$$\rho^0 = \frac{(r^0, g^0)}{(Ad^0, d^0)}$$

$$X^1 = X^0 - \rho^0 d^0$$

Iterations:

$$r^m = r^{m-1} - \rho^{m-1} Ad^{m-1}$$

$$\text{resolution of } Cg^m = r^m$$

$$d^m = g^m + \frac{(r^m, g^m)}{(r^{m-1}, g^{m-1})} d^{m-1}$$

$$\rho^m = \frac{(r^m, d^m)}{(d^m, Ad^m)}$$

$$X^{m+1} = X^m - \rho^m d^m$$

7.6.2 Conjugate residual method

The principle of this method is exactly the same as the principle of the conjugate gradient method, but it aims at minimizing a different functional, namely:

$$J(X) = (AX - B, AX - B)$$

Convergence is reached if A is a symmetric, positive definitive matrix.

Conjugate residual algorithm
Initialization:

$$r^0 = AX^0 - B$$

$$\text{resolution of } Cg^0 = r^0$$

$$d^0 = g^0$$

$$\text{resolution of } Cd'^0 = Ad^0$$

$$\rho^0 = \frac{(g^0, Ad^0)}{(d'^0, Ad^0)}$$

$$X^1 = X^0 - \rho^0 d^0$$

Iterations:

$$r^m = r^{m-1} - \rho^{m-1} Ad^{m-1}$$

$$g^m = g^{m-1} - \rho^{m-1} d'^{m-1}$$

$$d^m = g^m - \frac{(Ag^m, d'^{m-1})}{(Ad^{m-1}, d'^{m-1})} d^{m-1}$$

$$Ad^m = Ag^m - \frac{(Ag^m, d'^{m-1})}{(Ad^{m-1}, d'^{m-1})} Ad^{m-1}$$

$$\text{resolution of } Cd'^m = Ad^m$$

$$\rho^m = \frac{(Ad^m, g^m)}{(Ad^m, d'^m)}$$

$$X^{m+1} = X^m - \rho^m d^m$$

7.6.3 Conjugate gradient on the normal equation

The previous methods should, in principle, be restricted to symmetric, positive definite matrices A. They can, however, be transposed easily to non-symmetric matrices. It suffices to substitute A by ${}^tAA$ in every algorithm, then to multiply the right-hand side by ${}^tA$ to get an equivalent system with a symmetric matrix. The only requirement is the regularity of A.

This principle used together with the conjugate gradient method results in the normal equation method.

However, by substituting A by ${}^tA$, the gradient $g = AX - B$ would be changed in ${}^tA(AX - B)$. In fact, the definition of g is maintained, and the new algorithm is as follows:

Normal equation algorithm
Initialization:

$$r^0 = AX^0 - B$$

$$\text{resolution of } Cg^0 = r^0$$

$$\text{resolution of } {}^tCg'^0 = g^0$$

$$d^0 = {}^tA\, g'^0$$

$$\text{resolution of } Cd'^0 = Ad^0$$

$$\rho_0 = \frac{(d^0, d^0)}{(d'^0, d'^0)}$$

$$X^1 = X^0 - \rho^0 d^0$$

Iterations:

$$r^m = r^{m-1} - \rho^{m-1} A d^{m-1}$$

$$g^m = g^{m-1} - \rho^{m-1} d'^{m-1}$$

$$\text{resolution of } {}^tCg'^m = g^m$$

$$d^m = {}^tAg'^m + \frac{({}^tAg'^m, {}^tAg'^m)}{({}^tAg'^{m-1}, {}^tAg'^{m-1})} d^{m-1}$$

$$\text{resolution of } Cd'^m = Ad^m$$

$$\rho^m = \frac{({}^tAg'^m, {}^tAg'^m)}{(d'^m, d'^m)}$$

$$X^{m+1} = X^m - \rho^m d^m$$

7.6.4 Minimum error method

For the purpose of elaborating an iterative method for a simply regular matrix, the following functional is chosen:

$$J(X^m) = (A^{-1}(AX^m - B), A^{-1}(AX^m - B))$$

Minimum error algorithm
Initialization:

$$r^0 = AX^0 - B$$

$$\text{resolution of } Cg^0 = r^0$$

$$\text{resolution of } {}^tCg'^0 = g^0$$

$$d^0 = {}^tAg'^0$$

$$\rho^0 = \frac{(g^0, g^0)}{(d^0, d^0)}$$

$$X^1 = X^0 - \rho^0 d^0$$

Iterations:

$$r^m = r^{m-1} - \rho^{m-1} Ad^{m-1}$$

$$\text{resolution of } Cg^m = r^m$$

$$\text{resolution of } {}^tCg'^m = g^m$$

$$d^m = {}^tAg'^m + \frac{(g^m, g^m)}{(g^{m-1}, g^{m-1})} d^{m-1}$$

$$\rho^m = \frac{(g^m, g^m)}{(d^m, d^m)}$$

$$X^{m+1} = X^m - \rho^m d^m$$

7.6.5 GMRES method

The GMRES method (Generalized Minimum RESidual method), published in 1983 [168] is a worthy innovation, especially useful for poorly conditioned systems. We shall explain the fundamental principles before detailing the algorithm whose programming must be handled carefully.

This algorithm is iterative and, at each iteration, a minimization of $\|AX - B\|$ is sought. But X should be a vector of the "Krylov sub-space", generated by the vectors $r = AX^m - B$, Ar, $A^2r, \ldots,$ $A^{k-1}r$, where k is the dimension of the Krylov sub-space: this dimension should be a parameter of the method.

Thus, the solution has to be found in a restricted space. The idea originates from Lanczos' methods, which will not be presented here (see [57]).

We shall detail below the techniques for building the Krylov sub-space, and the minimization techniques on this sub-space.

First, an orthonormal base in this sub-space will be built, using the modified Gram–Schmidt method:

Let r be the starting residual, equal to $AX^m - B$. Then $u_1 = \frac{r}{\|r\|}$, and $\widetilde{u_2} = Au_1$ can then be defined. With $\beta_{21} = (\widetilde{u_2}, u_1)$ then $u_2 = \frac{\widetilde{u_2} - \beta_{21}u_1}{\|\widetilde{u_2} - \beta_{21}u_1\|}$. u_1 and u_2 are both unit and orthogonal vectors.

This idea of subtracting from each vector its projections on the previous vectors of the base continues for all the vectors of the base. The algorithm of orthogonalization is thus the following:

from i=1 to k
    $\widetilde{u}_{i+1} = Au_i$
    from j=1 to i
        $\beta_{i+1,j} = (\widetilde{u}_{i+1}, u_j)$ and $\widetilde{u}_{i+1}$ replaced by $\widetilde{u}_{i+1} - \beta_{i+1,j}u_j$
    $u_{i+1} = \frac{\widetilde{u}_{i+1}}{\|\widetilde{u}_{i+1}\|}$

Remark: the vector u_{k+1} is built. Though it is useful for the presentation, only $\|\widetilde{u}_{k+1}\|$ is needed and can be easily deduced from the formulae $\widetilde{u}_{i+1}$ replaced by $\widetilde{u}_{i+1} - \beta_{i+1,j}u_j$. As the base is orthogonal, we get:

$$\|\widetilde{u}_{k+1}\|^2 = \|Au_k\|^2 - \sum_{j=1}^{k} \beta_{k+1,j}^2$$

This algorithm of orthogonalization can be summarized by defining three matrices:

$$U_k = [u_1, u_2, \ldots, u_k] \text{ (matrix } n \times k)$$

$$U_{k+1} = [u_1, u_2, \ldots, u_{k+1}] \text{ (matrix } n \times (k+1))$$

and:

$$H_k = \begin{pmatrix} \beta_{2,1} & \beta_{3,1} & \ldots & \beta_{k+1,1} \\ \|\widetilde{u}_2\| & \beta_{3,2} & \ldots & \beta_{k+1,2} \\ 0 & \|\widetilde{u}_3\| & \ldots & \ldots \\ \ldots & \ldots & \ldots & \ldots \\ 0 & 0 & \beta_{k,k-1} & \ldots \\ 0 & 0 & \|\widetilde{u}_k\| & \beta_{k+1,k} \\ 0 & 0 & 0 & \|\widetilde{u}_{k+1}\| \end{pmatrix} \tag{7.35}$$

The matrix H_k has $k+1$ rows and k columns. The algorithm can be summarized by $AU_k = U_{k+1}H_k$. H_k is almost an upper triangular matrix; only the diagonal containing the terms $\|\widetilde{u}_2\|$, $\|\widetilde{u}_3\|\ldots$ overlaps the lower half. This matrix is called the upper Hessenberg matrix.

If X^m and $r^m = AX^m - B$ result from the immediately previous iteration, a better approximation will be sought, like $X^m + Z$, or by minimizing:

$$\|AX^m + AZ - B\| = \|r^m + AZ\| \tag{7.36}$$

As indicated earlier, Z should be sought in the Krylov sub-space, in the form:

$$Z = \sum_{j=1}^{k} y_j u_j \tag{7.37}$$

If Y is a vector whose components are denoted y_j, Z is the matrix–vector product $U_k Y$. If E is the vector whose components are $(\|r^m\|, 0, 0, 0, \ldots)$ (dimension $k+1$), r^m is written as $U_{k+1}E$. We then have:

$$r^m + AZ = U_{k+1}E + AU_kY = U_{k+1}(E + H_kY) \tag{7.38}$$

As U_{k+1} is made up of orthogonal and unit vectors, the norm of $U_{k+1}(E + H_kY)$ equals the norm of $E + H_kY$. We recall that the norm of a matrix is given by $\|A\| = \max(\frac{\|AX\|}{\|X\|})$, where $\|X\| \neq 0$.

The last step is to find Y, the vector which minimizes $\|E + H_kY\|$.

The technique used is a factorization technique called QR. It transforms a matrix M whose dimensions are (m,n) into a product QR, where Q is an orthogonal matrix whose dimensions are (m,m) and R is an upper triangular matrix, its dimensions being (m,n). Different variants may be found in reference [57].

Factorization will take place as $\overline{H} = RH_k$, where $\overline{H}$ is an upper triangular matrix of dimension $(k+1,k)$ and R an orthogonal matrix of dimension $(k+1,k+1)$. The process will then go on with $\overline{H}$, whose inversion is easier than for the Hessenberg matrix H_k. Before defining R, it can be noted at this point that, if $\overline{E} = RE$, we should minimize:

$$\|E + H_kY\| = \left\|R^{-1}\overline{E} + R^{-1}\overline{H}Y\right\| = \left\|\overline{E} + \overline{H}Y\right\|$$

As $\overline{H}$ is an upper triangular matrix of dimension $(k+1,k)$, the $(k+1)$th component of $\overline{H}Y$ is zero and, so, the $(k+1)$th component of $\overline{E} + \overline{H}Y$ is $\overline{e}_{k+1}$. The other components of $\overline{E}+\overline{H}Y$ can be made zero by a judicious choice of Y, and by inverting $\overline{H}$. It follows that the minimum of $\left\|\overline{E} + \overline{H}Y\right\|$ obtained with the best choice of Y is equal to $|\overline{e}_{k+1}|$.

We thus find that Y is the solution of the system below:

$$\begin{pmatrix} \overline{h}_{11} & \overline{h}_{12} & \ldots & \ldots & \overline{h}_{1k} \\ 0 & \overline{h}_{22} & \ldots & \ldots & \ldots \\ \ldots & \ldots & \ldots & \overline{h}_{k-1\,k-1} & \overline{h}_{k-1\,k} \\ 0 & 0 & \ldots & \ldots & \overline{h}_{k\;k} \end{pmatrix} \begin{pmatrix} y_1 \\ y_2 \\ \ldots \\ y_k \end{pmatrix} = \begin{pmatrix} \overline{e}_1 \\ \overline{e}_2 \\ \ldots \\ \overline{e}_k \end{pmatrix} \tag{7.39}$$

and the residual will be equal to $|\overline{e}_{k+1}|$.

Let us now specify how the matrix R is built. We have to define first Givens' rotations, whose size is $(k+1,k+1)$ and which are equal to the identity matrix, except for one rotation relative to two rows and two columns. Let R_j be the rotation:

$$R_j = \begin{pmatrix} 1 & 0 & 0 & 0 & 0 & 0 & 0 \\ 0 & 1 & 0 & 0 & 0 & 0 & 0 \\ 0 & 0 & 1 & 0 & 0 & 0 & 0 \\ 0 & 0 & 0 & c_j & s_j & 0 & 0 \\ 0 & 0 & 0 & -s_j & c_j & 0 & 0 \\ 0 & 0 & 0 & 0 & 0 & 0 & 0 \\ 0 & 0 & 0 & 0 & 0 & 0 & 1 \end{pmatrix} \tag{7.40}$$

The example above is given for $k+1$=7 and j=4. Coefficients c_j are located in the position (j,j) and $(j+1,j+1)$ with $c_j^2 + s_j^2 = 1$. The matrix R is equal to the product of k matrices, called Givens' rotations:

$$R_k \; R_{k-1} \; \dots \; R_2 \; R_1$$

Givens' rotations will cancel the problem-causing diagonal (with terms $\|\widetilde{u}_2\|$, $\|\widetilde{u}_3\|$, etc.) of the Hessenberg matrix H_k. To form $\overline{H}$, H_k will be transformed column by column. For example, the first column of H_k is:

$$\begin{pmatrix} \beta_{21} \\ \|\widetilde{u}_2\| \\ 0 \\ \dots \\ 0 \end{pmatrix} \quad \text{denoted as:} \quad \begin{pmatrix} h_{11} \\ h_{21} \\ 0 \\ \dots \\ 0 \end{pmatrix}$$

This column will be modified by rotation R_1, with:

$$c_1 = \frac{h_{11}}{\sqrt{h_{11}^2 + h_{21}^2}}$$

and:

$$s_1 = \frac{h_{21}}{\sqrt{h_{11}^2 + h_{21}^2}}$$

The first column of H_k is modified into:

$$\begin{pmatrix} \sqrt{h_{11}^2 + h_{21}^2} \\ 0 \\ 0 \\ \dots \\ 0 \end{pmatrix}$$

It will not be affected by the following rotations R_2 to R_k which will act, with regards to it, as the identity. The jth column of H_k will then only be transformed by the product R_j, R_{j-1}, ..., R_1.

The product R_{j-1}, R_{j-2}, ..., R_1 leads to a column:

$$\begin{pmatrix} \overline{h_{1j}} \\ \overline{h_{2j}} \\ \dots \\ \widehat{h}_{jj} \\ \widehat{h}_{j+1\;j} \\ 0 \\ \dots \\ 0 \end{pmatrix}$$

where $\widehat{h}_{jj}$ and $\widehat{h}_{j+1\;j}$ are intermediate results. As for the first column, let us choose:

$$c_j = \frac{\widehat{h}_{jj}}{\sqrt{\widehat{h}_{jj}^2 + \widehat{h}_{j+1\;j}^2}}$$

and:

$$s_j = \frac{\widehat{h}_{j+1\;j}}{\sqrt{\widehat{h}_{jj}^2 + \widehat{h}_{j+1\;j}^2}}$$

The jth column of $\overline{H}$ is changed, after being multiplied by R_j, into:

$$\begin{pmatrix} \overline{h}_{1j} \\ \overline{h}_{2j} \\ \cdots \\ \sqrt{\widehat{h}_{jj}^2 + \widehat{h}_{j+1\;j}^2} \\ 0 \\ 0 \\ \cdots \\ 0 \end{pmatrix}$$

Though this transformation of H_k into $\overline{H}$ seems complex to write, it is easily translated into the FORTRAN language. The change is made "on the spot", without using different registers for H_k and $\overline{H}$ in the memory space.

The GMRES method is very efficient for non-symmetric and poorly conditioned matrices. One of its disadvantages is the parameter k that must be adjusted. However, the method is not very sensitive to it. Values ranging from 3 to 15, or even to 100, can be found in the literature. Of course, the higher the value of k, the greater the space required in the memory. One should notice also that the number of matrix–vector products per iteration is k. With $k = 10$, the GMRES method will have no chance to perform better than a conjugate gradient that converges in 10 iterations.

Knowledge of the GMRES method helps to understand its limits. The method will surely degenerate if the matrix of the system to be solved is diagonal. In this case, the vectors Ar, A^2r, A^3r generating the Krylov sub-space are all colinear.

7.7 Preconditioning

7.7.1 Diagonal preconditioning

Let $AX = B$ be a linear system to be solved. The "diagonal-point" preconditioning is a real one from an etymological point of view as it is applied before the resolution. We form the diagonal matrix D such as $D_{ii} = \frac{1}{\sqrt{A_{ii}}}$.

If A_{ii} can be negative, then D_{ii} can also be taken as $\frac{1}{\sqrt{|A_{ii}|}}$.

But, in any case, A_{ii} must be non-zero. The original system is then solved, but written as:

$$DAD\;D^{-1}X = DB$$

The following terms appear:
A new matrix: $A' = DAD$
A new unknown vector: $X' = D^{-1}X$
A new right-hand side: $B' = DB$

Table 7.3: result of c–u preconditioning

	Without c–u preconditioning	With c–u preconditioning
Normal equation	No convergence	151 s
GMRES	195 s	139 s

On account of this building, and when A_{ii} is always positive, the diagonal of A' is only made of ones (this can be useful for the purpose of optimization techniques). The result of preconditioning here is to give a comparable weight to all the equations.

Once the system $A'X' = B'$ has been solved, it is easy to find X that is equal to DX'.

This shows the great interest in assembling the diagonal of the matrix with the EBE technique of storage, thus enabling the use of D.

7.7.2 c–u preconditioning for the Saint-Venant equations

Another preconditioning, referred to as c–u preconditioning, is specific to the Saint-Venant equations. It is based on the fact that the linear systems derived from a variational formulation such as the one described in Section 4.5.6, but with the celerity and velocity variables, have an excellent conditioning. That can be explained by the fact that the three unknowns of the linear system are homogeneous to a velocity and thus a symmetry appears. Actually, one can write $CM1 = -{}^tBM1$ and $CM2 = -{}^tBM2$. Changing the sign of the continuity equation would lead to a symmetric matrix (but unfortunately not positive definite). The preconditioning has to be done at the level of the linear system. In the celerity–velocity option, the changes in the variables were made at the continuum level. The idea here is to substitute the unknown water depth by the wave celerity at the discrete level and try to get the symmetry of the matrix as far as possible.

Initially, the value of the celerity c on each node (denoted by $c^n(i)$) is computed using the results of the earlier time step or the boundary conditions. In the case of tidal flat areas, this value can be kept within bounds to avoid zero-division error. The unknown $h^{n+1}(i)$, of time t^{n+1}, is then substituted in the whole linear system with $h^{n+1}(i)\frac{g}{2c^n(i)}$, where g is the gravity. The continuity equation is then multiplied by $\frac{2g}{c^n(i)}$. Once the new system is solved, the value of $h^{n+1}(i)$ has simply to be expressed according to the new unknown of the system.

This operation is formally equivalent to a diagonal preconditioning: the matrix A of the system is substituted with $D1\ A\ D2$, where $D1$ and $D2$ are diagonal matrices, functions of c^n.

This preconditioning method was published in 1994 (see [89]). It has proved to be efficient for the computation of tidal currents over areas stretching beyond the continental shelf; that is, very deep areas. For a model of the French Atlantic coast and the Channel (see Section 10.2), the CPU times on a CRAY C90 are given in Table 7.3 (computation of four tides, mesh made of 9500 elements, time step 150 s and with an additional diagonal preconditioning).

7.7.3 Block diagonal preconditioning

This preconditioning makes sense only if the matrix M is a block of square matrices. The principle and detailed explanations will be given next for a four-matrix block.

four-matrix block

Let $AX = B$ be the system to solve, where:

$$A = \begin{pmatrix} A11 & A12 \\ A21 & A22 \end{pmatrix}, \; X = \begin{pmatrix} X1 \\ X2 \end{pmatrix} \text{ and } B = \begin{pmatrix} B1 \\ B2 \end{pmatrix}$$

The purpose is to apply the idea of diagonal preconditioning, not only to matrices $A11$ and $A22$, but also to the extra-diagonal matrices $A12$ and $A21$. As a result, the four matrices of the block would have a diagonal made of ones.

$D11$, $D12$, $D21$ and $D22$ will respectively denote the diagonals of $A11$, $A12$, $A21$, and $A22$.

The basic idea is to approximate A with the matrix:

$$\widetilde{A} = \begin{pmatrix} D11 & D12 \\ D21 & D22 \end{pmatrix}$$

and use a LDU factorization of $\widetilde{A}$ in the form $L\sqrt{D}\,\sqrt{D}U$. The initial system $AX = B$ can then be written as:

$$\left(L\sqrt{D}\right)^{-1} A \left(\sqrt{D}U\right)^{-1} \sqrt{D}U \; X = \left(L\sqrt{D}\right)^{-1} B \tag{7.41}$$

This system is equal to:

$$\frac{1}{\sqrt{D}} L^{-1} A U^{-1} \frac{1}{\sqrt{D}} \sqrt{D} U X = \frac{1}{\sqrt{D}} L^{-1} B \tag{7.42}$$

In this form, the system can be viewed as a diagonal preconditioning of the system $A'X' = B'$, with the preconditioning matrix D being given and with:

$$A' = L^{-1} A U^{-1}$$

$$X' = UX$$

$$B' = L^{-1} B$$

Once the system is solved, the unknown X is computed by the formula:

$$X = U^{-1} X'$$

The following operations have to be done successively:
1) Computation of L, D, and U by a LDU factorization of:

$$\widetilde{A} = \begin{pmatrix} D11 & D12 \\ D21 & D22 \end{pmatrix}$$

2) Computation of A', X' and B'.

3) Resolution of the system $AX' = B'$ with a simple diagonal preconditioning where the diagonal D is given (see Section 7.7.1 above).

4) Computation of X as a function of X'.

The details of operations 1, 2 and 4 are as follows:

1) *LDU* factorization of $\widetilde{A}$

$\widetilde{A}$ is decomposed as:

$$\begin{pmatrix} I & 0 \\ L21 & I \end{pmatrix} \begin{pmatrix} D'11 & 0 \\ 0 & D'22 \end{pmatrix} \begin{pmatrix} I & U12 \\ 0 & I \end{pmatrix}$$

By identification, one can find:

$$D11^{'} = D11$$

$$L21 = \frac{D21}{D11}$$

$$U12 = \frac{D12}{D11}$$

$$D22^{'} = D22 - L21\ D11\ U12$$

In practice, the programming will use the same memory space for diagonals $D11^{'}$ and $D11$, $D22^{'}$ and $D22$, $L21$ and $D21$, $U12$ and $D12$.

This requires the successive operations:

$D21$ is substituted with $D21/D11$

$D22$ is substituted with $D22 - D21\ D12$

$D12$ is substituted with $D12/D11$

$\widetilde{A}$ is then written in the form:

$$\begin{pmatrix} I & 0 \\ D21 & I \end{pmatrix} \begin{pmatrix} D11 & 0 \\ 0 & D22 \end{pmatrix} \begin{pmatrix} I & D12 \\ 0 & I \end{pmatrix}$$

The diagonals $D11$ and $D22$ are inverted, their square root is computed and they are kept aside for further diagonal preconditioning (operation 3). They are no longer necessary for operations 2 and 4.

2) Computation of A', X' and B'

One uses the following formulae:

$$\begin{pmatrix} I & 0 \\ D21 & I \end{pmatrix}^{-1} = \begin{pmatrix} I & 0 \\ -D21 & I \end{pmatrix}$$

and:

$$\begin{pmatrix} I & D12 \\ 0 & I \end{pmatrix}^{-1} = \begin{pmatrix} I & -D12 \\ 0 & I \end{pmatrix}$$

The product:

$$\begin{pmatrix} I & 0 \\ -D21 & I \end{pmatrix} \begin{pmatrix} A11 & A12 \\ A21 & A22 \end{pmatrix}$$

is equal to:

$$\begin{pmatrix} A11 & A12 \\ A21 - D21\ A11 & A22 - D21\ A12 \end{pmatrix}$$

As for the LDU factorization, the computation of A' is made "on the spot": that is, in the very memory space where A has been stored.

One starts with the operations:

$A21$ is substituted with $A21 - D21\ A11$

$A22$ is substituted with $A22 - D21\ A12$

Multiplying on the right side by U^{-1} then requires that:

$A12$ is substituted with $A12 - A11\ U12$

$A22$ is substituted with $A22 - A21\ U12$

Following these operations the matrix A' takes the memory space of the matrix A.

The computation of X' is also made "on the spot", $X1$ being substituted with $X1 + D12\ X2$ ($X2$ unchanged).

For the computation of B' , $B2$ is substituted with $B2 - D21\ B1$ ($B1$ unchanged).

4) Computation of X

This takes place by substituting $X1$ with $X1 - D12X2$ (X2 unchanged).

Nine-matrix block

Let $AX = B$ be the system to solve, where:

$$A = \begin{pmatrix} A11 & A12 & A13 \\ A21 & A22 & A23 \\ A31 & A32 & A33 \end{pmatrix}, \ X = \begin{pmatrix} X1 \\ X2 \\ X3 \end{pmatrix} \text{ and } B = \begin{pmatrix} B1 \\ B2 \\ B3 \end{pmatrix}$$

Dij denotes the diagonal of Aij. The preconditioning principle is the same as for a four-matrix block.

The following operations have to be done:

1) Computation of L, D and U by LDU factorization of:

$$\widetilde{A} = \begin{pmatrix} D11 & D12 & D13 \\ D21 & D22 & D23 \\ D31 & D32 & D33 \end{pmatrix}$$

2) Computation of A', X' and B'.

3) Resolution of the system $AX' = B'$ using a simple diagonal preconditioning, where the diagonal matrix D is given.

4) Computation of X as a function of X'.

Hereafter, we detail operations 1, 2 and 4 in order to show the mechanisms of matrix decomposition in the same memory space:

1) LDU factorization of $\widetilde{A}$

$\widetilde{A}$ is broken down as:

$$\begin{pmatrix} I & 0 & 0 \\ D'21 & I & 0 \\ D'31 & D'32 & I \end{pmatrix} \begin{pmatrix} D'11 & 0 & 0 \\ 0 & D'22 & 0 \\ 0 & 0 & D'33 \end{pmatrix} \begin{pmatrix} I & D'12 & D'13 \\ 0 & I & D'23 \\ 0 & 0 & I \end{pmatrix}$$

The decomposition during which D and D' occupy alternatively the same memory space takes place as follows:

$D11$ remains unchanged

$D21$ is substituted by $D21/D11$, $D31$ is substituted by $D31/D11$

$D22$ is substituted by $D22 - D21\,D11\,D12$

$D32$ is substituted by $(D32 - D31\,D12)/D22$

$D23$ is substituted by $D23 - D13\,D21$ (division by $D22$ is intentionally omitted)

$D33$ is substituted by $D33 - D31\,D13 - D32\,D23$ ($D23$ equals in fact $D22\,D23$)

$D12$ is substituted by $D12/D11$, $D13$ is substituted by $D13/D11$

$D23$ is substituted by $D23/D22$ (to correct the omission mentioned three lines above)

The divisions are, in fact, multiplications because of the previous inversion of the diagonals $D11$, $D22$ and $D33$ which will be used in this form later. Following the inversion, the square root is computed and kept aside for the diagonal preconditioning.

2) Computation of A', X' and B'

The following formulae are used:

$$\begin{pmatrix} I & 0 & 0 \\ D21 & I & 0 \\ D31 & D32 & I \end{pmatrix}^{-1} = \begin{pmatrix} I & 0 & 0 \\ 0 & I & 0 \\ -D31 & -D32 & I \end{pmatrix} \begin{pmatrix} I & 0 & 0 \\ -D21 & I & 0 \\ 0 & 0 & I \end{pmatrix}$$

and:

$$\begin{pmatrix} I & D12 & D13 \\ 0 & I & D23 \\ 0 & 0 & I \end{pmatrix}^{-1} = \begin{pmatrix} I & -D12 & 0 \\ 0 & I & 0 \\ 0 & 0 & I \end{pmatrix} \begin{pmatrix} I & 0 & -D13 \\ 0 & I & -D23 \\ 0 & 0 & I \end{pmatrix}$$

These two decompositions are used to compute A', keeping in mind that A is multiplied on thc left by the lower part and multiplied on the right by the upper part. The changes are always made “on the spot”.

The left multiplication is carried out with the following operations:

$A21$ is substituted by $A21 - D21\ A11$

$A22$ is substituted by $A22 - D21\ A12$

$A23$ is substituted by $A23 - D21\ A13$

$A31$ is substituted by $A31 - D31\ A11 - D32\ A21$

$A32$ is substituted by $A32 - D31\ A12 - D32\ A22$

$A33$ is substituted by $A33 - D31\ A13 - D32\ A23$

The right multiplication is carried out with the following operations:

$A12$ is substituted by $A12 - A11\ D12$

$A22$ is substituted by $A22 - A21\ D12$

$A32$ is substituted by $A32 - A31\ D12$
$A13$ is substituted by $A13 - A11\ D13 - A12\ D23$
$A23$ is substituted by $A23 - A21\ D13 - A22\ D23$
$A33$ is substituted by $A33 - A31\ D13 - A32\ D23$
After these operations, matrix A' is in the memory space previously used by matrix A.

The computation of X' is made with the same conditions, following the steps:

$$X1 = X1 + D12\ X2 + D13\ X3$$

$$X2 = X2 + D23\ X3$$

$X3$ remains unchanged.
The computation of B' is carried out by following the steps:
$B1$ remains unchanged and:

$$B2 = B2 - D21\ B1$$

$$B3 = B3 - D31\ B1 - D32\ B2$$

4) Computation of X
This takes place by following the operations:
$X3$ remains unchanged and:

$$X2 = X2 - D23\ X3$$

$$X1 = X1 - D12\ X2 - D13\ X3$$

7.7.4 Preconditioning with LU factorization

This kind of preconditioning is not appropriate for non-assembled matrices. However, its comprehension is necessary for introducing the Crout preconditioning technique. Let L and U respectively be a lower and an upper matrix, whose product LU is close to A. The efficiency of the preconditioning is determined by the choice of L and U. Some examples will be given in the following subsections. Let us suppose that L and U are known. Instead of the system $AX = B$, the following system will be solved:

$$L^{-1}AU^{-1}UX = L^{-1}B$$

As a result, new quantities appear:
A new matrix: $A' = L^{-1}AU^{-1}$
A new unknown vector: $X' = UX$
A new right-hand side: $B' = L^{-1}B$
Once the system is solved, X is given by the formula:

$$X = U^{-1}X'$$

If A is a symmetric matrix, it is recommended to choose a LU factorization where U is ${}^tL$. When $LU = L\,{}^tL$, the iterative methods, such as conjugate gradient, can be adapted in order to use only one inversion by $(L^tL)^{-1}$. See [173].

7.7.5 Crout preconditioning

Principle

Crout preconditioning (see [173]) uses the idea of the LU factorization, but adapts it to the matrices stored with the EBE technique. A is approximated by a decomposition in the form $N = LDU$, where L and U are respectively the lower and upper matrices, with ones on the diagonal.

Let us assume that A already has a unit diagonal (if not and if the diagonal makes it possible, a diagonal preconditioning would have to be performed before Crout preconditioning).

A can then be written using the extra-diagonal element matrices X_e (with the notations of Section 7.3.3):

$$A = I + \sum_{e=1}^{nelem} P_e X_e P_e^t \tag{7.43}$$

The idea for approximating A is to apply the identity:

$$1 + \sum_{i=1}^{n} \varepsilon_i \approx \prod_{i=1}^{n} (1 + \varepsilon_i) \tag{7.44}$$

The smaller the ε_i, the truer this identity. In the case of A, this approximation leads to:

$$A \approx \prod_{e=1}^{nelem} (I + P_e X_e P_e^t) \tag{7.45}$$

X_e is a matrix with zero diagonal. Let $\widetilde{X_e}$ be $X_e + I_e$ where I_e is the identity matrix, at the elementary level.

Let $\widetilde{I_e}$ be equal to the identity matrix I (over the whole mesh), but with zero values on the diagonal in the positions corresponding to the nodes of element e: $\widetilde{I_e} = I - P_e I_e P_e^t$.

The Crout decomposition is applied to $\widetilde{X_e}$, i.e. $\widetilde{X_e} = L_e D_e U_e$ where L_e is a lower triangular matrix with unit diagonal, D_e is a diagonal matrix and U_e is a upper triangular matrix with unit diagonal. A is then approximated by:

$$A \approx \prod_{e=1}^{nelem} (\widetilde{I_e} + P_e (L_e D_e U_e) P_e^t) \tag{7.46}$$

Eventually, N is given by a similar but symmetrical expression:

$$N = \prod_{e=1}^{nelem} (\widetilde{I_e} + P_e L_e P_e^t) \prod_{e=1}^{nelem} (\widetilde{I_e} + P_e D_e P_e^t) \prod_{e=nelem}^{1} (\widetilde{I_e} + P_e U_e P_e^t) \tag{7.47}$$

The second group of products, $\prod_{e=1}^{nelem}(\widetilde{I}_e + P_e D_e P_e^t)$, is a diagonal matrix, denoted D, which can be easily computed. The corresponding operation is formally equivalent to a vector assembly, but with multiplications instead of additions (hence the name "multiplicative assembly"). The other matrices in the product are either lower triangular, or upper triangular. Solving a $NX = B$ type of linear system is then a succession of descending and ascending computations, first using the upper triangular matrices U_e then, after inversion of the diagonal matrix D, using the lower triangular matrices L_e.

Examples using triangles with linear interpolation

The matrix A is defined by its unit diagonal $DA(NPOIN)$ and its extra-diagonal terms $XA(NELEM, 6)$. A is not assumed to be symmetric.

***LDU* factorization:** the element matrices are decomposed as LDU products using the Crout algorithm. The result is stored in $DN(NPOIN)$ for the D_e matrices, after multiplicative assembly, and in $XN(NELEM, 6)$ for matrices L_e and U_e.

The Crout algorithm applied to the matrix A with elements a_{ij} is as follows:

For j=1 to n, apply:

For i=1 to n, apply

$\beta_{ij} = a_{ij} - \sum_{k=1}^{i-1} \alpha_{ik}\beta_{kj}$ (if i=1, the sum is zero).

then:

For i=j+1 to n, apply:

$\alpha_{ij} = \frac{1}{\beta_{ij}}(a_{ij} - \sum_{k=1}^{j-1} \alpha_{ik}\beta_{kj})$ (if j=1, the sum is zero).

This results in a LU factorization (L consists of α_{ij} and U of β_{ij}). The LDU factorization is eventually obtained by dividing each β_{ij} by β_{ii}.

L is finally the lower triangular matrix defined by $L_{ij} = \alpha_{ij}$ $(j < i)$, $L_{ii} = 1$, $L_{ij} = 0$ $(i < j)$.

D is the diagonal matrix defined by:

$$D_{ii} = \beta_{ii}$$

U is the upper triangular matrix:

$$U_{ij} = \beta_{ij}(j < i),\ U_{ii} = 1,\ U_{ij} = \frac{\beta_{ij}}{\beta_{ii}}(i < j)$$

In FORTRAN and with EBE storage, this becomes:

```
C
DO IELEM=1,NELEM
C
C Matrix to be decomposed
C (after diagonal preconditioning; that is, with a unit diagonal)
C
C Row 1 of element matrix
A11 = 1.D0
A12 = XM(IELEM,1)
```

```
A13 = XM(IELEM,2)
C Row 2 of element matrix
A21 = XM(IELEM,4)
A22 = 1.D0
A23 = XM(IELEM,3)
C Row 3 of element matrix
A31 = XM(IELEM,5)
A32 = XM(IELEM,6)
A33 = 1.D0
C
C Crout LU factorization
C
C Column 1 (BETA11=1 is omitted)
ALFA21 = A21
ALFA31 = A31
C
C Column 2
BETA12 = A12
BETA22 = A22 - ALFA21*BETA12
ALFA32 = (A32 - ALFA31*BETA12)/BETA22
C
C Column 3
BETA13 = A13
BETA23 = A23 - ALFA21*BETA13
BETA33 = A33 - ALFA31*BETA13 - ALFA32*BETA23
C
C LDU factorization
C Extra-diagonal terms are stored in XN
C Diagonal terms are stored in W2 and W3
C (W1 is not necessary because BETA11=1)
C
XN(IELEM,1) = BETA12
XN(IELEM,2) = BETA13
XN(IELEM,3) = BETA23/BETA22
C
XN(IELEM,4) = ALFA21
XN(IELEM,5) = ALFA31
XN(IELEM,6) = ALFA32
C
W2(IELEM) = BETA22
W3(IELEM) = BETA33
C
ENDDO
C
C Matrix L is stored in:
C XN(IELEM,4), XN(IELEM,5) and XN(IELEM,6)
C Matrix U is stored in XN(IELEM,1), XN(IELEM,2) and XN(IELEM,3)
```

```
C Arrays W2 and W3 are then multiplicatively assembled
C in DN (previously initialized to 1) as follows:
C------------------------------------------------------------
C LOOP WITH IMPOSED VECTORIZATION (Cray command)
C------------------------------------------------------------
C
CDIR$ IVDEP
DO 10 IELEM = 1 , NELEM
     DN(IKLE(IELEM,2)) = DN(IKLE(IELEM,2)) * W2(IELEM)
10 CONTINUE
C
CDIR$ IVDEP
DO 20 IELEM = 1 , NELEM
     DN(IKLE(IELEM,3)) = DN(IKLE(IELEM,3)) * W3(IELEM)
20 CONTINUE
C
```

The loop in the multiplicative assembly can be vectorized for the same reasons as the classical assembly, if the elements have been numbered appropriately.

Vector DN is then inverted because only the term $(DN)^{-1}$ is necessary:

```
C
C DN inversion
C
DO 30 I = 1 , NPOIN
     DN(I) = 1.D0/DN(I)
30 CONTINUE
C
```

System $NX = B$: solving this system is a succession of inversions of lower triangular matrices.

```
C
C INITIALIZATION: X = right-hand SIDE
C
DO I = 1 , NPOIN
     X(I) = B(I)
ENDDO
C
C INVERSIONS OF LOWER TRIANGULAR MATRICES
C
CDIR$ IVDEP
DO IELEM = 1 , NELEM
     X(IKLE(IELEM,2)) = X(IKLE(IELEM,2))
*    - XN(IELEM,4) * X(IKLE(IELEM,1))
     X(IKLE(IELEM,3)) = X(IKLE(IELEM,3))
*    - XN(IELEM,5) * X(IKLE(IELEM,1))
*    - XN(IELEM,6) * X(IKLE(IELEM,2))
ENDDO
C
```

```
C MULTIPLICATION BY DN (PREVIOUSLY INVERTED)
DO I = 1 , NPOIN
     X(I) = X(I)*DN(I)
ENDDO
C
C INVERSIONS OF UPPER TRIANGULAR MATRICES
C
CDIR$ IVDEP
DO IELEM = NELEM , 1 , -1
     X(IKLE(IELEM,2)) = X(IKLE(IELEM,2))
*    - XN(IELEM,3) * X(IKLE(IELEM,3))
     X(IKLE(IELEM,1)) = X(IKLE(IELEM,1))
*    - XN(IELEM,1) * X(IKLE(IELEM,2))
*    - XN(IELEM,2) * X(IKLE(IELEM,3))
ENDDO
C
```

Inversion loops cannot be vectorized even though one takes the same precautions as for vectorizing the assembly of vectors. It would be necessary to verify:

$$\text{IKLE(IELEM1,I)} \neq \text{IKLE(IELEM2,J)} \tag{7.48}$$

for I,J=1,2,3 and for every different element IELEM1 and IELEM2, chosen in a block of 64 elements (in the case of a CRAY computer). This statement is true only if I=J, considering our algorithm of element numbering (see Section 7.4.3). However, vectorization can be forced. It results in an error made by the computer and the result will not be a solution of the system $NX = B$. However, the aim of building N was to approximate M; that is, to approximate the solution $MX = B$. So, the result given by forced vectorization can be used as well; it will just give a different preconditioning. Furthermore, at the end of any iterative method, a stop test is performed to guarantee the adequacy of the solution. The tests performed have proved that Crout preconditioning, when appropriate, is extremely efficient even with a forced vectorization. In the case of diffusion matrices, the computation cost can be decreased to 50% whatever the time necessary to build the preconditioner.

7.7.6 Gauss–Seidel EBE preconditioning

The principle is similar to Crout preconditioning. Assuming a unit diagonal for M, L is defined as the lower part of M with unit diagonal, whereas U is defined as the upper part of M with a unit diagonal. Thus $L + U = M + I$.

Then, the successive descending and ascending computations follow the same principle as for Crout preconditioning.

7.7.7 Crout and Gauss–Seidel preconditioning for edge-based storage

All the reasoning made earlier for the Crout and Gauss–Seidel EBE preconditioning can be applied with the same efficiency to the edge-based storage (see for example [31]), except for two differences: first, extra-diagonal terms are assembled values, then the matrices to be decomposed have a dimension (2,2) whatever the finite element under consideration.

Assuming a previous diagonal preconditioning, we now consider the edge-based decomposition of the matrices:

$$\begin{pmatrix} 1 & a_{12} \\ a_{21} & 1 \end{pmatrix}$$

in:

$$\begin{pmatrix} 1 & a_{12} \\ a_{21} & 1 \end{pmatrix} = \begin{pmatrix} 1 & 0 \\ a_{21} & 1 \end{pmatrix} \begin{pmatrix} 1 & 0 \\ 0 & 1 - a_{12} * a_{21} \end{pmatrix} \begin{pmatrix} 1 & a_{12} \\ 0 & 1 \end{pmatrix}$$

This is Crout preconditioning, or:

$$\begin{pmatrix} 1 & a_{12} \\ a_{21} & 1 \end{pmatrix} \approx \begin{pmatrix} 1 & 0 \\ a_{21} & 1 \end{pmatrix} \begin{pmatrix} 1 & 0 \\ 0 & 1 \end{pmatrix} \begin{pmatrix} 1 & a_{12} \\ 0 & 1 \end{pmatrix}$$

This is Gauss–Seidel preconditioning. The latter is, generally, more efficient, as it avoids division by the diagonals. However, the saving of time resulting from such preconditioning is often quite low. The number of iterations of methods such as the conjugate gradient is divided by 2, but the cost of preconditioning is itself equivalent to one iteration.

Chapter 8

Parallelism

8.1 Introduction

Parallelism consists of using simultaneously a cluster of computers, or a cluster of processors in the same computer, to solve a single problem. Using n processors would ideally divide by n the time necessary to solve the same problem with only one processor.

The task would be easy if the problem could be broken down into sub-tasks, independent of each other. It becomes much more difficult when each processor needs the results from the others. We shall focus hereafter on parallelism performed with processors each having its own memory, and communicating with the others via message transmission, a clumsy way when compared with direct access into the RAM. This is the case with networks of workstations or PCs and is known as "shared-memory parallelism".

The emergence of vector computers, as we have seen earlier, led to changes in some algorithms. However, the possibility of vectorization is now detected automatically, and this is true for more and more complex operations due to the progress in computer technology. The arrival of parallel computers, with clusters of processors, initially brought the same hope. However, in the field of fluid mechanics software, this hope has not been fulfilled. Many experiences in automatic parallelism, where compilers themselves perform the task of optimizing the program, showed very poor improvement in CPU time. Furthermore, vectorization and parallelism appeared to be contradictory. Vectorization requires simple loops to perform sub-tasks whereas shared-memory parallelism hands over complex tasks to each processor so as to optimize communication time and data transmission time. Efficiently vectorized software would naturally be poorly parallelized. As an example, the assembly loops resulting from an EBE matrix–vector product (see Section 7.4.3) cannot be accelerated by more than a ratio of two even if we overlook the communication time. This is due to the fact that the results of each processor have to be assembled. That leads to a new cost which limits drastically the overall time to be gained. This is known as a "granularity" problem, the size of the tasks to be parallelized being too small. It is unlikely that progress in the algorithm would help in solving this theoretical bottleneck. From that, the idea emerges to break up the problem in another way: not by isolating "small" tasks, but in a kind of geographical way by decomposing the domain of computation.

This idea of domain decomposition aims at assigning to each processor one part of the domain over which it would solve the fluid mechanics problem. The results of the other processors would help in defining artificial boundary conditions arising from the partition. The assignments given to each processor are the same, but the data differ. According to the Flynn classification (see [51]), this corresponds to SMPD (Single Program Multiple Data) parallelism. The implementation is quite easy in the case of explicit algorithms: each equation is only a function of the variables related to the nodes in the immediate vicinity, computed at the earlier time step. Each processor is then in charge of the equations related to a group of nodes, and of the data concerning the neighbours of these nodes, resulting from the previous time step. This is executed merely by a partition of the domain, with an overlap of one element, and data transmission at the end of each time step.

In the case of our implicit algorithms, and especially with linear systems solved on the whole domain, the implementation is more complex, though possible. In fact, we can guarantee that the results of a parallel computation will be the same as a scalar computation, except for truncation errors because computations will not be performed in the same order. Parallelism (initially implemented on our algorithms by Reinhard Hinkelmann at the University of Hanover [101]) can be limited to the following problems:

- Partition of the domain.
- Communication between processors.
- Implementation of some basic algorithms: vector assembly, dot product, computation of normal vector and so on.

We shall examine these three points hereafter.

8.2 Partition of the domain

In the case of finite element meshes and implicit algorithms, it is better to partition without overlapping, edge to edge. Though the techniques of partition are beyond the scope of this book, we shall mention some of them and give the original references. Then we shall specify the kind of additional data structures necessary to describe the domain in order to implement parallelization.

The specifications for an efficient domain decomposer would be the following:

- Realizing a partition of meshes of triangles (the sub-domains of three-dimensional meshes of prisms will be made from sub-domains of triangles).
- Partitioning the domain into blocks in order to ensure a balance between the processors. Because the main algorithms are made of loops over the elements

and nodes, the partition will guarantee a balance between the numbers of elements, or similarly between the number of nodes within each sub-domain.

- Minimizing the number of nodes shared by different sub-domains and for which communications between processors will be necessary.

Other criteria found in the literature, aimed at simplifying the resolution of a linear system over each sub-domain, are not relevant to our approach (see [48]).

The simplest technique is known as recursive coordinate bisection. The domain is partitioned into two areas, by a vertical or horizontal line, whose position is determined in order to balance the number of nodes on either side. The partition is repeated several times on each part, resulting in a number of sub-domains equal to a power of 2. Eventually, the sub-domains have saw-toothed interfaces which do not minimize the number of nodes along the interfaces. It is then necessary to apply the technique of Kernighan and Lin [112], which moves some elements from one sub-domain to another, for minimizing the interfacial nodes, without upsetting the balance in the number of elements between the sub-domains.

All the other techniques are based mainly on graph theory and use information relative to the connectivity of the mesh. Some variants also use geometric information.

Greedy algorithms (see [46]) build sub-domains initiated from a node and extending over the areas of the mesh, not yet included in another sub-domain. Other domain decomposers artfully use algorithms of minimization of a matrix band. The resulting partition minimizes the links between sub-domains; that is, the number of distinct boundaries of each sub-domain. Unfortunately, the number of interfacial nodes is not optimized.

One of the most popular methods is Simon's recursive spectral bisection (see [174]). The domain is divided into two parts and the number of interfacial segments is minimized. The method looks for the eigenvalues of the connectivity table of the corresponding graph and provides a balance of the number of nodes. In order to get an equal number of elements, it is necessary to deal with a "dual" graph, built from the central nodes of the elements. In this case, Kernighan and Lin's method must be applied to minimize the number of interfacial segments.

Today, we have free decomposers on the software market, which can divide a domain into any number of sub-domains. We recommend:

- METIS [110] available on the website:

  http://www-users.cs.umm.edu/~karypis/metis

- CHACO [67]

- TOP/DOMDEC [47]

The website:

http://www.erc.msstate.edu/~biderba/research/other_sites.html
provides several addresses in the field of domain partitioning.

8.3 Requested data

Each sub-domain is assigned to a processor together with the information of a standard domain (connectivity table IKLE, node coordinates, and so on) but also with additional information, to help in assembling the results over the whole domain at the end of the computation. This information is as follows:

- An array KNOLG storing the global numbers of the nodes in the whole mesh. The inverse array, KNOGL, is defined within a loop:

  ```
  DO I=1,NPOIN
      KNOGL(KNOLG(I))=I
  ENDDO
  ```

- An array NACHB whose dimension is (5,NPTIR), where NPTIR is the number of interfacial nodes between the sub-domain and the others. These interfacial nodes are numbered from 1 to NPTIR. NACHB(1,IR) is the global number of the interfacial node IR. This array gathers the information to be transmitted to the other processors. The integers NACHB(2,IR) to NACHB(5,IR) are the numbers of the other sub-domains to which the nodes belong. They may have a -1 value if these sub-domains are less than four. The dimension five must be increased if a node belongs to more than five sub-domains.

Some additional information has to be defined on the actual boundary nodes of a two-dimensional domain, namely: ISEG, XSEG, YSEG and NUMLIQ.

If ISEG > 0: the boundary node just after belongs to another sub-domain, its global number is ISEG, and its real coordinates are XSEG and YSEG.

If ISEG < 0: the boundary node just before belongs to another sub-domain, its global number is $-$ISEG, and its real coordinates are XSEG and YSEG.

If ISEG $=-999999$: the node verifies the two conditions above (this pathological case is not considered by the software).

NUMLIQ is a specific numbering of the liquid boundaries, which allows association of boundary conditions to them, e.g. an imposed discharge. The numbering is easy to define when the whole domain is known. A possible convention is to start the numbering of these liquid boundaries (inflow and outflow) from the extreme south-west node of the domain and proceed in the anticlockwise direction along the edge. In the case of a sub-domain for which it is not possible to go over the whole contour, the numbering must be specified because it cannot be simply recomputed.

8.4 Communication between processors

Programming by communicative process requires the use of a library of communication functions, such as PVM [55], or MPI [19]. These libraries perform different operations:

- Setting up the parallel machine; that is, organizing the communication between various computers or processors.

- Running programs on all the processors (spawn order with PVM). In some cases, one processor has the function of a master, the others being the slaves. The program is first run on the master processor, and includes a start order for the slaves. A "group" name is given to the computation. Apart from this initialization step, there is then no more hierarchy between the processors.
- Various kinds of communication: point to point or collective.

8.4.1 Point-to-point communication

This is the basic communication between two processors, the receptor and the transmitter. The "send" function and the "receive" function can differ depending on the types of data (integers, real numbers, etc.). The functions' arguments are:

- identifier of receptor or transmitter,
- address of the data to transmit or receive,
- a message flag.

Transmissions can be made in two ways. Synchronous transmission occurs when waiting for the receptor to be ready; this avoids a copy in the buffer. Asynchronous transmission occurs whatever the state of the reception.

8.4.2 Collective communication

Collective communications use simultaneously all the processors. Three types may be distinguished:

- Broadcast: a processor (generally the master) sends the same data to all the slaves.
- Synchronization: refers to a barrier which all processors must reach before going onto the computation.
- Conditional transmission: each processor sends data and receives in return the sum, the maximum or the minimum of all the data transmitted.

8.4.3 Parallelism library

To make the source programs independent of the choice of the communication functions library, it is worth writing an interface library to deal with all the communications between the processors. A few functions are necessary:

P_DMAX(X): maximum of X over all the processors. X is a double-precision real.

P_DMIN(X): minimum of X over all the processors. X is a double-precision real.

P_DSUM(X): sum of all the values of X provided by all the processors. X is a double-precision real.

P_IMAX(I): maximum of I over all the processors. I is an integer.

P_IMIN(I): minimum of I over all the processors. I is an integer.

P_ISUM(I): sum of all the values of I provided by all the processors. I is an integer.

P_MAIL(CHAINE,NCAR): broadcasting a character string CHAINE, with a length NCAR, to all the processors.

P_READ(BUFFER,NBYTES,SOURCE,TYPE): reading of NBYTES bytes of data whose type is TYPE, to be stored at the address BUFFER and sent by the processor SOURCE.

P_SYNC: stops the processors till all of them call this function.

P_WRIT(BUFFER,NBYTES,DEST,TYPE): writing of NBYTES bytes of data, whose type is TYPE, located at the address BUFFER and intended for the processor DEST.

8.5 Adaptation of the algorithms

The data structure specific to parallelism will be described next, as well as the modifications made to the main algorithms.

8.5.1 Data structure specific to parallelism

From the basic information provided by the domain decomposer for every sub-domain, as mentioned in Section 8.2, the following data must be defined:

- An array INDPU, inverse of NACHB(1,*), whose task is to help store the data received from the other processors: INDPU(NACHB(1,I))=I. Given the number of an interfacial node, INDPU sends back its global number within the sub-domain.

- An array FAC, providing for each node the inverse of the number of sub-domains to which it belongs (see the algorithm for dot product below).

8.5.2 Scalar product

The dot product of two vectors over the whole domain implies local computing first, then summing up over all the sub-domains. While summing, the interface nodes are taken into account several times. The array FAC is then used to correct this error. The parallel dot product P_DOT of the vectors X and Y is written as follows:

```
P_DOT = 0.D0
DO I=1,NPOIN
    P_DOT=P_DOT+X(I)*Y(I)*FAC(I)
ENDDO
P_DOT=P_DSUM(P_DOT)
```

On a vector computer, multiplying by FAC(I) does not affect the time of computation. On the contrary, the call to the P_SUM routine causes a synchronization of the processors.

8.5.3 Matrix–vector product

First, building the matrices and computing the product are done locally, independently of the other sub-domains. The result is a vector without any contribution from the neighbouring sub-domains, on the interfacial nodes. The missing contributions result from the computation of the vector on the other sub-domains. For example, if the node I belongs to two different sub-domains, for which we get X(I)=3 and X(I)=5 respectively, the actual value of X(I) will be the sum 3+5. This can be achieved if processor 1 sends the value 3 to processor 2 and processor 2 sends the value 5 to processor 1. Although it appears simple, this operation is quite complex to achieve because it can lead to a fatal risk: both processors waiting endlessly for the contribution of the other before sending their own. The transmission is thus split into four different tasks, depending on the rank of the processors:

- transmission of the data to the higher rank processors,
- reception by the higher rank processors,
- transmission of the data to the lower rank processors,
- reception by the lower rank processors.

New data are required for every processor. Each processor prepares its own numbering of higher rank and lower rank processors as well as a numbering of nodes interfacing with each of these processors. Four new arrays IKP, NHP, IKM and NHM are necessary to navigate from one numbering to another:

- IKP(10,2): IKP(IZH,1) refers to the processor which is the IZHth local higher rank processor. IKP(IZH,2) is the number of interfacial nodes shared with this processor.
- IKM(10,2): IKM(IZH,1) refers to the processor which is the IZHth local lower rank processor. IKM(IZH,2) is the number of interfacial nodes shared with this processor.
- NHP(10,NPTIR): NHP(IZH,IR) is the global number in the sub-domain of a point whose number is IR in the interface with the IZHth higher rank processor.
- NHM(10,NPTIR): NHM(IZH,IR) is the global number in the sub-domain of a point whose number is IR in the interface with the IZHth lower rank processor.

ILMAX is the maximum distance in order between the processor and its neighbours. It will restrict the size of the loops over these neighbours.

Considering this information, the data transmission for vector V (defined over the whole domain), and for processor IPID, is written as follows:

1) storage of the data to be sent in a first buffer:

```
DO I=1,NPTIR
    BUF(I)=V(NACHB(1,I))
ENDDO
```

2) transmission to higher rank processors, e.g. processor ILP

2.1) storage of the data relative to the processor in a second buffer

```
        DO I=1,IKP(ILP,2)
            ERGBUF(I)=BUF(INDPU(NHP(ILP,I)))
        ENDDO
        2.2) sending
        CALL P_WRIT(ERGBUF,8*IKP(ILP,2),
                          IKP(ILP,1),ABS(IKP(ILP,1)-IPID))
    3) reception from lower rank processors, e.g. processor ILM
        3.1) reception
        CALL P_READ(ERGBUF,8*IKM(ILM,2),
                          IKM(ILM,1),ABS(IKM(ILM,1)-IPID))
        3.2) storage in vector V
        DO I=1,IKM(ILM,2)
            V(NHM(ILM,I))=V(NHM(ILM,I))+ERGBUF(I)
        ENDDO
    4) transmission to lower rank processors, e.g. processor ILM
        4.1) storage of the data relative to the processor in a second buffer
        DO I=1,IKM(ILM,2)
            ERGBUF(I)=BUF(INDPU(NHM(ILM,I)))
        ENDDO
        4.2) sending
        CALL P_WRIT(ERGBUF,8*IKM(ILM,2),
                          IKM(ILM,1),ABS(IKM(ILM,1)-IPID))
    5) reception from upper rank processors, e.g. processor ILP
        5.1) reception
        CALL P_READ(ERGBUF,8*IKP(ILP,2),
                          IKP(ILP,1),ABS(IKP(ILP,1)-IPID))
        5.2) summing into vector V
        DO I=1,IKP(ILP,2)
            V(NHP(ILP,I))=V(NHP(ILP,I))+ERGBUF(I)
        ENDDO
```

This algorithm can be made more complex to process several vectors at the same time or for tasks different from summing.

8.5.4 Specific cases

It is worth noting that just a few alterations of the scalar and matrix–vector products are sufficient to solve a linear equation with partial derivatives in shared memory. Furthermore, the contributions of the interfacial terms can be postponed until the final solving. They would be necessary before only if real nodal values, dot product, matrix–vector product or final results are needed. For a linear equation, these operations can be confined in the resolution of the linear system. Once the tools have been set up to communicate between processors, parallelization of fluid mechanics software is quite easy. However, specific cases, arising mainly from non-linearity, have to be considered. The basic idea is to get identical results on vector or parallel computers, except for truncation errors. Some examples follow:

- Diagonal preconditioning: the diagonal used for preconditioning has to be completed at the interfaces.

- Computation of nodal values after projection onto a basis: let us consider the gradient of a function f. A mean nodal value of $\overrightarrow{grad}(f)$ is given by:

$$\overrightarrow{grad}(f)_i = \frac{\int_\Omega \overrightarrow{grad}(f)\Psi_i \, d\Omega}{\int_\Omega \Psi_i \, d\Omega} \tag{8.1}$$

issued from the mean-value theorem. In this case the computation of the general term $\int_\Omega \Psi_i \, d\Omega$ has to be completed at the interfaces. The vector including:

$$\int_\Omega \overrightarrow{grad}(f)\Psi_i \, d\Omega$$

is unchanged when added to the right-hand side of a linear system (the solver itself will complete this term). However, it must be completed when used to compute another term such as a turbulent production.

- Any global concept: maximum Courant number, maximum Froude number, etc., have to be transmitted between processors.

- The method of characteristics: for high values of the Courant number, a characteristic curve may leave a sub-domain and enter others. Programming this situation within the parallel architecture would require a huge number of random transmissions, depending on the flow. If the Courant number is kept under 1, the transmissions would concern only the interfacial nodes. The program will then consider the result of the domain reached by the characteristic curve. A zero result can also be assigned to the points whose characteristic curve leaves the sub-domain. In this case, only the result with higher absolute value is considered, which is a process very similar to a complement at the interfaces. However, limiting the time–step is not fully satisfactory for an unconditionally stable method.

8.6 Results

Some examples of parallel computations using the previous techniques are given below. In all cases, the partitioning of the domain has been achieved with Simon's recursive spectral bisection.

8.6.1 Cluster of workstations HP9000/712/80

Communication library: PVM.

Wave propagation in a square tank using Telemac-3D software: see Table 8.1.

Table 8.1: parallelism on a cluster of stations HP9000

	1 processor	2 processors	4 processors
CPU time	418 s	279 s	222 s
Speed-up factor	1	1.5	1.88

8.6.2 CRAY T3E

Communication library: PVM.

20 time steps of the Malpasset dam-break wave propagation with Telemac-2D software: see Table 8.2.

Table 8.2: parallelism on CRAY T3E

	1 proc.	2 proc.	4 proc.	8 proc.
CPU time	171 s	87 s	46 s	27.5 s
Speed-up factor	1	1.97	3.73	6.2

This and the previous case outline the interest of a dedicated computer, independent of the exterior network, and its low transmission speed.

8.6.3 Origin 2400, 64-bit IRIX 6.5.11m

Machine consists of 32 processors of 300 MHz.

Communication library: MPI.

Study of a reach of the River Rhine (between km 664 and km 678) using Telemac-2D software. Mesh made of 49173 elements, computation of 10000 time steps of 0.5 s: refer to Table 8.3, whose data have been graciously provided by Jacek Jankowski, from Bundesanstalt für Wasserbau in Karlsruhe.

Table 8.3: parallelism on Origin 2400

	1 proc.	2 proc.	4 proc.	8 proc.	16 proc.	32 proc.
CPU time	75429 s	36026 s	12730 s	5417 s	3271 s	1822 s
Speed-up factor	1	2.1	5.9	13.9	23.1	41.4

This amazing result (the speed-up factor is higher than the number of processors) needs commenting on. In this case, a high number of elements in the whole mesh would inhibit the efficiency of one processor because of the memory size of the buffer assigned to data: 32 Mbytes. So efficiency increases when the sub-domain size decreases.

Other results are given in the chapter relative to applications (Section 10.1).

Chapter 9

Parameter estimation

9.1 Introduction

Parameter estimation, also called the inverse problem, consists of retrieving data of a problem from its solution. For free surface flow hydraulics, it involves assessing hidden or difficult-to-access parameters, such as friction coefficients on the bottom, depending on the observations and measurements (series of historical floods, tidal range in ports, etc.). The works on parameter estimation began in the Telemac system in 1994 with Barros's thesis [4]. The adjoint method, developed by Lions and Chavent (see [27] and [128]), has been chosen. It is based on the construction of a Lagrangian to which optimality conditions are applied, i.e. zero derivatives with respect to parameters, state variables and adjoint variables. We try to minimize a cost function J, differences between measurements and the result of a numerical simulation of an event, by a gradient method which consequently requires that we know how to calculate the gradient of the function J, with respect to all the parameters that are looked for. The adjoint system and adjoint variables are used to get this gradient.

One of the important choices of our procedure is to work rather on the discretized equations than on the original equations written in the continuum, which will provide a very precise validation. First of all, we shall explain the principle of the method applied to a linear system (or a series of systems in the case of unsteady flows), then we shall apply it to the linear systems obtained from the Saint-Venant equations. Recent tests also showed the efficiency of totally different techniques, based on surrogate models, neural networks and genetic algorithms; an example will be given in the next chapter in Section 10.2.

9.2 Principle of the adjoint method

We deal here with the primitive shallow water equations, solved as explained in Chapter 4, from Section 4.2 on. Our "direct" problem is a series of linear systems of the form $A^{it}X^{it} = B^{it}$, where the integer it varies between 1 and nit, the total number of iterations or time steps. A is a matrix, X is the vector of unknowns (the depth and the two components of the velocity) and B the vector of right-hand sides.

We shall denote X_{ip} and B_{ip} as the elements of the vectors X and B (ip ranging from 1 to np, the number of unknowns, which is also the rank of A). On a mesh of triangles of n nodes with a linear discretization, we would thus have $3n$ unknown variables in our case at each time step.

We have seen that our linear systems depend on the result of the previous iteration. A^{it} and B^{it} are therefore *a priori* functions of X^{it-1}. A^{it} and B^{it} also depend on parameters (water elevation, source terms, boundary conditions, friction coefficients, etc.) and we shall consider one among them, marked S, as a vector of rank ns. ns can be the number of nodes of the mesh, as well as the number of boundary points. S will be, for example, the value of the friction coefficient at each degree of freedom.

Note: when unknown variables are given on a boundary (e.g. in the case of a boundary condition of Dirichlet type), A^{it} and B^{it} are also functions of certain values of vector X^{it}. This fact will be considered later (Section 9.3) and would only make the presentation tedious.

In short and for now, A^{it} and B^{it} are functions of S and X^{it-1}.

Moreover, a vector of measurements marked M^{it} is known. However, the measurements of all the variables are certainly not available in all the points of the mesh, for all the time steps of calculation. When measurements are missing, this will be taken into account with the help of a "weight" vector marked W. A zero weight will be assigned to missing measurements.

Therefore, we try to find a vector S that minimizes the cost function J defined by:

$$J = \sum_{it=1}^{nit} \sum_{ip=1}^{np} W_{ip}(X_{ip}^{it} - M_{ip}^{it})^2 \tag{9.1}$$

that is, the sum over time and space of the squares of differences between the results of the calculation and the measurements.

A first straightforward idea for the weights W_{ip} would be to take them equal to 1 if one measurement is known and to 0 otherwise. However, the values of W_{ip} can be used for giving more weight to certain points, e.g. depending on the extent of their basis function, or, more prosaically, according to the reliability of the measurement.

The general principle of the search is a gradient method that will consist of a series of iterations of the form:

$$S_{is}^{m+1} = S_{is}^{m} - \rho \frac{\partial J^m}{\partial S_{is}^m}$$

starting from an initial value S_0, a set of parameters evaluated *a priori*; is varies from 1 to ns and m is the current iteration of the search process. We do not specify here how the coefficient ρ is chosen; this is another subject and several gradient methods are possible. The search is stopped when a local minimum of J is found for a set of parameters S.

This procedure requires knowledge of gradients $\frac{\partial J^m}{\partial S_{is}^m}$ which are ns in number. This could be done in a very clumsy way by carrying out several calculations where we would use each time the definition of the derivative by slightly varying one of the parameters. We shall use this idea for calibration purposes, but it is the adjoint system that is going to yield the necessary information more simply on the J gradients.

From the definition of J, we find that:

$$\frac{\partial J^m}{\partial S_{is}^m} = 2 \sum_{it=1}^{nit} \sum_{ip=1}^{np} W_{ip}(X_{ip}^{it} - M_{ip}^{it}) \frac{\partial X_{ip}^{it}}{\partial S_{is}^m} \tag{9.2}$$

which can also be written in the form of dot products:

$$\frac{\partial J^m}{\partial S_{is}^m} = 2 \sum_{it=1}^{nit} \left(W(X^{it} - M^{it}) \ , \ \frac{\partial X^{it}}{\partial S_{is}^m} \right) \tag{9.3}$$

where (X, Y) represents $\sum_{ip=1}^{np} X_{ip} Y_{ip}$. We shall also use the fact that:

$$(AX, Y) = (X,^t AY) \tag{9.4}$$

If the choice of the cost function is more complex, of the form $J = f(X_{ip}^{it})$, where J is a function of all the components of the vector X, for all the time steps, we shall have in a more general way:

$$\frac{\partial J^m}{\partial S_{is}^m} = \sum_{it=1}^{nit} \left(\frac{\partial f}{\partial X^{it}}, \frac{\partial X^{it}}{\partial S_{is}^m} \right) \tag{9.5}$$

The term $2\,W(X^{it} - M^{it})$ is therefore simply replaced by $\frac{\partial f}{\partial X^{it}}$. This will be the case when we try to find tidal ranges measured in ports, which are the result of Fourier analysis of variations of the free surface.

In any case, the main difficulty remains the number of gradients of the form $\frac{\partial X_{ip}^{it}}{\partial S_{is}^m}$ that have to be calculated since they are the number of $ns \times np \times nit$. Thanks to the technique proposed by Chavent [27], this will not be necessary and it will suffice for us to calculate a limited number of linear combinations of these gradients. We shall now explain the general principle, in Lagrangian terms, before presenting calculation details.

The Lagrangian is formed as:

$$L(X, S, P) = J(X, S) + (P, AX - B)$$

where P is the vector of Lagrange multipliers (arbitrarily introduced variables that will be specified afterwards). Here it is considered that X, S and P are independent variables. A and B depend on S and X. We try to minimize $J(X, S)$ under constraint, the constraint being that we always must have $AX = B$. The minimum looked for corresponds to the triplet X, S, P which satisfies the optimality conditions:

$$\frac{\partial L(X, S, P)}{\partial P} = 0 \tag{9.6}$$

which gives $AX - B = 0$, i.e. the direct system.

$$\frac{\partial L(X, S, P)}{\partial X} = 0 \tag{9.7}$$

which gives:

$$\frac{\partial J}{\partial X} + \left(P, \frac{\partial (AX - B)}{\partial X}\right) = 0$$

or, if A does not depend on X:

$$\frac{\partial J}{\partial X} + \left(P, A + \frac{\partial A}{\partial X}X - \frac{\partial B}{\partial X}\right) = 0$$

which is the "adjoint" system.

$$\frac{\partial L(X, S, P)}{\partial S} = 0 \tag{9.8}$$

which gives:

$$\frac{\partial J}{\partial S} = -\left(P, \frac{\partial (AX - B)}{\partial S}\right)$$

which is the equation of the J gradient given by Lagrange multipliers or "adjoint variables".

Application to an unsteady problem

We now consider that our problem is a succession of linear systems, the matrices and right-hand sides of which depend on the result of previously resolved systems.

By starting from the linear system of iteration it, $A^{it}X^{it} = B^{it}$, we get:

$$\frac{d}{dS_{is}^m}(A^{it}X^{it} - B^{it}) = 0 \tag{9.9}$$

which gives ns different equations. Each of these equations is transformed into an equality between a sum of two matrix–vector products and a right-hand side:

$$\frac{dA^{it}}{dS_{is}^m}X^{it} + A^{it}\frac{dX^{it}}{dS_{is}^m} = \frac{dB^{it}}{dS_{is}^m} \tag{9.10}$$

As A depends on X^{it-1}, we then write that:

$$\frac{dA^{it}}{dS_{is}^m} = \frac{\partial A^{it}}{\partial S_{is}^m} + \sum_{ip=1}^{np} \frac{\partial A^{it}}{\partial X_{ip}^{it-1}} \frac{\partial X_{ip}^{it-1}}{\partial S_{is}^m} \tag{9.11}$$

This is the same for B:

$$\frac{dB^{it}}{dS_{is}^m} = \frac{\partial B^{it}}{\partial S_{is}^m} + \sum_{ip=1}^{np} \frac{\partial B^{it}}{\partial X_{ip}^{it-1}} \frac{\partial X_{ip}^{it-1}}{\partial S_{is}^m} \tag{9.12}$$

and $\frac{dX^{it}}{dS_{is}^m} = \frac{\partial X^{it}}{\partial S_{is}^m}$ (no formal dependence between the components of X^{it} and those of X^{it-1}).

These equalities have a meaning even if $it = 1$, but in this case $\frac{\partial X_{ip}^{it-1}}{\partial S_{is}^m} = 0$ because the initial conditions do not depend on the choice of S.

Equation 9.9 is finally written as:

$$A^{it}\frac{\partial X^{it}}{\partial S_{is}^{m}} + \left(\sum_{ip=1}^{np} \frac{\partial A^{it}}{\partial X_{ip}^{it-1}} \frac{\partial X_{ip}^{it-1}}{\partial S_{is}^{m}}\right) X^{it} - \left(\sum_{ip=1}^{np} \frac{\partial B^{it}}{\partial X_{ip}^{it-1}} \frac{\partial X_{ip}^{it-1}}{\partial S_{is}^{m}}\right) = \frac{\partial B^{it}}{\partial S_{is}^{m}} - \frac{\partial A^{it}}{\partial S_{is}^{m}} X^{it} \tag{9.13}$$

Now we are going to construct the adjoint variables, marked P, of the same rank as X, our objective being that the gradient of J can finally be written as:

$$\frac{\partial J^{m}}{\partial S_{is}^{m}} = \sum_{it=1}^{nit} \sum_{ip=1}^{np} P_{ip}^{it} \left(\frac{\partial B^{it}}{\partial S_{is}^{m}} - \frac{\partial A^{it}}{\partial S_{is}^{m}} X^{it}\right)_{ip} \tag{9.14}$$

This equation will be true only if the adjoint vector P fulfils certain conditions which we are going to lay down, and we shall see that these conditions depend neither on the choice of S nor on its nature. We recognize on the right of the equals sign the right-hand side of Equation 9.13, which we are going to replace by the left-hand side of the same equation in order to get:

$$\begin{aligned} \frac{\partial J^{m}}{\partial S_{is}^{m}} = \sum_{it=1}^{nit} \sum_{ip=1}^{np} P_{ip}^{it} \left[A^{it}\frac{\partial X^{it}}{\partial S_{is}^{m}} + \left(\sum_{jp=1}^{np} \frac{\partial A^{it}}{\partial X_{jp}^{it-1}} \frac{\partial X_{jp}^{it-1}}{\partial S_{is}^{m}}\right) X^{it}\right]_{ip} \\ - \sum_{it=1}^{nit} \sum_{ip=1}^{np} P_{ip}^{it} \left[\left(\sum_{jp=1}^{np} \frac{\partial B^{it}}{\partial X_{jp}^{it-1}} \frac{\partial X_{jp}^{it-1}}{\partial S_{is}^{m}}\right)\right]_{ip} \end{aligned} \tag{9.15}$$

One of the two indices called ip had to be renamed as jp.

The new right-hand side is going to be split into three terms:

Term 1:

$$\sum_{it=1}^{nit} \sum_{ip=1}^{np} P_{ip}^{it} \left(A^{it}\frac{\partial X^{it}}{\partial S_{is}^{m}}\right)_{ip}$$

which is also $\sum_{it=1}^{nit} (P^{it}, A^{it}\frac{\partial X^{it}}{\partial S_{is}^{m}})$ or even $\sum_{it=1}^{nit} ({}^{t}A^{it}P^{it}, \frac{\partial X^{it}}{\partial S_{is}^{m}})$, or finally:

$$\sum_{it=1}^{nit} \sum_{ip=1}^{np} ({}^{t}A^{it}P^{it})_{ip} \frac{\partial X_{ip}^{it}}{\partial S_{is}^{m}}$$

The other two terms are:

Term 2:

$$\sum_{it=1}^{nit} \sum_{ip=1}^{np} P_{ip}^{it} \left(\left(\sum_{jp=1}^{np} \frac{\partial A^{it}}{\partial X_{jp}^{it-1}} \frac{\partial X_{jp}^{it-1}}{\partial S_{is}^{m}}\right) X^{it}\right)_{ip}$$

Term 3:

$$\sum_{it=1}^{nit}\sum_{ip=1}^{np} P_{ip}^{it}\left(-\sum_{jp=1}^{np}\frac{\partial B^{it}}{\partial X_{jp}^{it-1}}\frac{\partial X_{jp}^{it-1}}{\partial S_{is}^{m}}\right)_{ip}$$

The transformation of terms 2 and 3 is shown below in the example of term 2. Written in the form of dot products, it is equal to:

$$\sum_{it=1}^{nit}\left(P^{it},\left(\sum_{jp=1}^{np}\frac{\partial A^{it}}{\partial X_{jp}^{it-1}}\frac{\partial X_{jp}^{it-1}}{\partial S_{is}^{m}}\right)X^{it}\right) \tag{9.16}$$

or even, by putting P^{it} back into the sum over jp and factorizing the terms $\frac{\partial X_{jp}^{it-1}}{\partial S_{is}^{m}}$, to:

$$\sum_{it=1}^{nit}\sum_{jp=1}^{np}\left(P^{it},\frac{\partial A^{it}}{\partial X_{jp}^{it-1}}X^{it}\right)\frac{\partial X_{jp}^{it-1}}{\partial S_{is}^{m}} \tag{9.17}$$

also equal to:

$$\sum_{it=1}^{nit}\sum_{jp=1}^{np}\left(\frac{\partial^{t} A^{it}}{\partial X_{jp}^{it-1}}P^{it},X^{it}\right)\frac{\partial X_{jp}^{it-1}}{\partial S_{is}^{m}}$$

and finally, by renaming the mute index jp as ip, to:

$$\sum_{it=1}^{nit}\sum_{ip=1}^{np}\left(\frac{\partial^{t} A^{it}}{\partial X_{ip}^{it-1}}P^{it},X^{it}\right)\frac{\partial X_{ip}^{it-1}}{\partial S_{is}^{m}} \tag{9.18}$$

In the same way, we find for term 3:

$$\sum_{it=1}^{nit}\sum_{ip=1}^{np} P_{ip}^{it}\left(-\sum_{jp=1}^{np}\frac{\partial B^{it}}{\partial X_{jp}^{it-1}}\frac{\partial X_{jp}^{it-1}}{\partial S_{is}^{m}}\right)_{ip}=\sum_{it=1}^{nit}\sum_{ip=1}^{np}-\left(P^{it},\frac{\partial B^{it}}{\partial X_{ip}^{it-1}}\right)\frac{\partial X_{ip}^{it-1}}{\partial S_{is}^{m}} \tag{9.19}$$

Examination of terms 1, 2 and 3 gives the idea of factorizing $\frac{\partial X_{ip}^{it}}{\partial S_{is}^{m}}$, but this is $\frac{\partial X_{ip}^{it-1}}{\partial S_{is}^{m}}$ which appears in terms 2 and 3.

Let us note that this last term still has a meaning if $it = 1$ because $\frac{\partial X_{ip}^{0}}{\partial S_{is}^{m}} = 0$.
So an iteration $nit + 1$ is defined by selecting no adjoint variables: $P^{nit+1} = 0$.
For terms 2 and 3, we can therefore write:
Term 2:

$$\sum_{it=1}^{nit}\sum_{ip=1}^{np}\left(\frac{\partial^{t} A^{it+1}}{\partial X_{ip}^{it}}P^{it+1},X^{it+1}\right)\frac{\partial X_{ip}^{it}}{\partial S_{is}^{m}} \tag{9.20}$$

Term 3:

$$\sum_{it=1}^{nit}\sum_{ip=1}^{np} -\left(P^{it+1}, \frac{\partial B^{it+1}}{\partial X_{ip}^{it}}\right)\frac{\partial X_{ip}^{it}}{\partial S_{is}^{m}} \tag{9.21}$$

We have thus shown that our quantity:

$$\sum_{it=1}^{nit}\sum_{ip=1}^{np} P_{ip}^{it}\left(\frac{\partial B^{it}}{\partial S_{is}^{m}} - \frac{\partial A^{it}}{\partial S_{is}^{m}}X^{it}\right)_{ip}$$

is equal to:

$$\sum_{it=1}^{nit}\sum_{ip=1}^{np}\left[({}^{t}A^{it}P^{it})_{ip} + \left(X^{it+1}, \frac{\partial\, {}^{t}A^{it+1}}{\partial X_{ip}^{it}}P^{it+1}\right) - \left(P^{it+1}, \frac{\partial B^{it+1}}{\partial X_{ip}^{it}}\right)\right]\frac{\partial X_{ip}^{it}}{\partial S_{is}^{m}} \tag{9.22}$$

Now we recall that:

$$\frac{\partial J^{m}}{\partial S_{is}^{m}} = 2\sum_{it=1}^{nit}\sum_{ip=1}^{np} W_{ip}(X_{ip}^{it} - M_{ip}^{it})\frac{\partial X_{ip}^{it}}{\partial S_{is}^{m}}$$

which brings us to the conclusion that if we are capable of finding adjoint variables P such that:

$$({}^{t}A^{it}P^{it})_{ip} + \left(X^{it+1}, \frac{\partial\, {}^{t}A^{it+1}}{\partial X_{ip}^{it}}P^{it+1}\right) - \left(P^{it+1}, \frac{\partial B^{it+1}}{\partial X_{ip}^{it}}\right) = 2W_{ip}(X_{ip}^{it} - M_{ip}^{it}) \tag{9.23}$$

Then equation 9.14 will be solved.

Here we have a set of linear systems, the unknowns of which are the vectors of adjoint variables P^{it}, and these linear systems couple together P^{it} and P^{it+1}. We cannot find P^1 without having found P^2, but $P^{nit+1} = 0$, therefore we can find P^{nit}. This brings us to the conclusion that the linear systems giving adjoint variables will have to be resolved by going backwards in time. All the variables X^{it} can be known if a previous direct calculation is done; that is, a simulation test with the set of parameters S.

The adjoint variables will therefore be found by the series of the following linear systems:

$$({}^{t}A^{it}P^{it})_{ip} = 2W_{ip}(X_{ip}^{it} - M_{ip}^{it}) - \left(X^{it+1}, \frac{\partial\, {}^{t}A^{it+1}}{\partial X_{ip}^{it}}P^{it+1}\right) + \left(P^{it+1}, \frac{\partial B^{it+1}}{\partial X_{ip}^{it}}\right) \tag{9.24}$$

or even:

$$({}^{t}A^{it}P^{it})_{ip} = 2W_{ip}(X_{ip}^{it} - M_{ip}^{it}) - \left(\frac{\partial A^{it+1}}{\partial X_{ip}^{it}}X^{it+1} - \frac{\partial B^{it+1}}{\partial X_{ip}^{it}}, P^{it+1}\right) \tag{9.25}$$

The first system to be resolved is the one giving P^{nit}:

$$({}^tA^{nit}P^{nit})_{ip} = 2W_{ip}(X_{ip}^{nit} - M_{ip}^{nit}) \tag{9.26}$$

Resolving the series of adjoint systems formally represents the same effort as the direct calculation (same number of direct and adjoint systems, same number of unknowns, matrices of direct and adjoint systems being transpositions of each other).

These adjoint systems do not depend on the parameter which is to be identified; the dependence appears only when we calculate the gradient of J by using the adjoint variables (Equation 9.14).

9.3 Treatment of Dirichlet conditions

So far we have neglected the fact that A^{it} and B^{it} are also functions of X^{it} in the case of Dirichlet-type boundary conditions.

The previous calculations can be done again by adding this dependence, and would lead to the following adjoint linear systems:

$$\begin{aligned}({}^tA^{it}P^{it})_{ip} = 2W_{ip}(X_{ip}^{it} - M_{ip}^{it}) &- \left(\frac{\partial A^{it+1}}{\partial X_{ip}^{it}}X^{it+1} - \frac{\partial B^{it+1}}{\partial X_{ip}^{it}}, P^{it+1}\right)\\ &- \left(\frac{\partial A^{it}}{\partial X_{ip}^{it}}X^{it} - \frac{\partial B^{it}}{\partial X_{ip}^{it}}, P^{it}\right)\end{aligned} \tag{9.27}$$

In the Telemac system, Dirichlet conditions are thus processed (see Section 7.4.7) as follows.

If one unknown x has an imposed value x_0:

1) equation of x becomes $x = x_0$,

2) x is eliminated from all other equations, using its value x_0 (the right-hand sides of equations are changed).

With this principle, the numbering of equations remains unchanged, A rank too, but certain lines and columns of A are cancelled, apart from the diagonal term where 1 is placed.

The procedure can thus be formalized as follows.

Let D be a diagonal matrix, the terms of which are 1 for a normal point and 0 for the Dirichlet points. D has the following properties: $D^2 = D$, and $D(I - D) = 0$. I is a diagonal will all terms equal to 1.

$(I - D)X^{it}$ is a vector which contains only the imposed values of X. Taking into account known terms (Dirichlet) in X^{it} means changing B in:

$$B - A(I - D)X^{it}$$

The new right-hand side must therefore be equal to $B - A(I - D)X^{it}$ for the normal points and to $(I - D)X^{it}$ for Dirichlet points. This can be summarized in one formula: B is changed to B' with:

$$B' = D[B - A(I - D)X^{it}] + (I - D)X^{it}$$

Cancelling lines and columns of Dirichlet points and placing a 1 on the diagonal similarly means changing A to A', with:

$$A' = DAD + I - D$$

These last two equations show that A' of the iteration it does not depend on X^{it}, only B' depends on it.

Knowing that resolved direct systems take into account Dirichlet variables, and are in fact of the form $A'X = B'$, the question now is to know which adjoint system results from it and which condition must fulfil Dirichlet adjoint variables. All that has been done with A and B can be done again with A' and B', but to construct B' we must *a priori* know the Dirichlet values of adjoint variables. The matrix of an adjoint system will be ${}^tA'$. The equation of a Dirichlet adjoint variable of ip number will thus be reduced to P_{ip}^{it} = an unknown right-hand side. We are going to show that the value given to Dirichlet adjoint variables does not affect the others, and that it can be chosen in an arbitrary manner. For this, we are going to show that the right-hand side of the equation of a normal adjoint variable does not depend on these arbitrary values. The term:

$$-\left(\frac{\partial A^{it}}{\partial X_{ip}^{it}} X^{it} - \frac{\partial B^{it}}{\partial X_{ip}^{it}}, P^{it}\right)$$

that we have added to the right-hand side to take into account Dirichlet conditions is zero for a degree of freedom ip. The term:

$$-\left(\frac{\partial A^{it+1}}{\partial X_{ip}^{it}} X^{it+1} - \frac{\partial B^{it+1}}{\partial X_{ip}^{it}}, P^{it+1}\right)$$

is broken down into $-(\frac{\partial A^{it+1}}{\partial X_{ip}^{it}} X^{it+1}, P^{it+1})$ and $(\frac{\partial B^{it+1}}{\partial X_{ip}^{it}}, P^{it+1})$.

The term $(\frac{\partial A^{it+1}}{\partial X_{ip}^{it}} X^{it+1}, P^{it+1})$, by changing A to A', becomes:

$$\left(\frac{\partial DA^{it+1}D}{\partial X_{ip}^{it}} X^{it+1}, P^{it+1}\right)$$

(the term $I - D$ does not depend on X). This term is equal to:

$$\left(\frac{\partial A^{it+1}}{\partial X_{ip}^{it}} DX^{it+1}, DP^{it+1}\right)$$

So it is seen that Dirichlet arbitrary values of P^{it+1} will be cancelled by the product with D.

If we look at the term $(\frac{\partial B^{it}}{\partial X_{ip}^{it}}, P^{it})$, it now becomes:

$$\left(\frac{\partial\left[D(B^{it} - A^{it}(I - D)X^{it}) + (I - D)X^{it}\right]}{\partial X_{ip}^{it}}, P^{it}\right)$$

or even:

$$\left(\frac{\partial B^{it}}{\partial X^{it}_{ip}}, P^{it}\right) = \left(\frac{\partial D(B^{it} - A^{it}(I-D)X^{it})}{\partial X^{it}_{ip}}, P^{it}\right) \tag{9.28}$$

because:

$$\left(\frac{\partial (I-D)X^{it}}{\partial X^{it}_{ip}}, P^{it}\right) = \left(\frac{\partial X^{it}}{\partial X^{it}_{ip}}, (I-D)P^{it}\right) = 0$$

Actually $\frac{\partial X^{it}}{\partial X^{it}_{ip}}$ is a zero vector except if the point number ip is of Dirichlet type. In this case, the dot product is equal to the component ip of $(I-D)P^{it}$, which means P^{it}_{ip}, the adjoint variable of a Dirichlet value, hence zero. We then write:

$$\left(\frac{\partial D(B^{it} - A^{it}(I-D)X^{it})}{\partial X^{it}_{ip}}, P^{it}\right) = \left(\frac{\partial (B^{it} - A^{it}(I-D)X^{it})}{\partial X^{it}_{ip}}, DP^{it}\right) \tag{9.29}$$

and it is again found that Dirichlet arbitrary values of P^{it+1} will be cancelled by the product with D.

It remains to check that in the expression 9.14 selected arbitrary values do not come in either. If, in the expression, we change B to B' and A to A':

$$\left(\frac{\partial B^{it}}{\partial S^m_{is}} - \frac{\partial A^{it}}{\partial S^m_{is}} X^{it}\right)_{ip}$$

becomes:

$$\left[\frac{\partial \left[D(B^{it} - A^{it}(I-D)X^{it}) + (I-D)X^{it}\right]}{\partial S^m_{is}} - \frac{\partial \left[DA^{it}D + (I-D)\right]}{\partial S^m_{is}} X^{it}\right]_{ip} \tag{9.30}$$

All the vectors $(I-D)X^{it}$ have non-zero terms only for Dirichlet values, which do not depend on the parameter S, so these terms are eliminated. What remains is only:

$$\left[\frac{\partial \left[D(B^{it} - A^{it}(I-D)X^{it})\right]}{\partial S^m_{is}} - \frac{\partial \left[DA^{it}D\right]}{\partial S^m_{is}} X^{it}\right]_{ip}$$

in which there is always a product by D on the left that can be brought into P^{it} in order to obtain:

$$\frac{\partial J^m}{\partial S^m_{is}} = \sum_{it=1}^{nit} \left(DP^{it}, \frac{\partial \left(B^{it} - A^{it}(I-D)X^{it}\right)}{\partial S^m_{is}} - \frac{\partial \left(A^{it}D\right)}{\partial S^m_{is}} X^{it}\right) \tag{9.31}$$

in which, again, Dirichlet values of adjoint variables do not come in any longer for DP^{it} concerns only normal values.

In conclusion of this section, it is therefore seen that adjoint systems will have the same treatment of Dirichlet conditions as direct systems, and that Dirichlet adjoint variables can be selected arbitrarily.

9.4 Application to Telemac-2D algorithm

9.4.1 Friction coefficient estimation in Telemac-2D

In the case of Telemac-2D, friction comes into the momentum equation and is treated in an implicit formulation. The two components of friction force are respectively:

$$F_x = -\frac{u}{2h} C_f \sqrt{u^2 + v^2} \tag{9.32}$$

$$F_y = -\frac{v}{2h} C_f \sqrt{u^2 + v^2} \tag{9.33}$$

in which h is the water depth, C_f a dimensionless friction coefficient and u and v the two components of horizontal velocity. These forces are discretized respectively in the form of:

$$F_x = -\frac{u^{n+1}}{2h} C_f \sqrt{u^2 + v^2} \tag{9.34}$$

$$F_y = -\frac{v^{n+1}}{2h} C_f \sqrt{u^2 + v^2} \tag{9.35}$$

the magnitude of the velocity $\sqrt{u^2 + v^2}$ being taken at the time t^n. These forces written in an implicit form come to modify the matrix A of the final linear system, B remaining unchanged. We further choose to do a "mass lumping" of these friction forces which means that, for the part concerning velocities, A is changed to $A + F$, where $F = kI$, I is the identity matrix and k the vector equal to:

$$k = \frac{1}{2h} C_f \sqrt{u^2 + v^2} \int_\Omega \Psi_i \, d\Omega \tag{9.36}$$

where Ψ_i is the basis relative to the point with index i.

So we have:

$$\frac{\partial J^m}{\partial S_{is}^m} = \sum_{it=1}^{nit} \sum_{ip=1}^{np} P_{ip}^{it} \left(\frac{\partial B^{it}}{\partial S_{is}^m} - \frac{\partial A^{it}}{\partial S_{is}^m} X^{it} \right)_{ip} = -\sum_{it=1}^{nit} \sum_{ip=1}^{np} P_{ip}^{it} \left(\frac{\partial A^{it}}{\partial S_{is}^m} X^{it} \right)_{ip} \tag{9.37}$$

for B does not depend on S, or even:

$$\frac{\partial J^m}{\partial S_{is}^m} = -\sum_{it=1}^{nit} \left(P^{it}, \frac{\partial A^{it}}{\partial S_{is}^m} X^{it} \right) = -\sum_{it=1}^{nit} \left(P^{it}, \frac{\partial (k)}{\partial S_{is}^m} X^{it} \right) \tag{9.38}$$

It is therefore necessary to know the dependence of k in terms of S, thus specifying the law used. For a Chézy or Strickler formula, k scales on $1/S^2$ and $\frac{\partial(k)}{\partial S_{is}^m}$ thus has a value of $\frac{-2k}{S_{is}^m}$ for index $i = is$ and 0 for the others. Thus we get:

$$\frac{\partial J^m}{\partial S_{is}^m} = \sum_{it=1}^{nit} P_{is}^{it} \frac{2kX_{is}^{it}}{S_{is}^m} \tag{9.39}$$

Note: we have assumed so far that a friction coefficient was estimated per point. If we look for a limited number of coefficients by dividing the mesh into different zones, or arranging points class-wise, then the gradient becomes:

$$\frac{\partial J^m}{\partial S_{is}^m} = \sum_{it=1}^{nit} \sum_{i \in classe(is)} P_i^{it} \frac{2kX_i^{it}}{S_{is}^m} \tag{9.40}$$

in which index i covers all the points to which the same friction coefficient S_{is}^m is assigned.

9.4.2 Calculation of right-hand sides of adjoint systems

Principle

Let us recall that the right-hand side of the adjoint system has for component ip:

$$2W_{ip}(X_{ip}^{it} - M_{ip}^{it}) - \left(\frac{\partial A^{it+1}}{\partial X_{ip}^{it}} X^{it+1} - \frac{\partial B^{it+1}}{\partial X_{ip}^{it}}, P^{it+1} \right) - \left(\frac{\partial A^{it}}{\partial X_{ip}^{it}} X^{it} - \frac{\partial B^{it}}{\partial X_{ip}^{it}}, P^{it} \right) \tag{9.41}$$

With a cost function of the formulation $f(X_{ip}^{it})$, let us recall that the term $2W_{ip}$ $(X_{ip}^{it} - M_{ip}^{it})$ is replaced by $\frac{\partial f}{\partial X_{ip}^{it}}$. A and B originate from a variational formulation, which gives to each point a weight proportional to the integral of its basis $\int_\Omega \Psi_i \, d\Omega$. For the sake of homogeneity, it would be necessary to take a cost function J which is not $\sum_{it=1}^{nit} \sum_{ip=1}^{np} W_{ip}(X_{ip}^{it} - M_{ip}^{it})^2$ but rather:

$$\sum_{it=1}^{nit} \int_\Omega \Psi_i \, W_{ip}(X_{ip}^{it} - M_{ip}^{it})^2 d\Omega$$

which complicates the calculation of its gradient. Without changing the calculation of the gradient, it can also be considered that W_{ip} is equal not to 1 or 0 but to $\int_\Omega \Psi_i \, d\Omega$ or 0, which restores the homogeneity of the right-hand side. In any case, it is easy to calculate $2W_{ip}(X_{ip}^{it} - M_{ip}^{it})$.

In the case of Telemac-2D, the matrix A is composed of blocks and vector X contains the water depth and the two components of the velocity. The three blocks of the right-hand sides are marked $CV1$, $CV2$ and $CV3$:

$$A = \begin{pmatrix} AM1 & BM1 & BM2 \\ CM1 & AM2 & 0 \\ CM2 & 0 & AM3 \end{pmatrix}, \quad X = \begin{pmatrix} H \\ U \\ V \end{pmatrix}, \quad B = \begin{pmatrix} CV1 \\ CV2 \\ CV3 \end{pmatrix}$$

Note: strictly speaking, the unknown H is in fact in Telemac-2D: $\delta H = H^{n+1} - H^n$; we consider here for more clarity and symmetry that the unknown is H^{n+1}. This changes slightly the right-hand sides which become (with the notations of Section 4.5.6):

$$CV1 = \frac{M^h}{\Delta t} H^n + (\theta_h - 1)(T_x^h + T_y^h) H^n + (\theta_u - 1)(B_x^{uh} U^n + B_y^{uh} V^n) + M^h SCE + TB1 \tag{9.42}$$

where $\frac{M^h}{\Delta t} H^n$ appears in $CV1$.

$$CV2 = \frac{M^u}{\Delta t} U^n + (\theta_u - 1)(T_x^u + T_y^u) U^n - g(1 - \theta_h) C_x^{uh} Z_s^n + M^u F_x \tag{9.43}$$

where $-gC_x^{uh} Z_s^n$ becomes $-g(1 - \theta_h) C_x^{uh} Z_s^n$ in $CV2$.

$$CV3 = \frac{M^u}{\Delta t} V^n + (\theta_u - 1)(T_x^u + T_y^u) V^n - g(1 - \theta_h) C_y^{uh} Z_s^n + M^u F_y \tag{9.44}$$

where $-gC_y^{uh} Z_s^n$ becomes $-g(1 - \theta_h) C_y^{uh} Z_s^n$ in $CV3$.

If the method of characteristics is used for advection terms, these three right-hand sides are written as:

$$CV1 = \frac{M^h}{\Delta t} \widetilde{H} + (\theta_u - 1)(B_x^{uh} U^n + B_y^{uh} V^n) + M^h SCE + TB1 \tag{9.45}$$

$$CV2 = \frac{M^u}{\Delta t} \widetilde{U} - g(1 - \theta_h) C_x^{uh} Z_s^n + M^u F_x \tag{9.46}$$

$$CV3 = \frac{M^u}{\Delta t} \widetilde{V} - g(1 - \theta_h) C_y^{uh} Z_s^n + M^u F_y \tag{9.47}$$

For the adjoint system, the following notations will be adopted:

$$^tA = \begin{pmatrix} ^tAM1 & ^tCM1 & ^tCM2 \\ ^tBM1 & ^tAM2 & 0 \\ ^tBM2 & 0 & ^tAM3 \end{pmatrix} = \begin{pmatrix} AM1_{adj} & BM1_{adj} & BM2_{adj} \\ CM1_{adj} & AM2_{adj} & 0 \\ CM2_{adj} & 0 & AM3_{adj} \end{pmatrix} \tag{9.48}$$

$$P = \begin{pmatrix} H_{adj} \\ U_{adj} \\ V_{adj} \end{pmatrix} \qquad B_{adj} = \begin{pmatrix} CV1_{adj} \\ CV2_{adj} \\ CV3_{adj} \end{pmatrix} \tag{9.49}$$

This adjoint system is split into three blocks:

$$^tAM1\, H_{adj} + {}^tCM1\, U_{adj} + {}^tCM2\, V_{adj} = CV1_{adj} \tag{9.50}$$

$$^tBM1\, H_{adj} + {}^tAM2\, U_{adj} = CV2_{adj} \tag{9.51}$$

$$^tBM2\, H_{adj} + {}^tAM3\, V_{adj} = CV3_{adj} \tag{9.52}$$

Let us recall that the right-hand side of the adjoint system is (Equation 9.25):

$$2W_{ip}(X_{ip}^{it} - M_{ip}^{it}) - \left(X^{it+1}, \frac{\partial^t A^{it+1}}{\partial X_{ip}^{it}} P^{it+1} \right) + \left(P^{it+1}, \frac{\partial B^{it+1}}{\partial X_{ip}^{it}} \right)$$

This gives, for each point ip of the mesh:

$$(CV1_{adj})_{ip} = 2W_{ip}(H^{it} - H^{it}_{mes})_{ip}$$

$$+\left(H^{it+1}_{adj}, \frac{\partial CV1^{it+1}}{\partial H^{it}_{ip}}\right) + \left(U^{it+1}_{adj}, \frac{\partial CV2^{it+1}}{\partial H^{it}_{ip}}\right) + \left(V^{it+1}_{adj}, \frac{\partial CV3^{it+1}}{\partial H^{it}_{ip}}\right)$$

$$-\left(H^{it+1}, \frac{\partial^t AM1^{it+1}}{\partial H^{it}_{ip}} H^{it+1}_{adj}\right)$$

$$-\left(H^{it+1}, \frac{\partial^t CM1^{it+1}}{\partial H^{it}_{ip}} U^{it+1}_{adj}\right) - \left(H^{it+1}, \frac{\partial^t CM2^{it+1}}{\partial H^{it}_{ip}} V^{it+1}_{adj}\right)$$

$$-\left(U^{it+1}, \frac{\partial^t BM1^{it+1}}{\partial H^{it}_{ip}} H^{it+1}_{adj}\right) - \left(U^{it+1}, \frac{\partial^t AM2^{it+1}}{\partial H^{it}_{ip}} U^{it+1}_{adj}\right)$$

$$-\left(V^{it+1}, \frac{\partial^t BM2^{it+1}}{\partial H^{it}_{ip}} H^{it+1}_{adj}\right) - \left(V^{it+1}, \frac{\partial^t AM3^{it+1}}{\partial H^{it}_{ip}} V^{it+1}_{adj}\right) \tag{9.53}$$

$$(CV2_{adj})_{ip} = 2W_{ip}(U^{it} - U^{it}_{mes})_{ip}$$

$$+\left(H^{it+1}_{adj}, \frac{\partial CV1^{it+1}}{\partial U^{it}_{ip}}\right) + \left(U^{it+1}_{adj}, \frac{\partial CV2^{it+1}}{\partial U^{it}_{ip}}\right) + \left(V^{it+1}_{adj}, \frac{\partial CV3^{it+1}}{\partial U^{it}_{ip}}\right)$$

$$-\left(H^{it+1}, \frac{\partial^t AM1^{it+1}}{\partial U^{it}_{ip}} H^{it+1}_{adj}\right)$$

$$-\left(H^{it+1}, \frac{\partial^t CM1^{it+1}}{\partial U^{it}_{ip}} U^{it+1}_{adj}\right) - \left(H^{it+1}, \frac{\partial^t CM2^{it+1}}{\partial U^{it}_{ip}} V^{it+1}_{adj}\right)$$

$$-\left(U^{it+1}, \frac{\partial^t BM1^{it+1}}{\partial U^{it}_{ip}} H^{it+1}_{adj}\right) - \left(U^{it+1}, \frac{\partial^t AM2^{it+1}}{\partial U^{it}_{ip}} U^{it+1}_{adj}\right)$$

$$-\left(V^{it+1}, \frac{\partial^t BM2^{it+1}}{\partial U^{it}_{ip}} H^{it+1}_{adj}\right) - \left(V^{it+1}, \frac{\partial^t AM3^{it+1}}{\partial U^{it}_{ip}} V^{it+1}_{adj}\right) \tag{9.54}$$

$$(CV3_{adj})_{ip} = 2W_{ip}(V^{it} - V^{it}_{mes})_{ip}$$

$$+\left(H^{it+1}_{adj}, \frac{\partial CV1^{it+1}}{\partial V^{it}_{ip}}\right) + \left(U^{it+1}_{adj}, \frac{\partial CV2^{it+1}}{\partial V^{it}_{ip}}\right) + \left(V^{it+1}_{adj}, \frac{\partial CV3^{it+1}}{\partial V^{it}_{ip}}\right)$$

$$-\left(H^{it+1}, \frac{\partial^t AM1^{it+1}}{\partial V^{it}_{ip}} H^{it+1}_{adj}\right)$$

Plate 1: porosity in a rectangular channel: effect on free surface and velocity

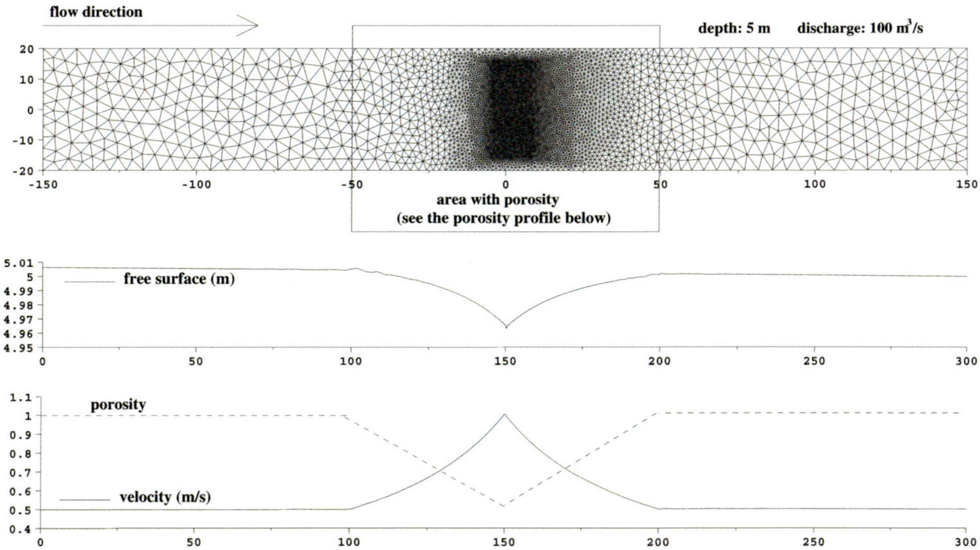

Plate 2: flow in a canal with bottom friction:
2- and 3-dimensional simulation with a Manning coefficient of 0.02

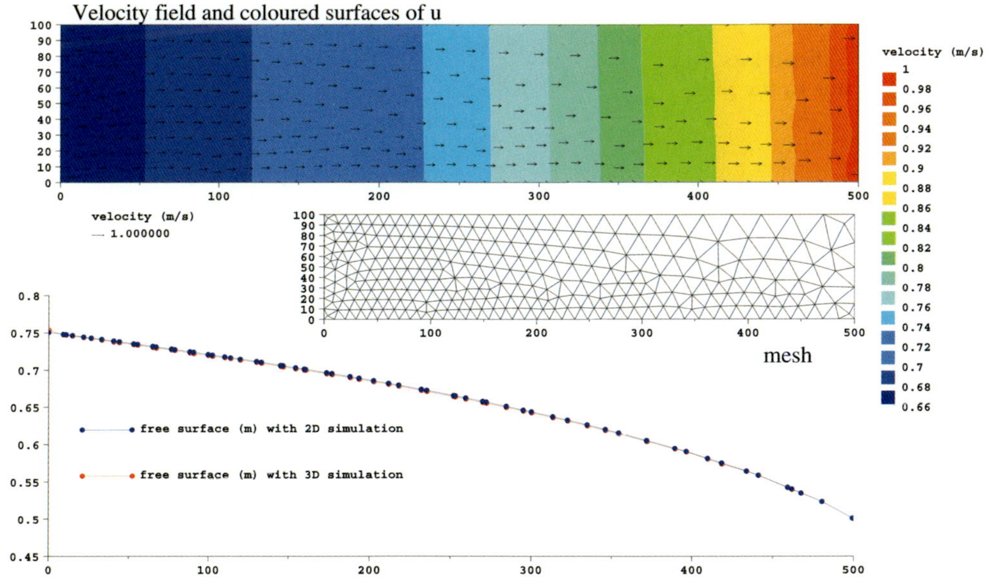

Plate 3: bridge pier in a canal: simulation with porosity and drag force

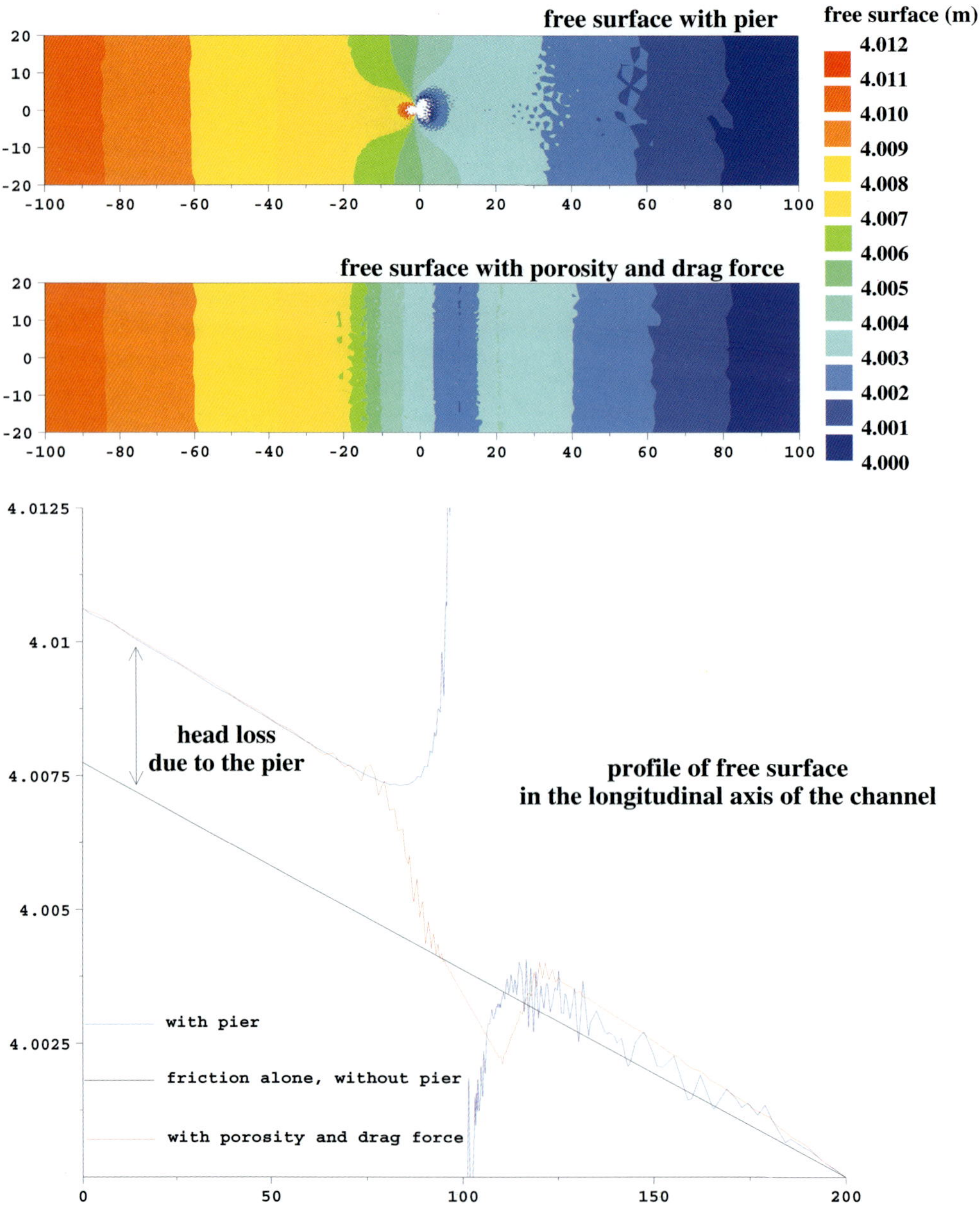

Plate 4: effect of 32 marine turbines (in white rectangle) on tidal currents: drag forces modelled in shallow water equations

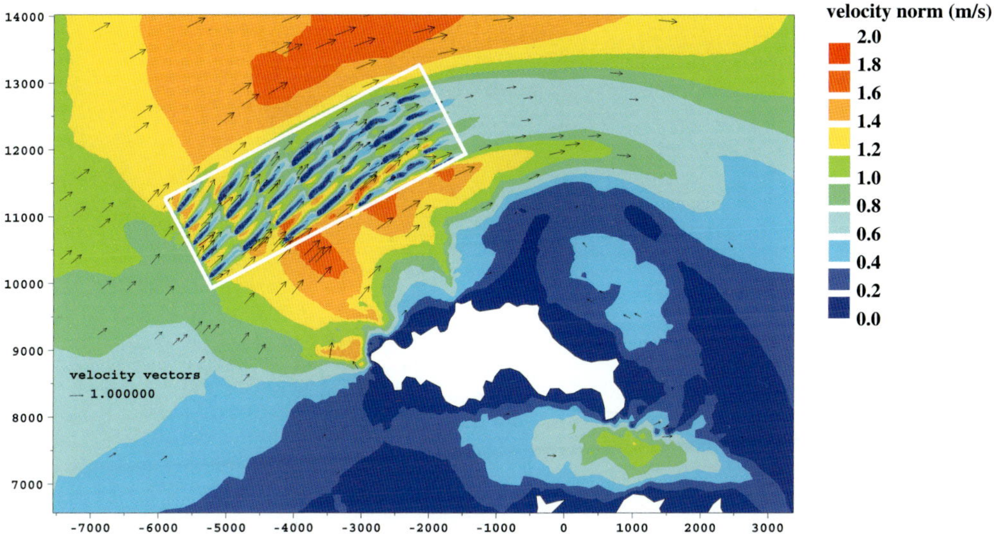

Plate 5: Lynch test case (tide on a quadratic slope): linearized shallow water equations without advection

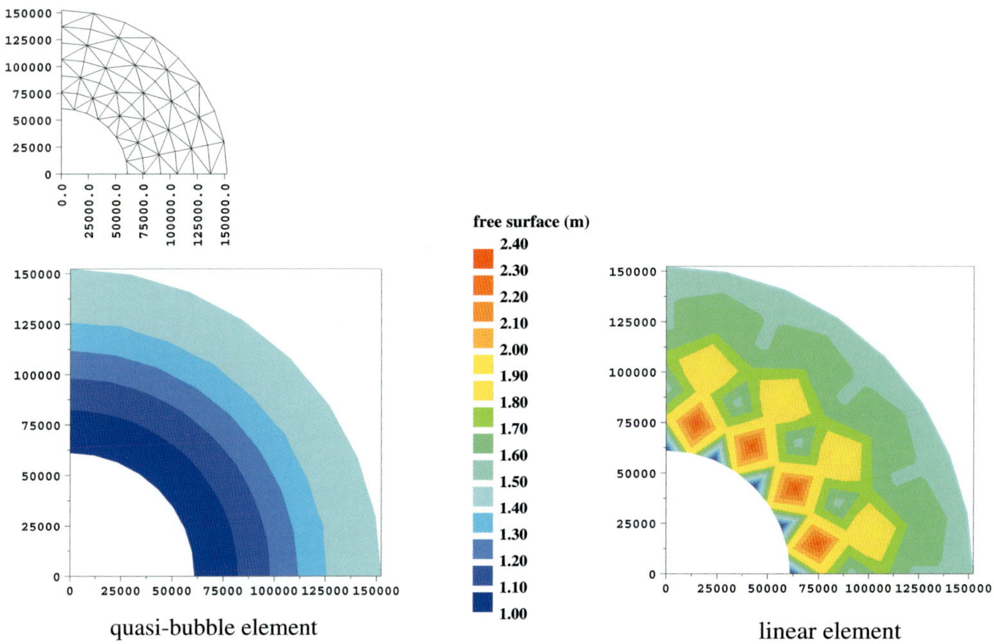

Plate 6: Saint-Venant equations: von Karman eddies behind bridge piers

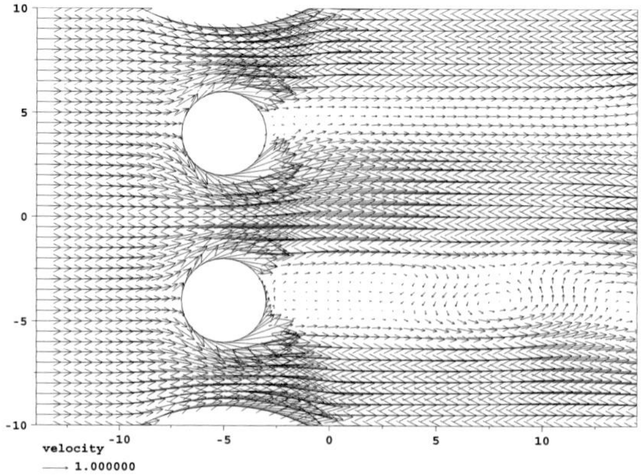

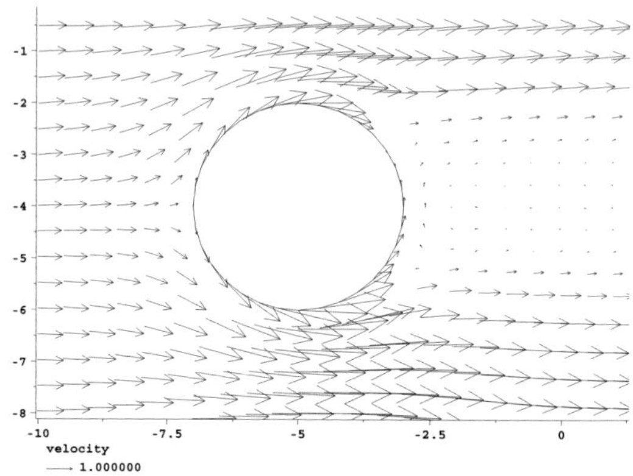

Plate 7: Saint-Venant equations: propagation of a wave, effect of boundary conditions

after 2 s, one iso-line every 1/6 m

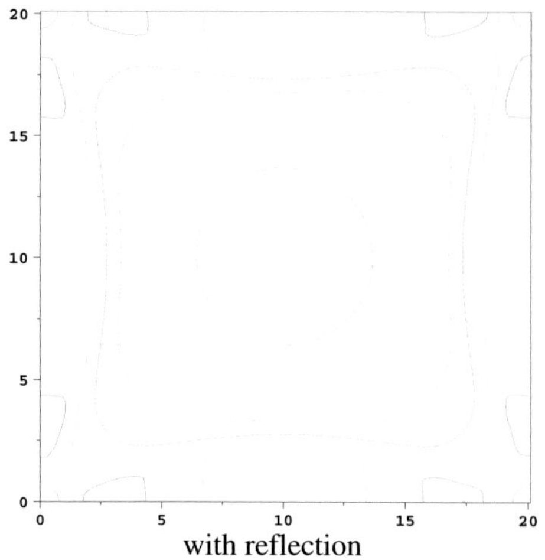

with reflection on the domain boundaries

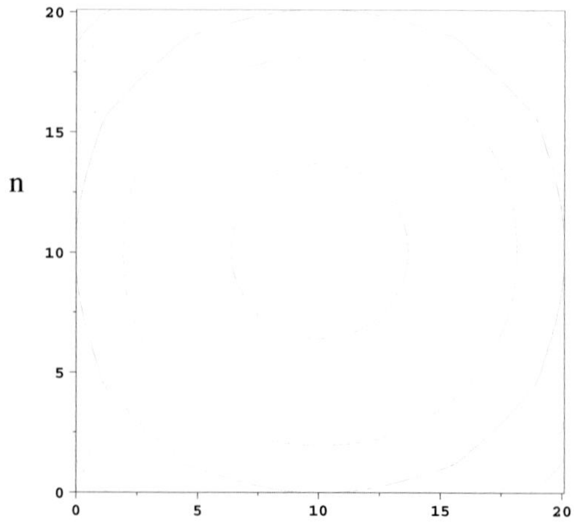

with Thompson radiation conditions

Plate 8: transcritical flow over a bump in a channel with a hydraulic jump

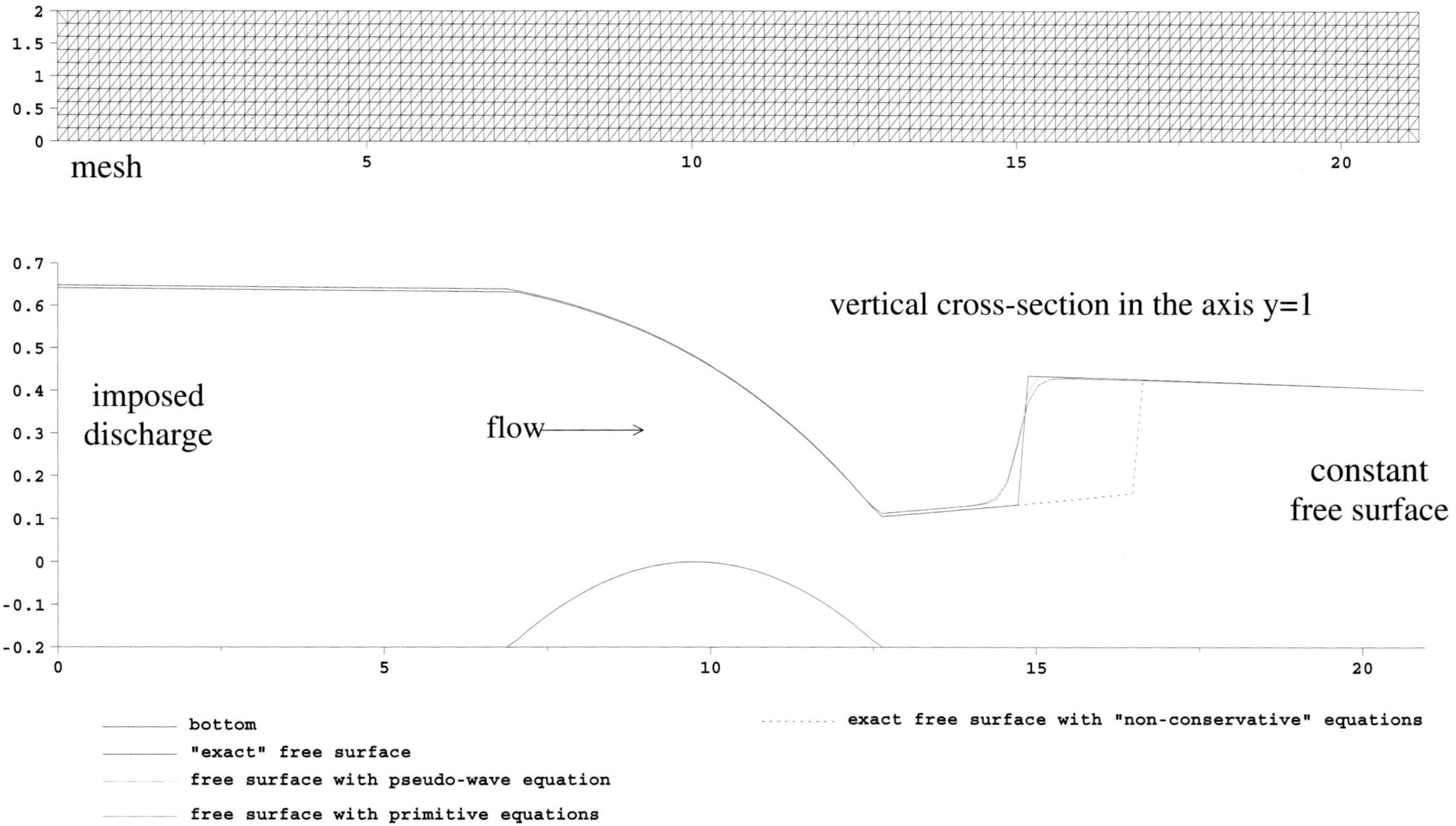

Plate 9: ebb tide on a beach: free surface at different times
with imposed sinusoidal elevation on the left

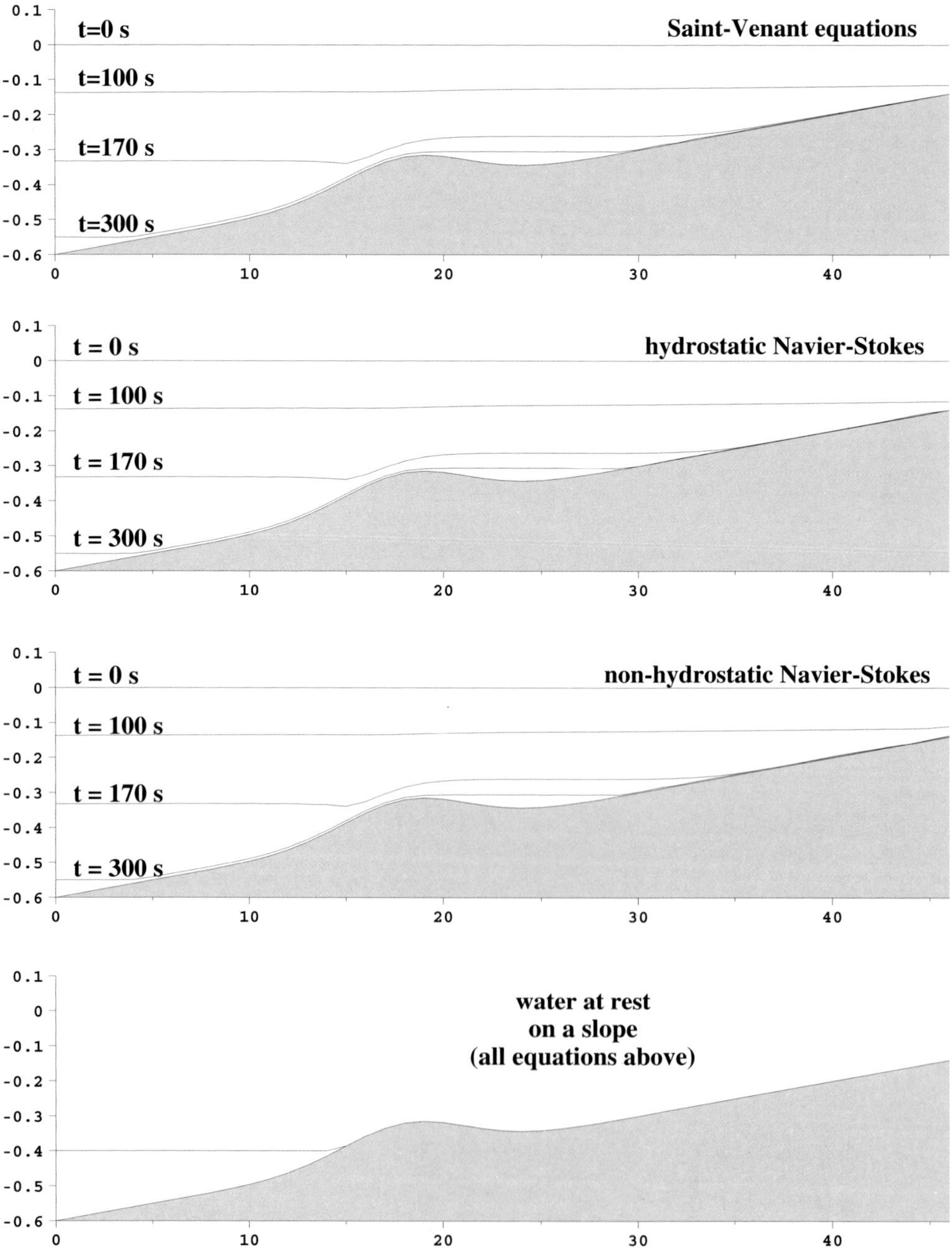

Plate 10: Saint-Venant equations and von Karman eddies behind bridge piers: effect of discretization on the free surface

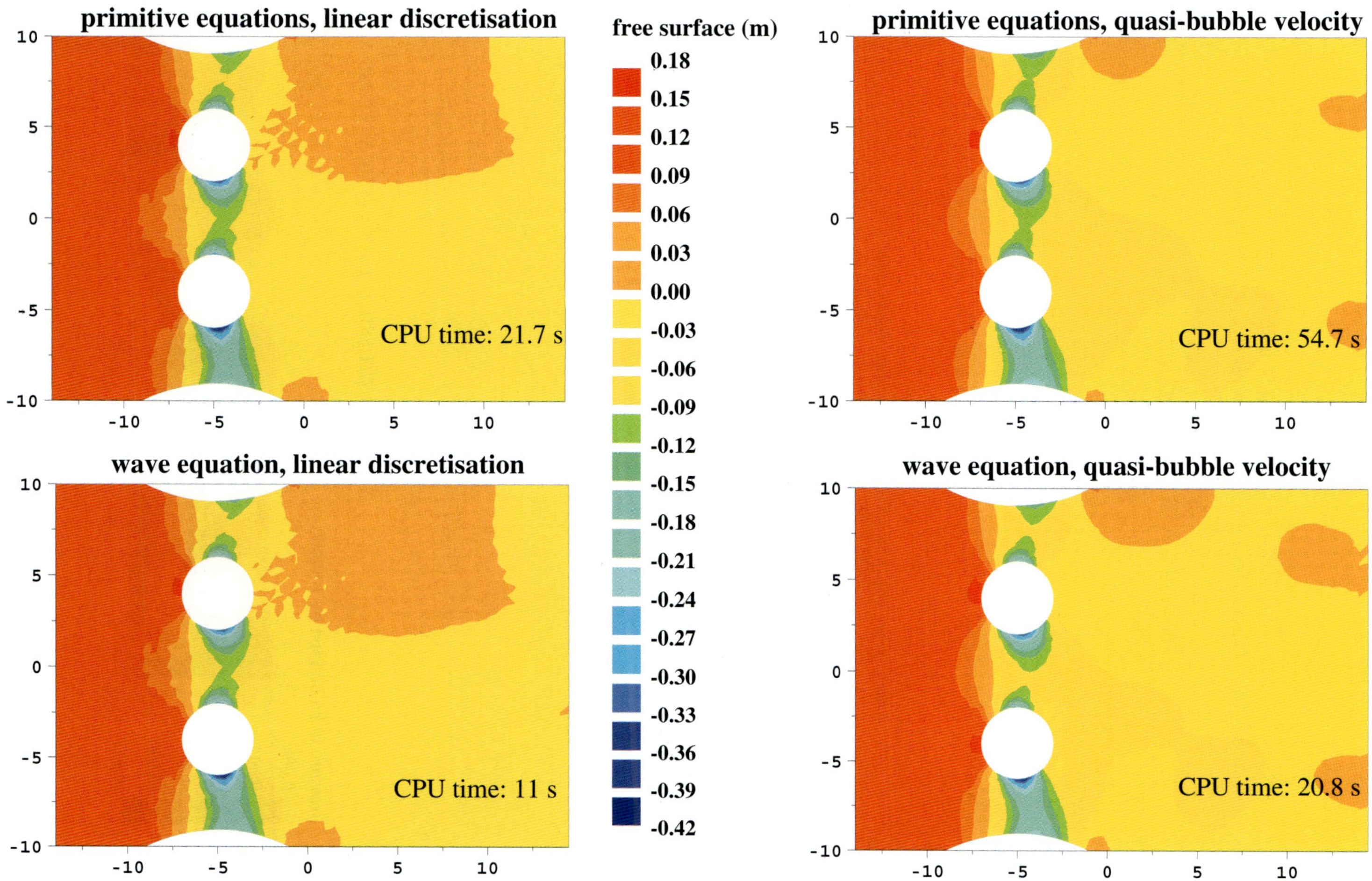

Plate 11: transcritical flow over a bump in a canal with a hydraulic jump:
non-hydrostatic Navier-Stokes equations with five planes vertically in the mesh

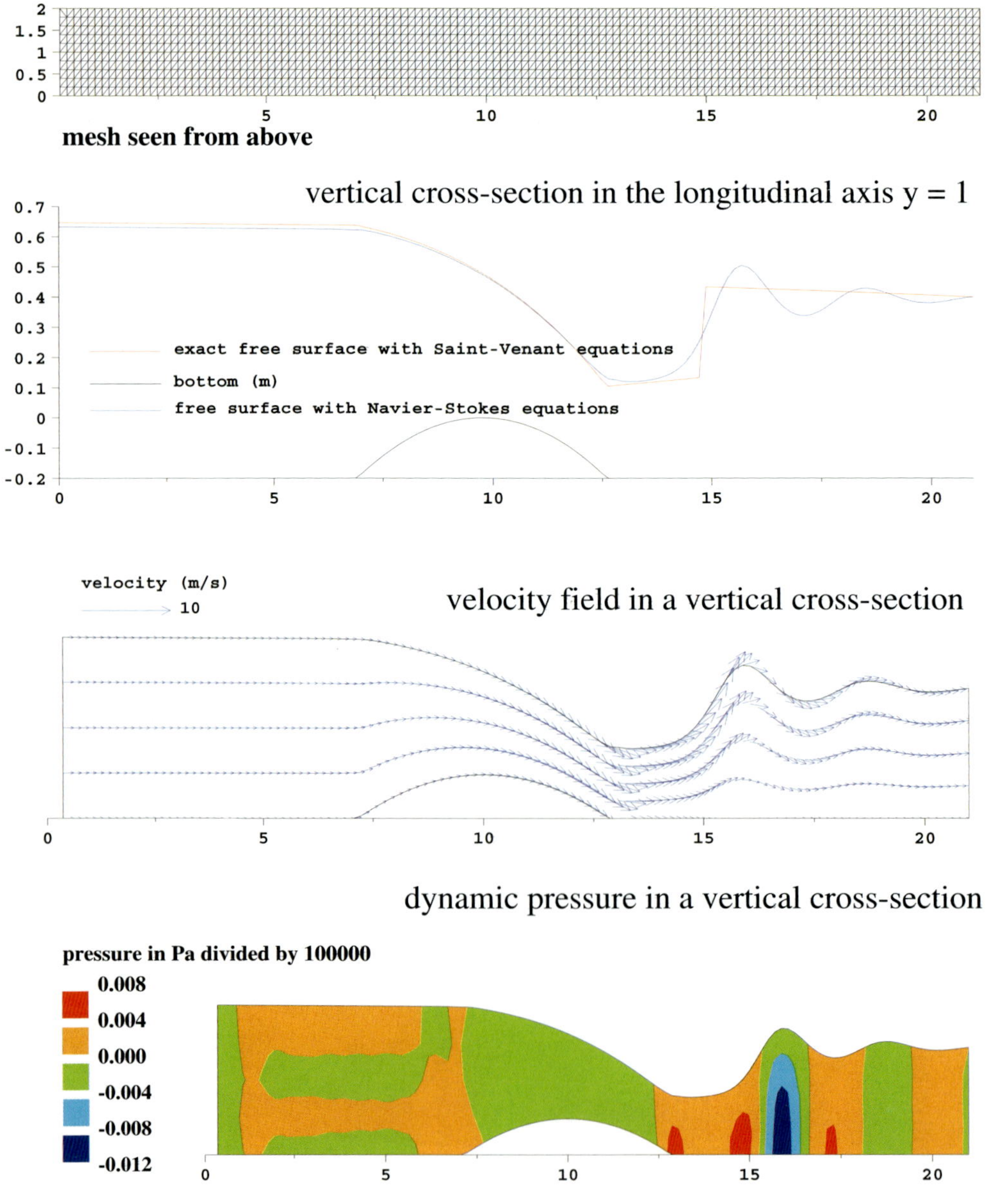

Plate 12: propagation of a solitary wave in a channel

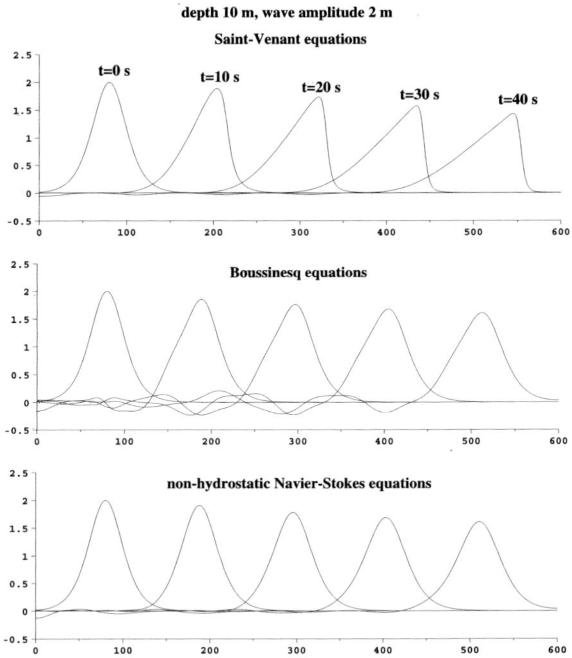

Plate 13: Saint-Venant equations: propagation of a circular wave

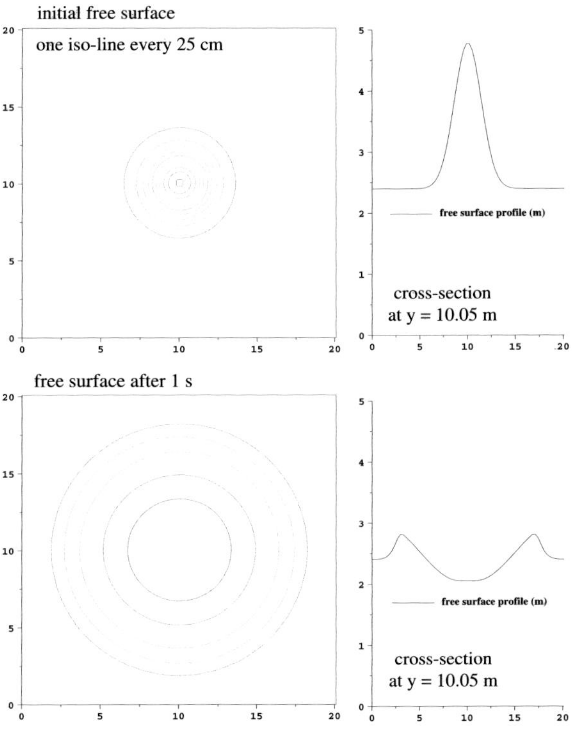

Plate 14: dam break in a river, Stoker's solution

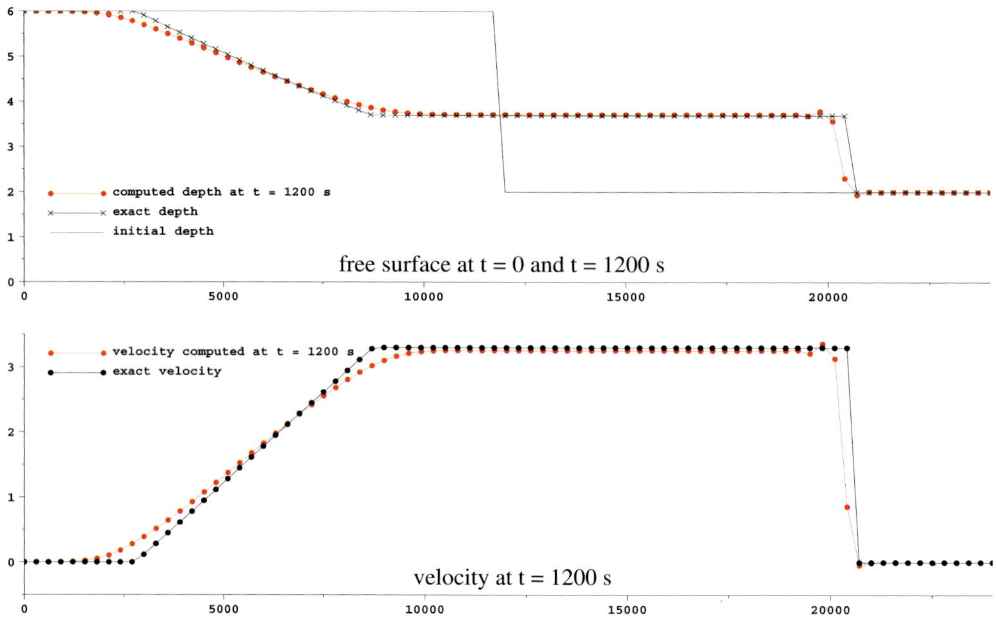

Plate 15: Malpasset site: contour of computational domain, position of measurement probes (6 to 14) and electric transformers (A,B,C)

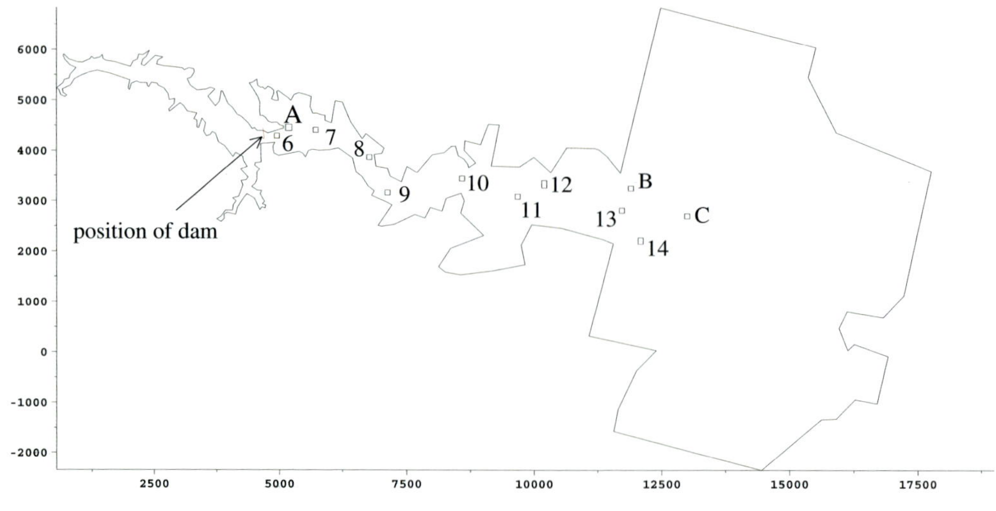

Plate 16: Malpasset site: bathymetry and details of the mesh

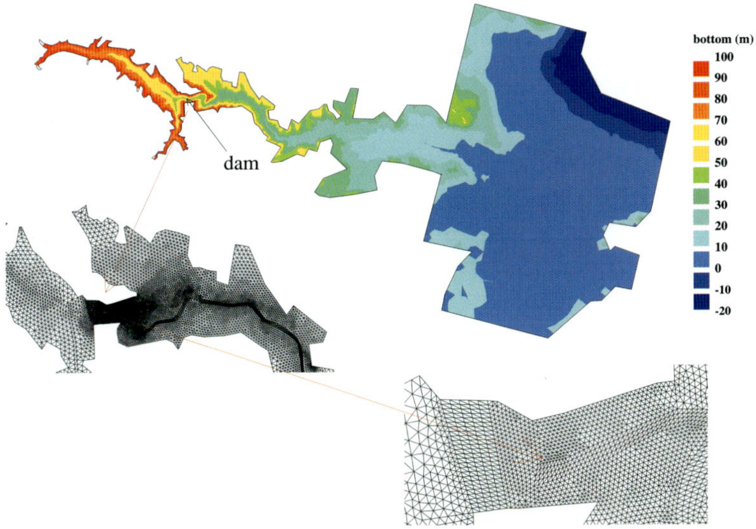

Plate 17: Malpasset dam break: velocity field after 100 s

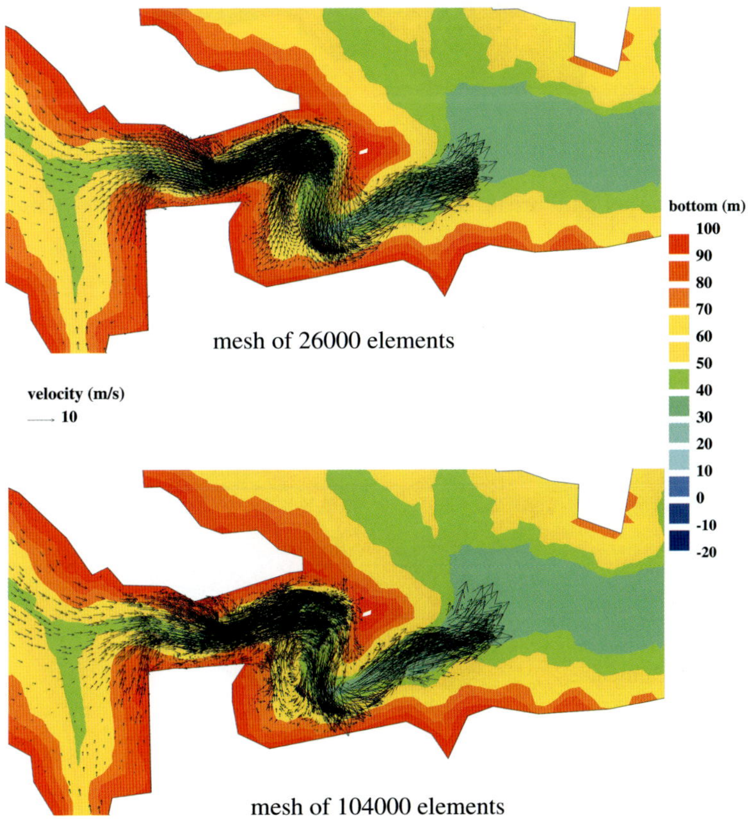

mesh of 26000 elements

mesh of 104000 elements
(1 vector every 4 is shown)

Plate 18: Malpasset dam break: depth and velocity after 20 minutes

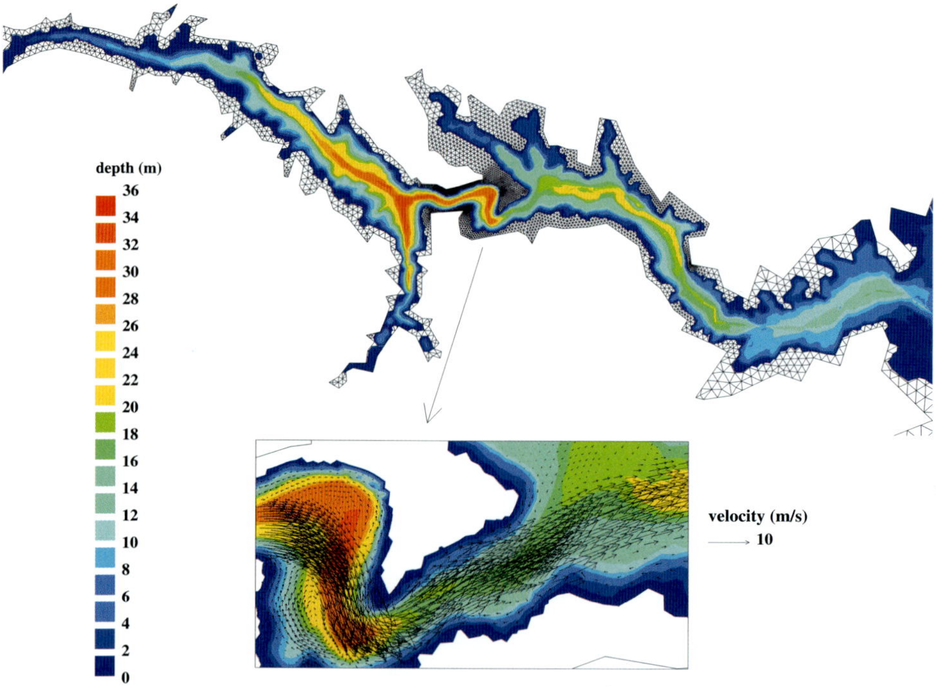

Plate 19: Malpasset dam break: the flood wave reaches the sea after 36 minutes

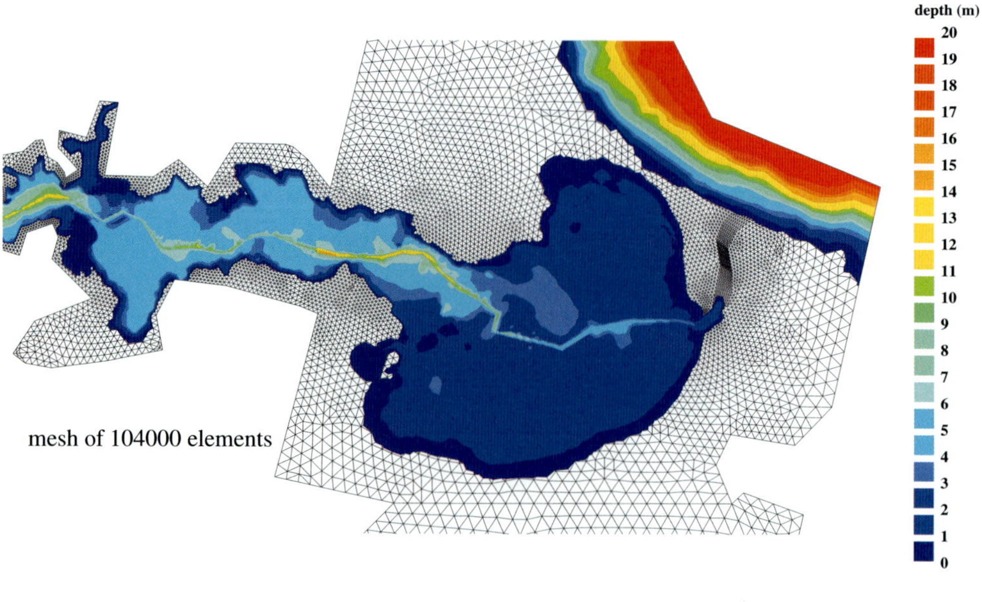

Plate 20: Malpasset dam break: flood after 2200 s

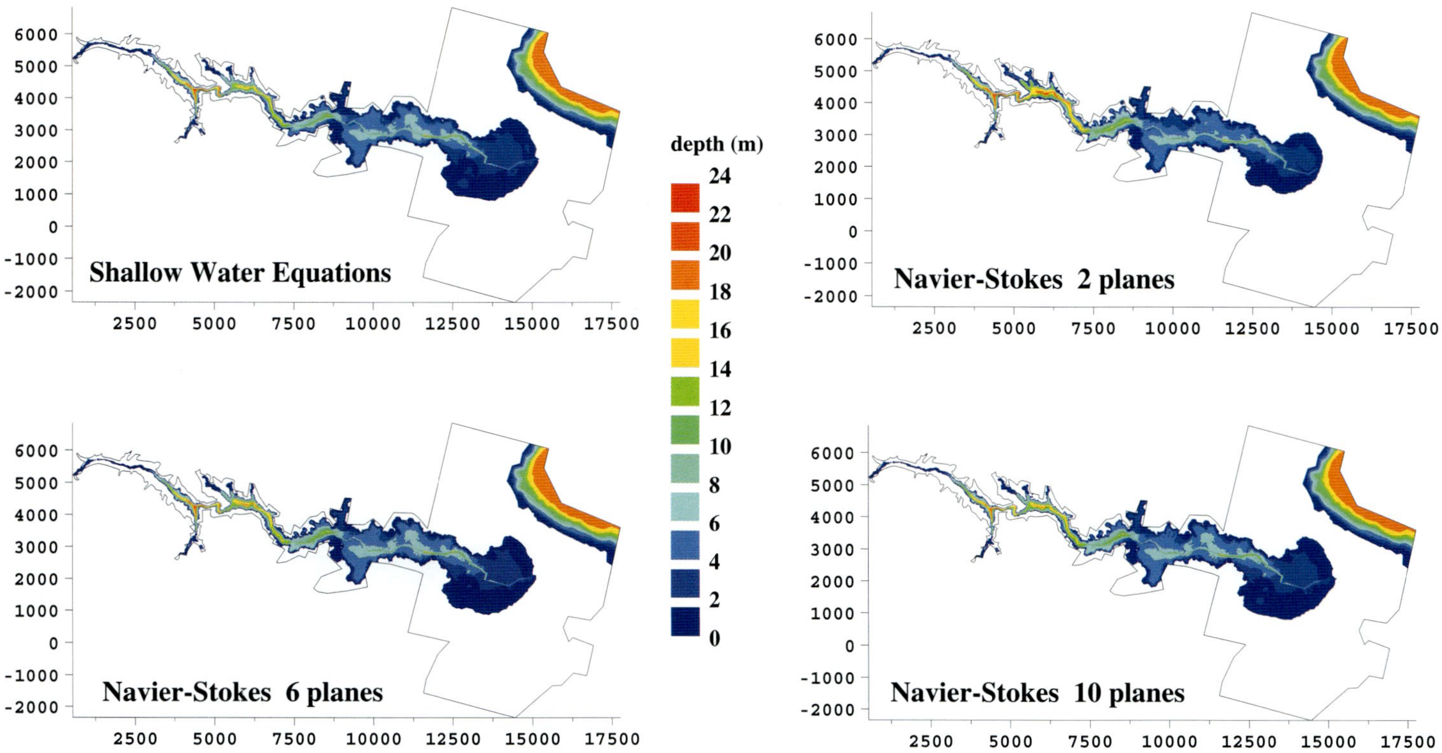

Plate 21: semi-diurnal tide in the Channel: tidal range and phase

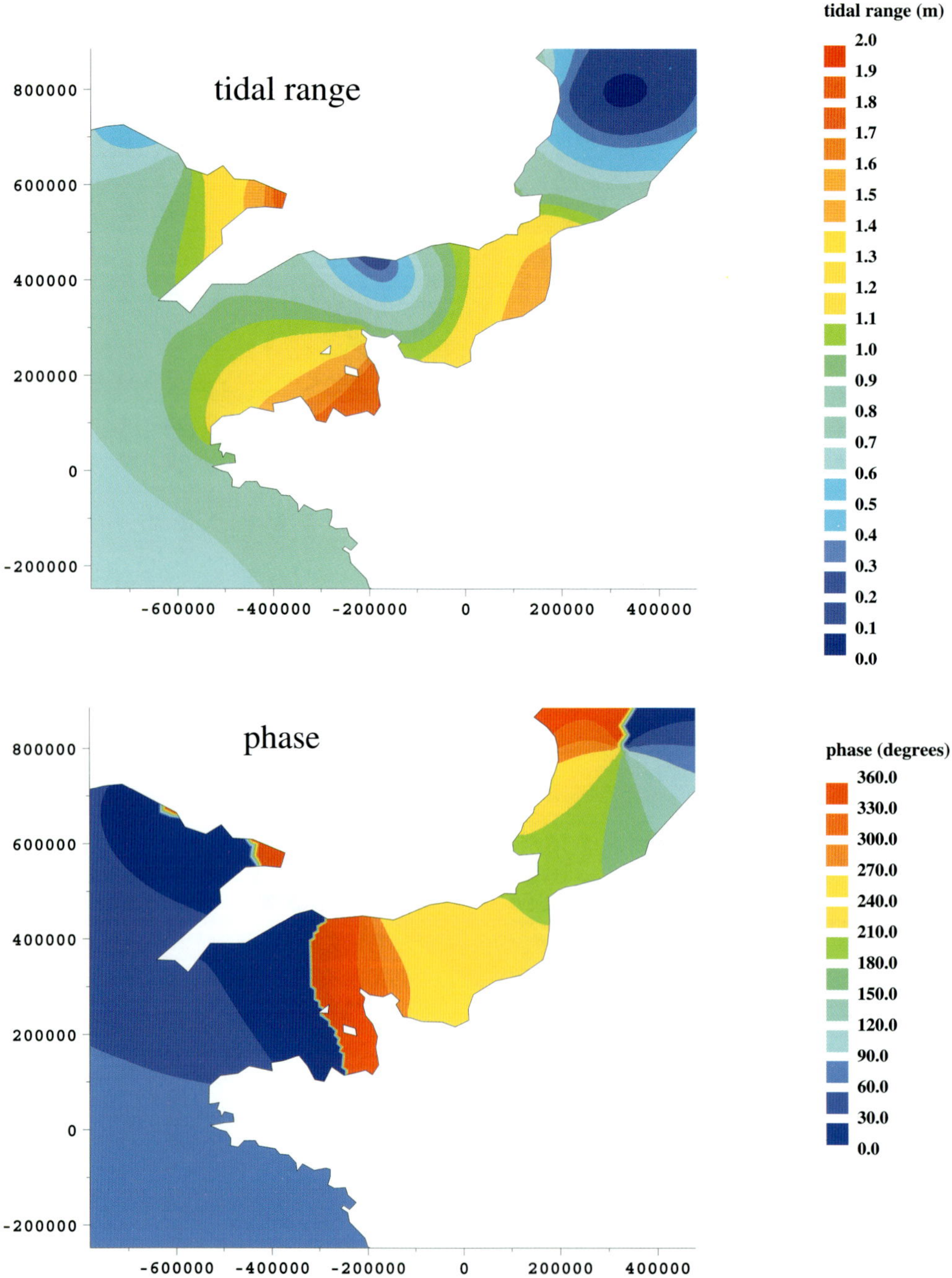

Plate 22: semi-diurnal tide in the Channel: bathymetry, mesh and friction coefficients

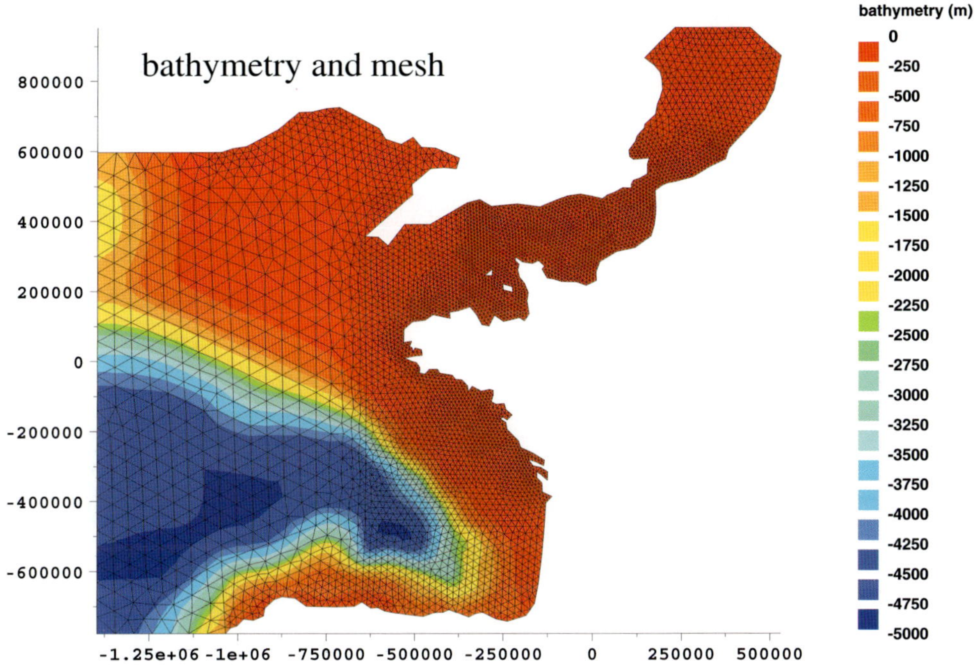

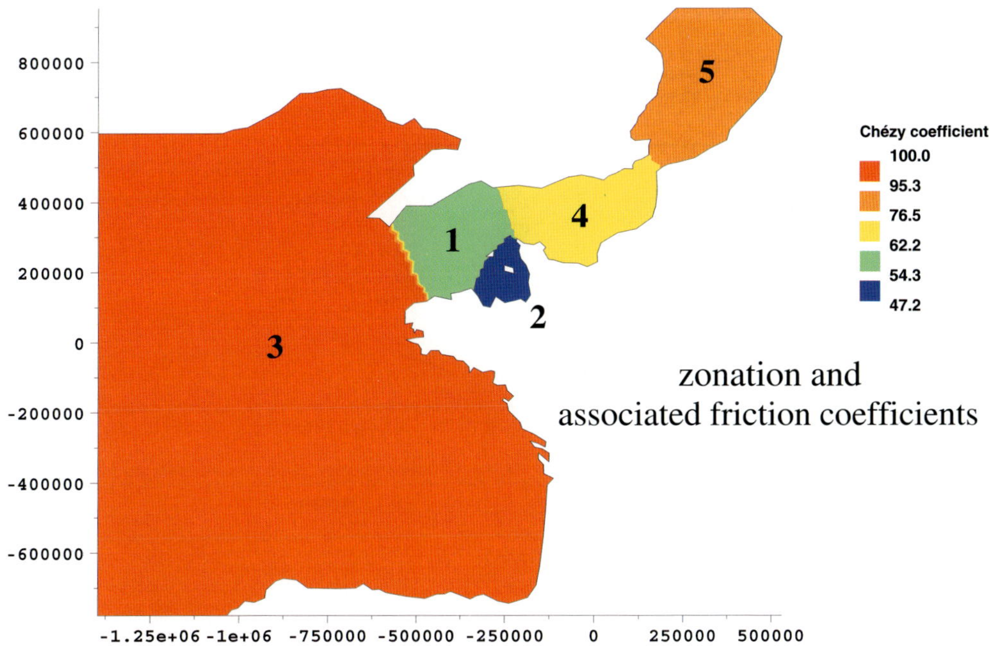

Plate 23: mixing of two fluids with different density:
vertical cross-section in the longitudinal axis of the canal

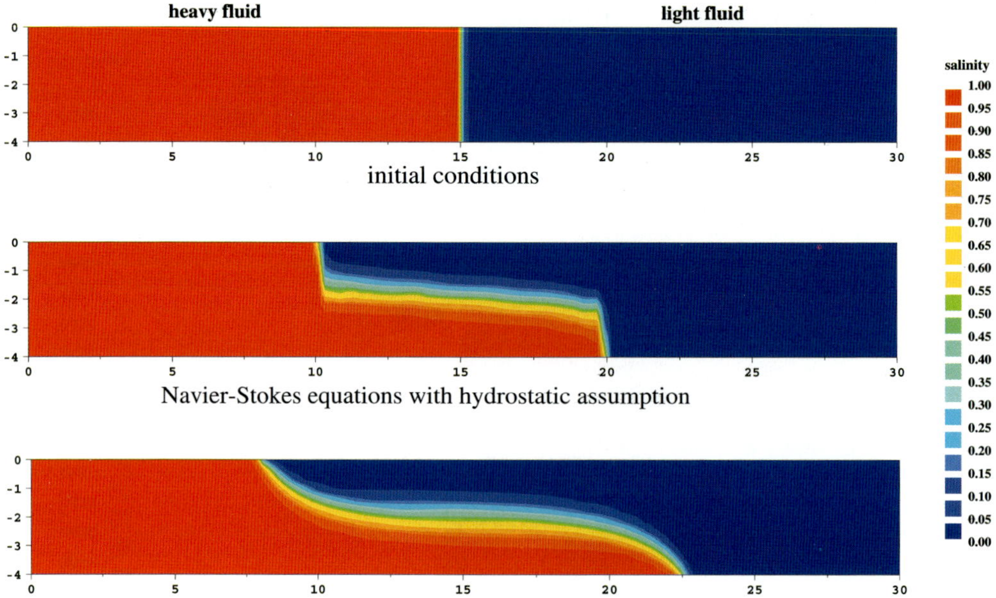

Plate 24: automatic estimation of the friction coefficient in a canal
(data simulated with a Strickler coefficient of 35, initial guess 5)

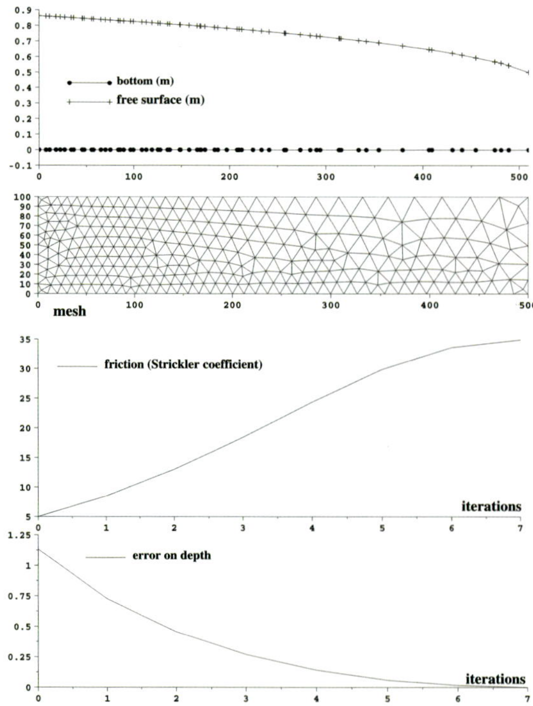

$$-\left(H^{it+1}, \frac{\partial^t CM1^{it+1}}{\partial V_{ip}^{it}} U_{adj}^{it+1}\right) - \left(H^{it+1}, \frac{\partial^t CM2^{it+1}}{\partial V_{ip}^{it}} V_{adj}^{it+1}\right)$$

$$-\left(U^{it+1}, \frac{\partial^t BM1^{it+1}}{\partial V_{ip}^{it}} H_{adj}^{it+1}\right) - \left(U^{it+1}, \frac{\partial^t AM2^{it+1}}{\partial V_{ip}^{it}} U_{adj}^{it+1}\right)$$

$$-\left(V^{it+1}, \frac{\partial^t BM2^{it+1}}{\partial V_{ip}^{it}} H_{adj}^{it+1}\right) - \left(V^{it+1}, \frac{\partial^t AM3^{it+1}}{\partial V_{ip}^{it}} V_{adj}^{it+1}\right) \tag{9.55}$$

Besides the three differences from the measurements, there are 30 terms depending on the matrices and right-hand sides of the direct system which we have to calculate. A first analysis allows us to eliminate 13 of them, in fact:

- $CM1$ and $CM2$ depend neither on H, nor on U, and V (six terms).
- $BM1$ and $BM2$ depend only on H (four terms).
- $AM1$ does not depend on H (one term).
- $CV2$ does not depend on V (one term).
- $CV3$ does not depend on U (one term).

We are left with:

$$(CV1_{adj})_{ip} = 2W_{ip}(H^{it} - H_{mes}^{it})_{ip} + T11 + T12 + T13 + T14 + T15 + T16 + T17 \tag{9.56}$$

$$(CV2_{adj})_{ip} = 2W_{ip}(U^{it} - U_{mes}^{it})_{ip} + T21 + T22 + T23 + T24 + T25 \tag{9.57}$$

$$(CV3_{adj})_{ip} = 2W_{ip}(V^{it} - V_{mes}^{it})_{ip} + T31 + T32 + T33 + T34 + T35 \tag{9.58}$$

by denoting:

$$T11 = \left(H_{adj}^{it+1}, \frac{\partial CV1^{it+1}}{\partial H_{ip}^{it}}\right)$$

$$T12 = \left(U_{adj}^{it+1}, \frac{\partial CV2^{it+1}}{\partial H_{ip}^{it}}\right)$$

$$T13 = \left(V_{adj}^{it+1}, \frac{\partial CV3^{it+1}}{\partial H_{ip}^{it}}\right)$$

$$T14 = -\left(U^{it+1}, \frac{\partial^t BM1^{it+1}}{\partial H_{ip}^{it}} H_{adj}^{it+1}\right)$$

$$T15 = -\left(U^{it+1}, \frac{\partial^t AM2^{it+1}}{\partial H_{ip}^{it}} U_{adj}^{it+1}\right)$$

$$T16 = -\left(V^{it+1}, \frac{\partial^t BM2^{it+1}}{\partial H_{ip}^{it}} H_{adj}^{it+1}\right)$$

$$T17 = -\left(V^{it+1}, \frac{\partial^t AM3^{it+1}}{\partial H_{ip}^{it}} V_{adj}^{it+1}\right)$$

$$T21 = \left(H_{adj}^{it+1}, \frac{\partial CV1^{it+1}}{\partial U_{ip}^{it}}\right)$$

$$T22 = \left(U_{adj}^{it+1}, \frac{\partial CV2^{it+1}}{\partial U_{ip}^{it}}\right)$$

$$T23 = -\left(H^{it+1}, \frac{\partial^t AM1^{it+1}}{\partial U_{ip}^{it}} H_{adj}^{it+1}\right)$$

$$T24 = -\left(U^{it+1}, \frac{\partial^t AM2^{it+1}}{\partial U_{ip}^{it}} U_{adj}^{it+1}\right)$$

$$T25 = -\left(V^{it+1}, \frac{\partial^t AM3^{it+1}}{\partial U_{ip}^{it}} V_{adj}^{it+1}\right)$$

$$T31 = \left(H_{adj}^{it+1}, \frac{\partial CV1^{it+1}}{\partial V_{ip}^{it}}\right)$$

$$T32 = \left(V_{adj}^{it+1}, \frac{\partial CV3^{it+1}}{\partial V_{ip}^{it}}\right)$$

$$T33 = -\left(H^{it+1}, \frac{\partial^t AM1^{it+1}}{\partial V_{ip}^{it}} H_{adj}^{it+1}\right)$$

$$T34 = -\left(U^{it+1}, \frac{\partial^t AM2^{it+1}}{\partial V_{ip}^{it}} U_{adj}^{it+1}\right)$$

$$T35 = -\left(V^{it+1}, \frac{\partial^t AM3^{it+1}}{\partial V_{ip}^{it}} V_{adj}^{it+1}\right)$$

Calculation rules

All these terms have the form: $(\frac{\partial B}{\partial X_{ip}}, P)$ (category 1) or $(\frac{\partial A}{\partial X_{ip}} X, P)$ (category 2).

We shall deal with these two categories successively.

Category 1: $(\frac{\partial B}{\partial X_{ip}}, P^{it+1})$

1.1 Case where B is of the form MX^{it}: the commonest case is that where M is a matrix which does not depend on X^{it} (e.g. a mass or gradient matrix). So we have:

$$\left(\frac{\partial B}{\partial X_{ip}}, P^{it+1}\right) = \left(\frac{\partial(MX^{it})}{\partial X_{ip}}, P^{it+1}\right) = \left(\frac{M\,\partial(X^{it})}{\partial X_{ip}}, P^{it+1}\right) = \left(\frac{\partial(X^{it})}{\partial X_{ip}}, {}^tMP^{it+1}\right)$$

Now $\frac{\partial(X^{it})}{\partial X_{ip}}$ is a vector, all components of which are zero except component ip which is equal to 1. $(\frac{\partial(X^{it})}{\partial X_{ip}}, {}^tMP^{it+1})$ is therefore equal to $({}^tMP^{it+1})_{ip}$. The vector of the right-hand side, whose component ip is $(\frac{\partial B}{\partial X_{ip}^{it}}, P^{it+1})$, will therefore be ${}^tM\,P^{it+1}$.

Conclusion: if M does not depend on X^{it}, a right-hand side MX^{it} of the direct system corresponds to a right-hand side ${}^tM\,P^{it+1}$ of the adjoint system.

Category 2: $(\frac{\partial A}{\partial X_{ip}}X, P)$

2.1 Case where A is a diagonal of general term $f(X_i)$: in this case, which is the case of friction matrices:

$$\left(\frac{\partial A}{\partial X_{ip}}X, P\right) = \frac{\partial f}{\partial X_{ip}}X_{ip}P_{ip}$$

2.2 Case where A is a linear function of X: if we break down X into the form $\sum_{i=1}^{n} X_i\Psi_i$, we notice that the derivative with respect to X_{ip} eliminates all the terms except $X_{ip}\Psi_{ip}$ which after differentiation with respect to X_{ip} gives Ψ_{ip}. The matrix $\frac{\partial A}{\partial X_{ip}}$ is therefore a matrix obtained from A by replacing X by Ψ_{ip}.

Calculation of terms $T11$ to $T35$

Henceforth we consider that calculations are done without sub-iteration for treating non-linear terms, and therefore that $h_{prop} = h^n$, $u_{conv} = u^n$ and $v_{conv} = v^n$. Let us now calculate the 17 terms $T11$ to $T35$:

Term $T11$:

$$(T11)_{ip} = \left(H_{adj}^{it+1}, \frac{\partial CV1^{it+1}}{\partial H_{ip}^{it}}\right)$$

The dependences of h in $CV1$ are:
$\frac{M^h}{\Delta t}H^n$ (case 1.1) which gives:

$$T11\_1 = \frac{M^h}{\Delta t}H_{adj}^{it+1} \tag{9.59}$$

$(\theta_h - 1)(T_x^h + T_y^h)^{it}H^{it}$ (case 1.1) which gives:

$$T11\_2 = (\theta_h - 1)^t(T_x^h + T_y^h)^{it}H_{adj}^{it+1} \tag{9.60}$$

$(\theta_u - 1)(B_x^{uh} U^{it} + B_y^{uh} V^{it})$ (case 2.2) which gives:

$$T11\_3 = (\theta_u - 1)(B_{x\,adj}^{uh} U^{it} + B_{y\,adj}^{uh} V^{it}) \tag{9.61}$$

by denoting $B_{x\,adj}^{uh}$ as the matrix B_x^{uh} where the water depth has been replaced by the adjoint water depth, i.e.:

$$B_{x\,adj}^{uh}(i,j) = -\int_\Omega \Psi_j^u \frac{\partial}{\partial x}(\varphi_i^h\, h_{adj}^{it+1})\, d\Omega \tag{9.62}$$

and also:

$$B_{y\,adj}^{uh}(i,j) = -\int_\Omega \Psi_j^u \frac{\partial}{\partial y}(\varphi_i^h\, h_{adj}^{it+1})\, d\Omega \tag{9.63}$$

$TB1$, equal to $-\int_\Gamma h_{prop} \vec{u}\,.\,\vec{n}\,\varphi_i^{it} d\Gamma$ and simplified here in $-\int_\Gamma h^{it} \vec{u}\,.\,\vec{n}\,\varphi_i^n d\Gamma$, gives a term:

$$(T11\_4)_{ip} = \left(TB1_{adj}^h\right)_{ip} = -\int_\Gamma h_{adj}^{it+1} \vec{u}\,.\,\vec{n}\,\varphi_{ip}^n d\Gamma \tag{9.64}$$

So we have:

$$T11 = \frac{M^h}{\Delta t} H_{adj}^{it+1} + (\theta_h - 1)^t (T_x^h + T_y^h)^{it} H_{adj}^{it+1}$$

$$+(\theta_u - 1)(B_{x\,adj}^{uh} U^n + B_{y\,adj}^{uh} V^n) + TB1_{adj}^h \tag{9.65}$$

Terms $T12$ and $T13$:

$$T12 = \left(U_{adj}^{it+1}, \frac{\partial CV2^{it+1}}{\partial H_{ip}^{it}} \right)$$

The only dependence on h in $CV2$ is the term $-g(1-\theta_h) C_x^{uh} Z_s^n$, equivalent in sensitivity to $-g(1-\theta_h) C_x^{uh} H^n$. We are still in the case 1.1 which indicates that:

$$\begin{aligned} T12 &= -g(1-\theta_h)^t C_x^{uh} U_{adj}^{it+1} = -\frac{(1-\theta_h)^t}{\theta_h} CM1\, U_{adj}^{it+1} \\ &= -\frac{(1-\theta_h)}{\theta_h} BM1_{adj}\, U_{adj}^{it+1} \end{aligned} \tag{9.66}$$

$$\begin{aligned} T13 &= -g(1-\theta_h)^t C_y^{uh} V_{adj}^{it+1} = -\frac{(1-\theta_h)^t}{\theta_h} CM2\, V_{adj}^{it+1} \\ &= -\frac{(1-\theta_h)}{\theta_h} BM2_{adj}\, V_{adj}^{it+1} \end{aligned} \tag{9.67}$$

Terms $T14$ and $T16$:

$$(T14)_{ip} = -\left(U^{it+1}, \frac{\partial^t BM1^{it+1}}{\partial H^{it}_{ip}} H^{it+1}_{adj}\right) = -\left(\frac{\partial BM1^{it+1}}{\partial H^{it}_{ip}} U^{it+1}, H^{it+1}_{adj}\right)$$

We are in the case of category 2.2, where, knowing $BM1$, we have:

$$(T14)_{ip} = -\sum_{i=1}^{n}\sum_{j=1}^{n} -\theta_u \left[\frac{\partial}{\partial h_{ip}}\left(\int_\Omega \Psi^u_j \frac{\partial}{\partial x}(h^{it}\,\varphi^h_i)\,d\Omega\right) u^{it+1}_j h^{it+1}_{adj\ i}\right] \tag{9.68}$$

or even:

$$(T14)_{ip} = \sum_{i=1}^{n}\sum_{j=1}^{n} \theta_u \left[\left(\int_\Omega \Psi^u_j \frac{\partial}{\partial x}(\varphi^h_{ip}\,\varphi^h_i)\,d\Omega\right) u^{it+1}_j h^{it+1}_{adj\ i}\right] \tag{9.69}$$

which finally gives:

$$(T14)_{ip} = \theta_u \int_\Omega u^{it+1} \frac{\partial}{\partial x}(\varphi^h_{ip}\,h^{it+1}_{adj})\,d\Omega \tag{9.70}$$

that is to say:

$$T14 = -\theta_u B^{uh}_{x\,adj} U^{it+1} \tag{9.71}$$

and also:

$$T16 = -\theta_u B^{uh}_{y\,adj} V^{it+1} \tag{9.72}$$

The terms $T14$ and $T16$ can be regrouped with $T11\_3$ because they require the same matrices.

Terms $T15$ and $T17$:

$$(T15)_{ip} = -\left(U^{it+1}, \frac{\partial^t AM2^{it+1}}{\partial H^{it}_{ip}} U^{it+1}_{adj}\right)$$

The dependences of h in $AM2$ come only from friction and thus from the matrix already noted:

$$F^u_{ii} = -\frac{1}{\cos(\alpha)} \frac{C_f}{2h^{it}_i} \sqrt{(u^{it}_i)^2 + (v^{it}_i)^2} \int_\Omega \varphi^u_i d\Omega$$

We are in case 2.1, which gives:

$$(T15)_{ip} = \frac{\partial(F^u_{ii})}{\partial h^{it}_{ip}} u^{it+1}_{adj\ ip}\ u^{it+1}_{ip} = -\frac{1}{h^{it}_{ip}} F^u_{ip\,ip}\, u^{it+1}_{adj\ ip}\ u^{it+1}_{ip}$$

if there is no dependence of C_f in relation to h, which is not the case for the Strickler formula for which we would then have:

$$(T15)_{ip} = -\frac{4}{3h^{it}_{ip}} F^u_{ip\,ip}\, u^{it+1}_{adj\ ip}\ u^{it+1}_{ip}$$

So we have:

$$T15 = -\frac{1}{H^{it}} F^u\, U_{adj}^{it+1}\; U^{it+1} (\text{Chézy}) \text{ or } T15 = -\frac{4}{3H^{it}} F^u\, U_{adj}^{it+1}\; U^{it+1} \;(\text{Strickler}) \tag{9.73}$$

and:

$$T17 = -\frac{1}{H^{it}} F^u\, V_{adj}^{it+1}\; V^{it+1} (\text{Chézy}) \text{ or } T17 = -\frac{4}{3H^{it}} F^u\, V_{adj}^{it+1}\; V^{it+1} \;(\text{Strickler}) \tag{9.74}$$

Terms $T21$ and $T31$:

$$T21 = \left(H_{adj}^{it+1}, \frac{\partial CV1^{it+1}}{\partial U_{ip}^{it}} \right)$$

The dependences of $CV1$ in relation to the velocity are in the transport term $(\theta_h - 1)(T_x^h + T_y^h)^{it} H^{it}$, in the term $(\theta_u - 1)B_x^{uh} U^{it}$ and the boundary term $TB1$. By denoting T_{adj}^x as the matrix of the general term:

$$T_{adj}^x(i,j) = \int_\Omega \frac{\partial h^{it}}{\partial x} \Psi_i^u \varphi_j^h$$

and:

$$T_{adj}^y(i,j) = \int_\Omega \frac{\partial h^{it}}{\partial y} \Psi_i^u \varphi_j^h$$

we find:

$$T21 = (\theta_h - 1) T_{adj}^x H_{adj}^{it+1} + (\theta_u - 1)^t B_x^{uh} H_{adj}^{it+1} + TB1_{adj}^u \tag{9.75}$$

$$T31 = (\theta_h - 1) T_{adj}^y H_{adj}^{it+1} + (\theta_u - 1)^t B_y^{uh} H_{adj}^{it+1} + TB1_{adj}^v \tag{9.76}$$

by denoting:

$$\left(TB1_{adj}^u\right)_{ip} = -\int_\Gamma h_{adj}^{it+1} h^n \Psi_{ip}^u n_x d\Gamma \text{ and } \left(TB1_{adj}^v\right)_{ip} = -\int_\Gamma h_{adj}^{it+1} h^n \Psi_{ip}^u n_y d\Gamma$$

Terms $T22$ and $T32$:

$$T22 = \left(U_{adj}^{it+1}, \frac{\partial CV2^{it+1}}{\partial U_{ip}^{it}} \right)$$

The dependence in relation to the velocity u in $CV2$ originates from:

$$\frac{M^u}{\Delta t} U^{it} \text{ and } (\theta_u - 1)(T_x^u + T_y^u)^{it} U^{it}$$

where $\frac{M^u}{\Delta t} U^{it}$ gives $\frac{M^u}{\Delta t} U_{adj}^{it+1}$ and $(\theta_u - 1)(T_x^u + T_y^u)^{it} U^{it}$ gives a term $(\theta_u - 1)(U_{adj}^{it+1}, \frac{\partial((T_x^u + T_y^u)^{it} U^{it})}{\partial U_{ip}^{it}})$ equal to:

$$(\theta_u - 1)\left(U_{adj}^{it+1}, \frac{\partial(\int_\Omega \overrightarrow{u}^{it}.\overrightarrow{grad}(u^{it})\varphi_i^u d\Omega)}{\partial U_{ip}^{it}}\right)$$

Or even:

$$(\theta_u - 1)\int_\Omega \frac{\partial}{\partial x}(u^{it}\Psi_{ip}^u)U_{adj}^{it+1}d\Omega + (\theta_u - 1)\int_\Omega v^{it}\frac{\partial \Psi_{ip}^u}{\partial y}U_{adj}^{it+1}d\Omega$$

and therefore:

$$T22 = \frac{M^u}{\Delta t}U_{adj}^{it+1} + (\theta_u - 1)\int_\Omega \frac{\partial}{\partial x}(u^{it}\Psi_{ip}^u)U_{adj}^{it+1}d\Omega + (\theta_u - 1)\int_\Omega v^{it}\frac{\partial \Psi_{ip}^u}{\partial y}U_{adj}^{it+1}d\Omega \tag{9.77}$$

We find in the same way that:

$$T32 = \frac{M^u}{\Delta t}V_{adj}^{it+1} + (\theta_u - 1)\int_\Omega \frac{\partial}{\partial y}(v^{it}\Psi_{ip}^u)V_{adj}^{it+1}d\Omega + (\theta_u - 1)\int_\Omega u^{it}\frac{\partial \Psi_{ip}^u}{\partial x}V_{adj}^{it+1}d\Omega \tag{9.78}$$

Terms $T23$ and $T33$:

$$T23 = -\left(H^{it+1}, \frac{\partial^t AM1^{it+1}}{\partial U_{ip}^{it}}H_{adj}^{it+1}\right)$$

The dependence of $AM1$ in relation to u originates only from the transport term $\theta_h(T_x^h + T_y^h)$ and even only $\theta_h T_x^h$ which gives a term:

$$-\theta_h \int_\Omega \Psi_{ip}^u \frac{\partial h^{it+1}}{\partial x}h_{adj}^{it+1}d\Omega$$

equal to $-\theta_h T_{x\,adj}^h H_{adj}^{it+1}$ by denoting:

$T_{x\,adj}^h$ the matrix of general term $\int_\Omega \frac{\partial h^{it+1}}{\partial x}\Psi_i^u\varphi_j^h d\Omega$.

$T_{y\,adj}^h$ the matrix of general term $\int_\Omega \frac{\partial h^{it+1}}{\partial y}\Psi_i^u\varphi_j^h d\Omega$.

We have:

$$T23 = -\theta_h T_{x\,adj}^h H_{adj}^{it+1} \tag{9.79}$$

and:

$$T33 = -\theta_h T_{y\,adj}^h H_{adj}^{it+1} \tag{9.80}$$

Terms $T24$ and $T25$, $T34$ and $T35$:

$$T24 = -\left(U^{it+1}, \frac{\partial^t AM2^{it+1}}{\partial U_{ip}^{it}}U_{adj}^{it+1}\right)$$

The dependences of $AM2$ according to u are the transport matrix $\theta_u T_x^u$ and the friction. We denote:

$T^u_{x\,adj}$ the matrix of general term $\int_\Omega \frac{\partial u^{it+1}}{\partial x} \Psi_i^u \varphi_j^u d\Omega$.
$T^u_{y\,adj}$ the matrix of general term $\int_\Omega \frac{\partial u^{it+1}}{\partial y} \Psi_i^u \varphi_j^u d\Omega$.
$T^v_{x\,adj}$ the matrix of general term $\int_\Omega \frac{\partial v^{it+1}}{\partial x} \Psi_i^u \varphi_j^u d\Omega$.
$T^v_{y\,adj}$ the matrix of general term $\int_\Omega \frac{\partial v^{it+1}}{\partial y} \Psi_i^u \varphi_j^u d\Omega$.
We have:

$$T24 = -\theta_u T^u_{x\,adj} U^{it+1}_{adj} - \frac{U^{it} U^{it+1} U^{it+1}_{adj}}{\left(U^{it}\right)^2 + \left(V^{it}\right)^2} F^u \tag{9.81}$$

$$T25 = -\theta_u T^v_{x\,adj} V^{it+1}_{adj} - \frac{U^{it} V^{it+1} V^{it+1}_{adj}}{\left(U^{it}\right)^2 + \left(V^{it}\right)^2} F^u \tag{9.82}$$

$$T34 = -\theta_u T^u_{y\,adj} U^{it+1}_{adj} - \frac{V^{it} U^{it+1} U^{it+1}_{adj}}{\left(U^{it}\right)^2 + \left(V^{it}\right)^2} F^u \tag{9.83}$$

$$T35 = -\theta_u T^v_{y\,adj} V^{it+1}_{adj} - \frac{V^{it} V^{it+1} V^{it+1}_{adj}}{\left(U^{it}\right)^2 + \left(V^{it}\right)^2} F^u \tag{9.84}$$

9.5 Descent methods for minimizing the cost function

9.5.1 Conjugate gradient

We use the classic algorithm of the conjugate gradient, in which the gradient of the functional we want to minimize intervenes. In the classic case of the resolution of the linear system $AX = B$, this gradient is equal to $AX - B$ for $J = \frac{1}{2}(AX, X) - (B, X)$. Here this gradient is given by the adjoint variables. We have:

Initialization:

$$g_0 = \frac{\partial J}{\partial S^0_{ip}}$$

$$d_0 = g_0$$

$$\rho_0 = -\frac{J}{(d_0, g_0)}$$

Iteration m:

$$g_m = \frac{\partial J}{\partial S^m_{ip}}$$

$$d_m = g_m + \frac{(g_m, g_m)}{(g_{m-1}, g_{m-1})} d_{m-1}$$

$$\rho_m = -\frac{J}{(d_m, g_m)}$$

$$S^{m+1} = S^m - \rho_m d_m$$

9.6 Validation and results

In real cases, there is always a limited number of measurements, much less than the number of unknowns, therefore we cannot hope to identify in a unique way a friction coefficient per degree of freedom. That is why we are restricting ourselves to a limited number of zones (as many as the measurements) on which we assume that the friction coefficient is constant. Test cases with simulated data illustrate the procedure. Plate 24 reports a very simple case of a single Strickler coefficient in a canal with steady state flow. The initial guess is 5, whereas the value used to compute the simulated data is 35. The correct depth and coefficient are found after seven iterations. A more complex case is the flow around bridge piers (see Plate 13) with different Strickler coefficients in four rectangular zones as per Figure 9.1 below.

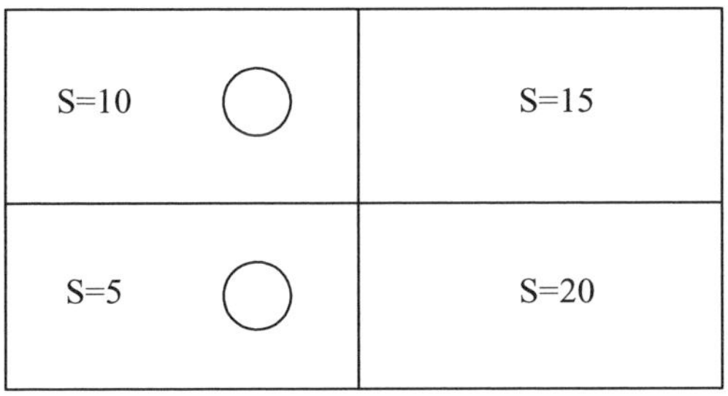

Figure 9.1: Strickler coefficient for each zone

Results of a simulation with such coefficients are kept aside and considered afterwards as data. Starting from a constant friction coefficient of 25, we then try to identify the values used for constituting the data. The four values are found at around 1% in 20 iterations. However, such a result is not a validation of a correct calculation of the gradient of the cost function. A mistake in the magnitude could allow a decrease of J; a deviation in the direction could go unnoticed, the solution being found by describing a spiral. We can approximate the gradient by varying the friction coefficient in one single point and comparing the obtained values of J. We write:

$$\nabla J \approx \frac{J(S + \delta S) - J(S)}{\delta S}$$

If δS is 0.05, the difference with the gradient obtained by the adjoint system is about 1%. A real application of the parameter estimation is presented in the following chapter (Section 10.2).

Chapter 10

Applications

Hofstadter's law: *it always takes longer than you expect, even when you take into account Hofstadter's law.*

10.1 Malpasset dam-break wave

10.1.1 Introduction

The numerical simulation of the Malpasset dam-break wave with Telemac-2D software served as a test case in the European project called PCECOWATER (Parallel Computing of Environment COastal and lake shallow WATER dynamics), in the framework of the ESPRIT programme, between 1996 and 1998. The partners in the project were EDF, its Italian equivalent ENEL, the research centre CRS4 in Sardinia, Reiter Research Bureau Ltd and the electricity producer Kemijoki Oy in Finland. We shall briefly recall the circumstances and consequences of the accident before presenting the numerical simulation, its validation and performances obtained in parallelism in this case. For more details and a list of reference documents, see [84].

10.1.2 The Malpasset dam accident

The Malpasset accident is a rare example of total and instantaneous collapse of an arch dam. The dam with a capacity of 55 million cubic metres had been constructed for irrigation and as a reservoir for drinking water. It was situated in a neck of the Reyran valley (Var department) in the south-east of France, about 12 km from Fréjus. Its height was 66.5 m, crown length 223 m, and thickness varied from 1.55 m to 6.77 m. On 2 December 1959 at 9:14 p.m., after a period of heavy rain that quickly filled the reservoir, the reason for which a fixed wheel gate had been opened, the dam gave way suddenly. There were no witnesses in the vicinity and no abnormal indication was noticed by the engineers who had visited the site some hours before the accident. About 1.5 km away, strong vibrations were felt in the ground and a brief rumbling was heard, followed by a strong wind. Then a first wave descended on the valley, followed by a wall of water. Electricity was cut, and the wave spread over the populated zones, claiming 433 victims. Some 1.5 km downstream, over 800 m of the Esterel highway was damaged and a bridge washed away. Huge boulders were swept away and deposited further downstream. There is almost nothing left of the arch on the site and the downstream morphology has been markedly changed. The river no longer flows on its bed but on an alluvial substratum.

The inquiry showed that the main cause of the accident was an underground fault downstream of the dam. Under the weight of water, the arch became detached from its foundation and turned in one piece on its right side, while the left side collapsed, then the central part. The legal proceedings that followed concluded that the cause of the accident had been unpredictable; however, this would not be the case today, with the geotechnical standards subsequently imposed.

After the accident, a Permanent Technical Committee for Dams, "Comité Technique Permanent des Barrages" (CTPB), was constituted in France. The May 1968 decree stipulates that each dam exceeding 20 m and with a capacity of more than 15 million cubic metres must be inspected by the CTPB. Amongst other measures, a predictive dam-break wave simulation along with a map of flooded zones in case of accident must be furnished. The numerical simulations are mostly carried out in dimension one with Saint-Venant equations. In some cases, when the wave spreads over wide plains or when there is an opening into the sea, a two-dimensional numerical simulation is necessary. This is the reason why the Telemac-2D software is used in EDF. The data collected at the time of the Malpasset accident, as well as a physical model, allow a validation of this kind of calculation.

10.1.3 Available data

The data available for our simulation are as follows:

- Photographs of the dam and the valley after the accident.
- An on-the-spot inquiry by the local police, showing elevations attained on the left and right banks.
- Propagation time of the break wave. In fact, the wave destroyed three transformers on its passage, which will be referred to as A, B and C, the breakdown time of which has been recorded exactly. Transformer A was struck 100 s after the break but this figure may be erroneous because the time of the dam break, around 9:14 p.m., is not known exactly. The transit time between A and B is much better known, 1140 s, and will serve for validation. The wave reached the sea at around 9:50 p.m.
- Topography: recent maps are no longer valid and it was necessary to resort to the IGN (Institut Géographique National) 1:20000 map of Saint-Tropez, number 3, dated 1931. The valley is very narrow downstream of the dam, with two right-angled bends. The valley then widens on draining a few small tributaries, and narrows down again before finally reaching the coastal plain.
- Physical model: in 1964, a physical model on the scale of 1:400 had been prepared at Laboratoire National d'Hydraulique (National Hydraulics Laboratory) to reproduce the accident, and a one-dimensional simulation was attempted. So

we have measurements that were carried out on this model, the calibration of which led to estimate that the Strickler coefficient in the valley was in the range of 30–40; 14 measurement points were collected on free surface elevation in this model, 5 of which were in the reservoir itself. At each measurement point, free surface maximum elevation was noted. These data are put together in Table 10.1, with one- and two-dimensional results. In the case of Telemac-2D, computation is done with a Strickler coefficient of 30, without initial discharge into the river. The x and y coordinates in metres of measurement points are given in brackets.

- One-dimensional calculations: these come from the RUPTUR software [59]. The cross-sections are set up every 100 m. Two Strickler coefficients have been used: 25 upstream of the dam and 35 elsewhere. The results give an average velocity of the wave as 33.3 km/h, to be compared with the observed 27 km/h. This difference is explained by the fact that one-dimensional simulations do not reproduce the slowing-down effect of bends (unless it is integrated artificially in the equations in the form of local head loss). An error of 11 m is noticed at measurement point 7. Points 13 and 14 were not included in the calculation domain.

Table 10.1: measurement of the maximum water level (Max. level) on the physical model and 1D and 2D numerical results

Measurement point	Distance to dam	Bottom	Max. level	1D	2D
6 (4947;4289)	336 m	43.9	84.2	84.49	87.97
7 (5717;4407)	1320 m	34.5	49.1	60.41	54.43
8 (6775;3869)	2160 m	30.0	54.0	52.40	53.25
9 (7128;3162)	3420 m	27.4	40.2	37.68	47.91
10 (8585;3443)	4840 m	23.1	34.9	36.41	36.51
11 (9674;3085)	6000 m	19.1	27.4	29.16	25.37
12 (10939;3044)	7220 m	11.4	21.5	20.46	19.13
13 (11724;2810)	7980 m	09.3	16.1	*	17.65
14 (12723;2485)	8960 m	07.5	12.9	*	12.76

10.1.4 Two-dimensional numerical modelling

The domain has been discretized into 26000 elements (see Plates 15 and 16). The dimensions are 17500 m × 9000 m. The elevation of the bottom goes from −20 m (sea bed) to 100 m, the estimated value of the free surface elevation in the lake, about which there is some uncertainty (however, the tests showed that an error of 50 cm had only 1% effect on the propagation time of the wave).

The dam is modelled by a straight line between the points of coordinates (4701.18 m, 4143.1 m) and (4655.5 m, 4392.10 m). The position of the three transformers A, B and C is (5550 m, 4400 m), (11900 m, 3250 m) and (13000 m, 2700 m) respectively.

The mesh size varies from 5 m (downstream of the dam) to 300 m (sea).

The reservoir is included in the calculations. The initial condition consists of a water body 100 m above sea level upstream of the dam, and water at rest at elevation 0 in the sea, defined by the area where the bottom elevation is negative. At time t=0, there is thus a 55 m high wall of water at the location of the dam. In the first configuration, there is no water in the river at the start of calculation. A series of tests has also been done with a constant discharge of 20 m^3/s and a slight acceleration of the wave is noticed. We know that a gate in the dam was opened but the exact value of discharge coming out of it, between 0 and 40 m^3/s, is not known.

Numerical parameters:

- The continuity equation is dealt with in conservative form, as indicated in Section 6.3. The loss of mass during calculation time is 0.937 litres over a total of 96841720 m^3, or a relative precision of 10^{-11}, to the order of precision of machines.

- The velocity advection stage is carried out with the method of characteristics for scalar machines, and is replaced by the distributive scheme N in parallelism.

- Discretization of the elevation is linear and that of the velocity is quasi-bubble.

- The event lasts 4000 s, and is reproduced by 8000 time steps of 0.5 s. The Courant number based on the velocity $\frac{\sqrt{u^2+v^2}\Delta t}{\Delta x}$, is to the order of 2. A time step of 1 s is also possible but the free surface then shows some oscillations immediately downstream of the dam in the first few seconds.

- The iterative solver chosen is GMRES, with a relative precision of 10^{-4}, a Krylov sub-space dimension of 3 and simple diagonal preconditioning.

Physical parameters:

- The turbulent diffusion coefficient, also including dispersion, has been taken as constant and equal to 1 m^2/s. A sensitivity analysis shows that it has little influence on transit times between the transformers (see Table 10.2). Only a value of 10 m^2/s entails a notable change.

- The friction coefficient is much more significant, especially for propagation time, as we can see in Table 10.3.

Table 10.2: influence of diffusion coefficient with a Strickler coefficient of 30

Diffusion coefficient	Arrival at A	Time from A to B	Time from A to C
Nature	100 s ?	1140.0 s	1320.0 s
0.001 m^2/s	109.5 s	1176.0 s	1318.5 s
0.01 m^2/s	109.5 s	1176.5 s	1319.5 s
0.1 m^2/s	109.5 s	1176.5 s	1320.0 s
1 m^2/s	110.5 s	1176.5 s	1325.0 s
10 m^2/s	110.0 s	1201.5 s	1368.0 s

Table 10.3: influence of Strickler coefficient

Strickler coefficient	Arrival at A	Time from A to B	Time from A to C
Nature	100 s ?	1140.0 s	1320.0 s
25	110.5 s	1286.5 s	1443.0 s
30	110.5 s	1176.5 s	1325.0 s
32,5	110.0 s	1133.5 s	1280.5 s
40	109.5 s	1037.0 s	1183.0 s
100	108.5 s	850.0 s	987.5 s

We observe variations from –25.4% to +12.8% of transit time between A and B. On the other hand, the choice has little influence on arrival time at A. Table 10.4 shows the results obtained with an initial discharge of 20 m^3/s in the river. The differences are 4.6% for the travel time between A and C, and 1% between A and B. This is evidently due to the fact that propagation on a wet bottom is faster. Moreover, certain troughs are already filled with water before the arrival of the wave.

Table 10.4: influence of Strickler coefficient with an initial discharge of 20 cubic metres

Strickler coefficient	Arrival at A	Time from A to B	Time from A to C
Nature	100 s ?	1140.0 s	1320.0 s
30	108.0 s	1165.0 s	1263.5 s
35	107.5 s	1084.0 s	1187.5 s

Still with an initial discharge in the river, the maximum elevations of measurement points are as given in Table 10.5.

Table 10.5: maximum water level at the measurement points

Measurement point	Strickler 30	Strickler 35
6	88.35	89.10
7	54.44	54.21
8	53.26	53.20
9	47.93	47.89
10	36.52	36.38
11	25.38	25.23
12	19.14	18.92
13	17.66	17.57
14	12.76	12.65

The influence of friction on the elevations remains weak, without marked tendency in one direction or the other. A more precise calibration would necessitate varying the Strickler coefficient in space. In the real predictive computations of dam-break waves, this coefficient is in fact fixed by the CTPB, according to a scale which takes into account the nature of the terrain crossed (towns, forests, etc.).

10.1.5 Results

Here we present a case without initial discharge in the river, with a Strickler coefficient of 30. The mesh has 26000 elements, but a finer mesh has also been used to show that a sufficient precision in space was attained. Each triangle of the initial mesh has been divided into four, and time step divided by 2. Then calculation takes 5 hours on a CRAY C98. The arrival time of the wave at A is decreased by 2%; the time between A and C remains unchanged.

Plate 17 shows the velocity field downstream of the dam, 1 minute after the break. The effect of the bend is already noticeable, with a start of recirculation on the right bank. Maximum velocity is 30 m/s. Compared with a theoretical one-dimensional dam break on a flat bottom, which would give 23.2 m/s, this shows the acceleration effect of the slope and the neck. The top of Plate 17 is the result obtained with the mesh of 26000 elements, the bottom with a mesh made of 104000 elements (one vector only out of four is drawn in this case, for clarity). There is no appreciable difference, but in the finer mesh some small eddies are more visible. The water depth and velocities after 20 minutes are shown in Plate 18. In this plate, the mesh is visible if the depth is less than 5 cm. At the place of the two right-angled bends, it is observed that flow does not occupy the whole wet section. For example, it is concentrated on the right bank for the first bend. This configuration is due to a significant transverse slope (4 m between two banks), also noted on the physical model.

Plate 19 represents the water depth and velocities at the time when the wave reaches the sea. For more clarity, only a small proportion of velocity vectors is shown. The model perhaps does not represent reality very well in the coastal zone, for the Nice–Marseilles railway track has not been included in the model, and undoubtedly played a role in concentrating the wave, before being destroyed.

10.1.6 Performances on various machines

Calculation with a mesh of 26000 elements, and 8000 time steps of 0.5 s, has been chosen as a test for evaluating performances of various machines, first in scalar mode, then in parallel mode with domain decomposition.

In scalar mode

Table 10.6 lists the CPU time for nine computers. It is seen that these times are roughly proportional to clock period, except for the CRAY which is a vector computer. We see here the interest in optimization of algorithms with our various matrix storage methods. The Silicon Graphics workstation is another exception because it is a 64-bit machine, which is about twice as fast as a 32-bit machine of the same clock period. PCs having Windows NT seem to perform less well, though it must be pointed out that these results also depend on the compilers used.

In parallel mode

Table 10.7 compares the runs in parallel of an IBM SP II, CRAY T3E, UNIX station-network and a PC network with 1, 2, 4 or 8 sub-domains processed in parallel. The language of communication is MPI on Origin and PVM in all other cases.

Table 10.6: CPU time in scalar mode

Machine	CPU time
CRAY C98 (vector computer, 64 bits)	00 h 37 min
CRAY T3E (1 processor)	04 h 59 min
IBM SP II (1 processor)	05 h 12 min
Silicon Graphics Octane (195 MHz, 64 bits)	05 h 41 min
SUN Sparc (200 MHz)	10 h 08 min
Hewlett Packard C180 (180 MHz)	10 h 10 min
PENTIUM II Windows NT (233 MHz)	12 h 15 min
Hewlett Packard B132 (133 MHz)	13 h 55 min
Hewlett Packard 9000 712/80 (80 MHz)	24 h 00 min

Table 10.7: CPU time in parallel mode

Machine	1 processor	2 processors	4 processors	8 processors
	Time	Speed-up	Speed-up	Speed-up
Origin 2400	11184 s	2.2	5.9	13.9
IBM SP II	18700 s	1.67	3.16	4.96
CRAY T3E	17955 s	1.97	3.73	6.2
HP9000 712/80	86400 s	1.6	2.3	
Pentium 233 MHz	44100 s	1.5		

The speed-up factor decreases slightly with the number of processors (except with the Origin 2400, as we have seen in Section 8.6), because the number of communications increases in a concomitant manner. It is remarkable that at least eight processors of the CRAY T3E are required for competing with the CRAY C98 vector computer. Let us recall here that the time gain due to vectorization of Telemac-2D is a factor of 11. Experiments on the workstations' network or PCs give quite random results (Table 10.7 shows only the best performances). It seems difficult to perform parallelism in an efficient manner if we do not use dedicated networks which remove all the competition from the other users. Higher gains can be obtained if the decrease in memory requirements of a sub-domain in relation to the entire domain removes the necessity of recourse to the virtual memory. A division by 3 of the calculation time could thus be obtained with two workstations.

10.1.7 More recent results

Since 1998, the wave equation has been implemented in Telemac-2D and allowed larger time steps (4 seconds) in this computation. Moreover, in 2002, the simulation could be done in three dimensions with non-hydrostatic Navier–Stokes equations, even with two planes over the vertical, which is to say with only one layer of prismatic elements, between the bottom and the free surface. Results are compared with the Saint-Venant equations in Plate 20. Though the flood is slightly delayed with two planes, there is a remarkable agreement between results in two and three dimensions. Computer times for various configurations are given in Table 10.8.

Table 10.8: recent results on a 750 MHz one-processor machine

Malpasset dam break, 1000 time steps of 4 s	Time on HP C 3700 (750 MHz)
Saint-Venant equations (wave equation)	166 s
Navier–Stokes equations, 2 planes	16 minutes
Navier–Stokes equations, 6 planes	1 hour 55 minutes

10.1.8 Conclusion

This simulation of the Malpasset dam-break wave shows that two-dimensional and even three-dimensional calculations of dam-break waves or propagation of floods over domains of several kilometres are now current industrial applications. The improvement in relation to one-dimensional simulations is obvious from the moment the geometries comprise wide plains or marked bends. The celerity of flood waves or dam breaks is also better respected.

10.2 Tides in the English Channel and the Atlantic Ocean

10.2.1 Estimation with adjoint method

This large tidal model (about 520 km × 950 km) is used to provide boundary conditions for more local studies. It extends from the Bay of Biscay and Iberian Peninsula up to just south of the North Sea. The western border of the domain is situated beyond the continental shelf, roughly towards the 12.5 degree west longitude. The mesh, made of 9414 elements and 5007 nodes, is refined near the coasts, with elements varying between 10 km and 50 km. The system of coordinates is a Mercator projection; therefore the procedure described in Section 2.3.7 is used here. Four tides (i.e. a duration of 181500 s) are simulated starting from a state of rest. On the open boundaries, water elevations of the M2 wave (semi-diurnal wave due to the moon) are imposed. These water elevations are obtained by a Fourier analysis of measurements, to extract the M2 component. The time step is 150 s. The Courant number $\frac{c\Delta t}{\Delta x}$ calculated with the wave celerity is therefore generally more than 1. With a poor relative precision (10^{-3}) some oscillations of the free surface are observed on crossing the continental shelf. These oscillations are removed either by a higher precision (10^{-6}) or by the selection of a quasi-bubble element for the velocity. The diffusion coefficient is set at 0 and the velocity advection is ignored. Plate 21 shows the magnitude and phase of the M2 wave in coloured surfaces. The effect of amplification of the Mont Saint-Michel bay in France and the Severn estuary in England is quite visible. The amphidromic point will be noticed in the North Sea where all the phases converge.

Verification of the model is done with measurements of tidal range (see Table 10.9) and phase of the wave (see Table 10.10) in ports or at the location of a navigation

Table 10.9: magnitude of wave M2

Measurement point	Measurement	Manual calibration	Estimation
Zeebrugge	1.59 m	1.50 m	1.57 m
Calais	2.46 m	2.35 m	2.34 m
Boulogne	2.93 m	3.04 m	2.89 m
Dieppe	3.08 m	3.17 m	3.00 m
Cherbourg	1.87 m	1.96 m	1.95 m
Saint-Malo	3.74 m	3.58 m	3.64 m
Buoy C1	1.29 m	1.29 m	1.29 m
Ouessant	2.06 m	1.96 m	1.97 m

Table 10.10: phase of wave M2 in degrees

Measurement point	Measurement	Manual calibration	Estimation
Zeebrugge	14	21	16
Calais	345	347	345
Boulogne	331	333	334
Dieppe	311	312	316
Cherbourg	228	227	228
Saint-Malo	178	179	181
Buoy C1	103	105	105
Ouessant	111	114	115

beacon. The Zeebrugge port is situated in the North Sea, the navigation beacon C1 in the Atlantic off the Brittany coast.

The Chézy formula is used for simulating friction, and five different coefficients are used and applied to as many distinct zones which form a tessellation of the domain. These zones are given in Plate 22 which also shows the bathymetry and entire meshing.

Two simulations, which differ only by friction coefficients, are compared with measurements: the first one is the result of a manual calibration, the second one uses a set of friction coefficients given by automatic parameter estimation. The two sets of coefficients are listed in Table 10.11. For manual calibration (a study which lasted for about six months) the maximum error is 16 cm (Saint-Malo port) on the magnitudes and 7 degrees on the phases (in Zeebrugge port, the furthest away from the boundary conditions). The mean square deviation is 10.25 cm for the magnitudes and 3 degrees for the phases. With automatic calibration (which requires a hundred calculations, now much less than a day on a workstation), the maximum error is 12 cm (Calais) for the magnitudes and 5 degrees for the phases. The mean square deviation for the magnitudes is 6.8 cm (3 cm gain) and 2.9 degrees for the phases. However, it is necessary to specify that the set of coefficients obtained by automatic calibration depends a lot on initial values, for the method can converge towards local minima. Here the initial value is taken from manual calibration and we find that the friction values are not very far from it, but they provide, however, a considerable improvement for magnitudes.

Table 10.11: friction coefficients in different areas

Zone	Manual parameter estimation	Automatic parameter estimation
1	60	54.3
2	45	47.2
3	90	95.3
4	73	62.2
5	60	76.5

A greater precision, of about 5 cm, can be obtained with a finer meshing of 30000 elements. However, such a precise calibration is just an illusion and a number of factors can be responsible for the error in magnitude varying from 5 to 10 cm. A sensitivity analysis has shown that a variation of 5% of boundary conditions is enough to produce such an error. It is also the order of magnitude of differences obtained when the tide-generating force is added inside the domain.

10.2.2 Estimation with surrogate models, neural networks and genetic algorithms

This section aims at testing techniques that are not based on an adjoint system. The main interest is that the details of the resolution and the process leading from the results (water velocities and depth) to the cost function may be considered as black boxes. Generally these techniques require a large number of iterations and cannot be contemplated in computational fluid dynamics. However, running the model thousands of times may not be necessary if we use a surrogate model that will be substituted for the simulation model. A further advantage of these techniques is that the optimization can be made with respect to several criteria, namely here the tide amplitude and phase at various measurement points. In real-life studies the different criteria may also correspond to several flood conditions. This study was done within the framework of the CLACIC project, dedicated to parameter estimation with knowledge-based methods, which ran from March 2003 to November 2005, and was funded by the French Ministry of Industry and Research as part of the programme "Réseau National des Technologies Logicielles". Other partners were the Centre d'Etudes Maritimes et Fluviales (CETMEF), Commissariat à l'Energie Atomique (CEA), Divergent from the University of Compiègne, Verne-Ader and LaRIA/UPJV from the University of Amiens, and Deltacad consultants. Jean-Marc Martinez, Gilles Arnaud and Michel Dumas from CEA have helped writing this section.

Surrogate models

A neuronal surrogate model is built prior to the optimization stage. We shall review here the major principles of the methodology in which the surrogate model supersedes the model in the search for the best set of coefficients minimizing the deviation between computations and measurements. Various optimization techniques that are inconsistent with the model computer times will thus become suitable. Both "raw force" (i.e. a random search) and genetic algorithms may be used. The surrogate model will provide (for any set of friction coefficients) a good approximation of the simulation results in a very short CPU time. An experimental protocol determines the set of points to which the surrogate model will be fitted so that it will represent the original model as accurately as possible. In the present study, the experimental protocol provides 120 sets of values for the five friction coefficients corresponding to the five areas in the Channel. For each set, the tide simulation provides 16 results, i.e. wave amplitude and phase values as measured in eight harbours. A neuronal surrogate model is derived from these 120 sets of results. Thus, 16 actual functions of the five friction coefficients are ultimately obtained and will make it possible to estimate the eight tidal ranges and the eight phases for other sets of friction coefficients. The simulation examples required for teaching the neuronal networks were specified by a Latin Hypercube Sampling (LHS), i.e. independent random variables, each being evenly distributed within the [20,100] range (corresponding to Strickler friction coefficients). The number of experiments was set to 120 (i.e. 120 Telemac-2D computations) so that the method can be compared with the adjoint method that had also involved 120 computations. The neuronal networks are of the MLP (Multi-Layer Perceptron) type with a single hidden layer. The learning operations were carried out on centred and reduced data (inputs and outputs). It should be pointed out that a "modulo π" processing is applied to the eight phases being computed by Telemac-2D (all the computed phases are then within the $[-\pi,\pi]$ range).

Single- or multiple-criterion problem

Two optimization problems have been solved:

- The first one is a single-criterion problem; the cost function $f1(x)$ is the sum of squared amplitude deviations at the eight measuring points.
- The second one is a dual-criterion problem, additionally aiming at minimizing the sum $f2(x)$ of squared phase deviations.

The first problem is solved with different methods: the former ("raw force") implements a random search that is focussed on the most promising areas in the search domain as the investigation proceeds; the latter uses genetic algorithms. Since the genetic algorithms induce changes in a population of individuals representing the potential solutions to the problem, not the minimum of function $f1$ but the set of points x meeting $f_1(x) < f_1^{\min} + \epsilon$ is searched for; ϵ may be interpreted as an order of magnitude of the error caused by the surrogate model. There will not be a single solution, but a set of points in the vicinity of the solution. From the results of that optimization, it will appear that the topology of that vicinity will provide very useful information about the behaviour of the solutions.

The second problem is solved by means of genetic algorithms. Examining the Pareto front will be of the utmost importance for analysing the problem. The Pareto front is composed of solutions which appear to be the best as regards one of the criteria, among sets of solutions which give the same performance on other criteria.

For the single-criterion case, the following problem is solved.

Find the points x meeting:

$$f_1(x) < f_1^{\min} + \epsilon \quad \text{with } f_1^{\min} = 0.01 \quad \text{and} \quad \epsilon = 0.004$$

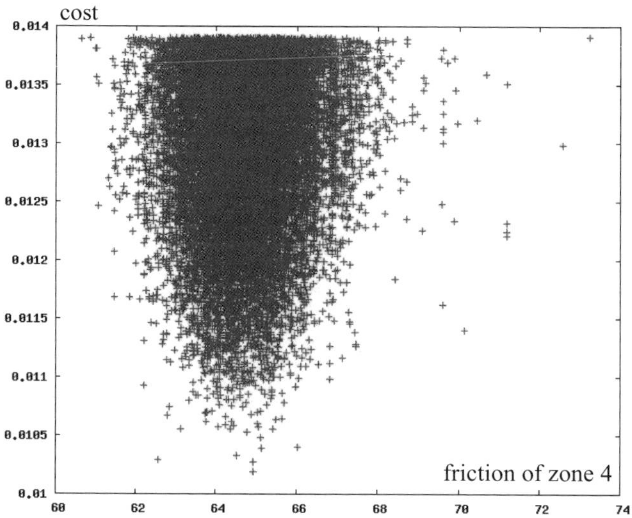

Figure 10.1: cluster of solutions for coefficient of area 4

That problem is classically formulated as an optimization problem under constraint. The solutions to that problem are displayed in the form of scatter plots.

In Figure 10.1, the friction coefficient of zone 4 is the abscissa and the cost is the ordinate. Here the friction coefficient of nearly all solutions is clearly in the range [62, 68]. Other cases, small zones for example, show very scattered results. In yet other cases, a bimodal solution appears, with two possible values. Plots where one coefficient is the abscissa and another the ordinate show possible correlations, as in Figure 10.2 with coefficients of areas 1 and 5. Here we have a negative correlation, high values of one coefficient corresponding to low values of the other.

For the dual-criterion problem, the purpose is now to find a set of friction coefficients which minimizes two functions $f1(x)$ and $f2(x)$, where $f1(x)$ denotes the sum of squared amplitude deviations at all eight measuring points and $f2(x)$ is the sum of squared phase deviations. Figure 10.3 shows the whole Pareto front. Errors on tidal range, in metres, are the abscissa, and errors on phases, in degrees, are the ordinate. The front is split into two parts by a kind of discontinuity. The part on the left, corresponding to the lower tidal range errors, seems to be the one most pertinent to the problem. The error on $f1(x)$ varies by a factor 5. The two criteria are contradictory and the minima of each function are obtained for quite different sets of friction coefficients. On other examples, it appears that the functions exhibit discontinuities and are not monotonic.

Results and analysis

The second and third columns of Table 10.12 give the Strickler friction coefficients obtained respectively with a random search and with genetic algorithms. For the latter, a population of 2000 individuals and 11 generations have been used, which

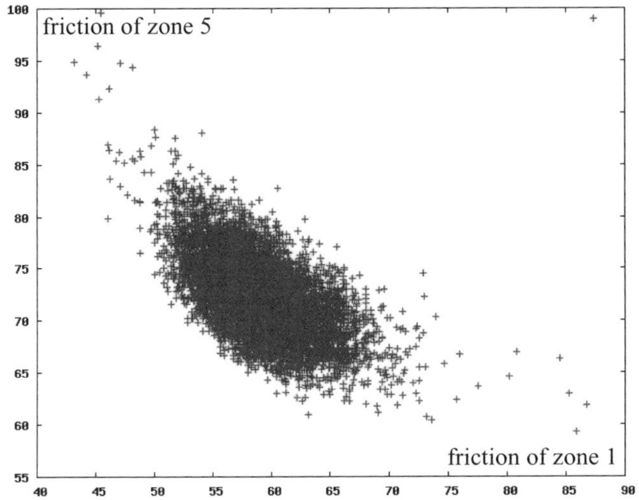

Figure 10.2: correlation between two friction coefficients

implied 29000 assessments.

Table 10.12: friction coefficients obtained with surrogate models.

Zone	Random search	Genetic algorithms
1	45	49.696
2	75	55.393
3	99	97.990
4	65	65.191
5	76	79.796

Table 10.13 and Table 10.14 give the new values obtained for the amplitudes and phases.

It is first observed that the sets of friction coefficients achieved through the various methods are rather different. Thus, it appears that several solutions can be found. We have seen that the standard deviations over amplitudes and phases with the manual setting were 10.25 cm and 3 degrees, respectively, and that the adjoint method with 120 computations brought an improvement, since the deviations shift down to 6.8 cm and 2.89 degrees. In comparison, the surrogate model-based techniques happen to be attractive. As a matter of fact we get:

Reference: expert estimation + adjoint method: 6.8 cm and 2.89 degrees

Random search (several million random selections): 5.9 cm and 2.14 degrees

Genetic algorithms: 5.8 cm and 2.15 degrees

All the results are better than the reference. It seems, after other tests with eight areas not reported here, that the differences among the various methods only exert a

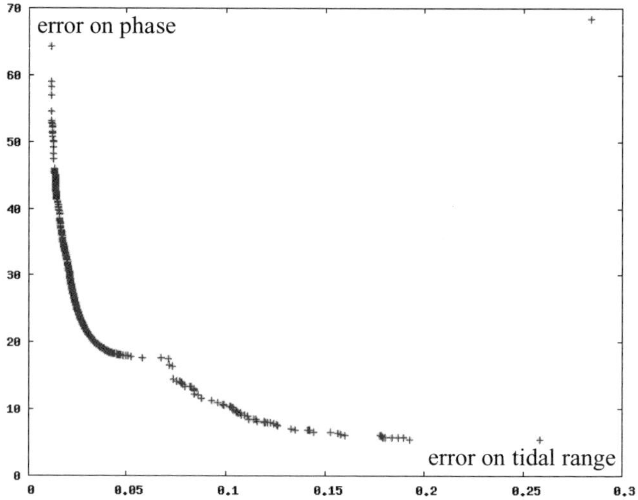

Figure 10.3: Pareto front in a dual-criterion problem

small effect on the accuracy being achieved, though they lead to rather different sets of coefficients, and this once again raises the question of local minima. Shifting up to eight areas or to dual-criterion optimization actually brings little improvement and requires more expertise.

Conclusion and prospects

Those methods based on a minimization from a response surface reveal here that, at the same computational cost, they can provide better results. Finding sets of five coefficients that are better than an optimization involving an expert for six months and a thesis is a very promising achievement. The primary advantage of the new methods being tested here is that they consider the evaluation of the criteria and the computational algorithms as black boxes, whereas the adjoint methods should follow these algorithms as closely as possible, thus heavily burdening both development and maintenance. These new methods, however, also imply a human skill, especially with several criteria, e.g. by displaying and analysing the Pareto fronts. The outcome of that approach will then depend on either the user-friendliness or the required knowledge level for implementing them in real-life conditions.

10.3 Solitary wave

Solitary waves propagating without distortion are noticed at times in the sea or in a canal after the sudden closing of a lock. In 1960, Laitone [120] proposed an approximation to the first order (compared with $\frac{A}{H_0}$, see below) of such a wave, which is a particular limit case of a cnoidal wave. In a canal with a constant depth, along axis

Table 10.13: magnitude of wave M2 obtained with surrogate models.

Measurement point	Measurement	Random search	Genetic algorithm
Zeebrugge	1.59 m	1.63 m	1.66 m
Calais	2.46 m	2.39 m	2.42 m
Boulogne	2.93 m	2.93 m	2.95 m
Dieppe	3.08 m	3.03 m	3.04 m
Cherbourg	1.87 m	1.94 m	1.93 m
Saint-Malo	3.74 m	3.65 m	3.65 m
Buoy C1	1.29 m	1.29 m	1.29 m
Ouessant	2.06 m	1.98 m	1.98 m

Table 10.14: phase of wave M2

Measurement point	Measurement (degrees)	Random search	Genetic algorithm
Zeebrugge	14	14.79	14.63
Calais	345	343.78	343.94
Boulogne	331	332.31	332.84
Dieppe	311	313.49	314.300
Cherbourg	228	226.56	226.95
Saint-Malo	178	175.70	178.18
Buoy C1	103	105.41	105.40
Ouessant	111	114.68	114.77

Ox, this solution is as follows:

$$U(x,t) = \sqrt{gH_0}\frac{A}{H_0}\frac{1}{\cosh(\sqrt{\frac{3A}{4H_0^3}}(x-ct))^2} \tag{10.1}$$

$$W(x,z,t) = \sqrt{3gH_0}\left(\frac{A}{H_0}\right)^{3/2}\frac{z}{H_0}\frac{\tanh(\sqrt{\frac{3A}{4H_0^3}}(x-ct))}{\cosh(\sqrt{\frac{3A}{4H_0^3}}(x-ct))^2} \tag{10.2}$$

$$h(x,t) = H_0 + \frac{A}{\cosh(\sqrt{\frac{3A}{4H_0^3}}(x-ct))^2} \tag{10.3}$$

where H_0 is the reference depth. The magnitude of the wave is A, and its celerity is:

$$c = \sqrt{g(H_0 + A)}$$

Second-order [20] and third-order approximations [180] have also been found since Laitone's first publication.

With a depth of 10 m and a magnitude A of 2 m, we are in the application domain of the Boussinesq equations. On the other hand, the Saint-Venant equations reveal non-physical behaviour and the solitary wave is deformed and tends to break. A series of tests has been carried out in a 600 m long canal. The initial condition represents a solitary wave whose maximum is already 80 m away from the left entry of the canal (i.e. at a distance such that the water depth at the entry is less than 1/100th of its magnitude). The calculation consists of 400 iterations of 0.1 s, after which the wave gets close to the outlet on the right of the canal. The size of elements is about 1 m, which gives a Courant number (calculated from the wave celerity) of the order of 1. The following have been tested:

- The Saint-Venant equations.
- The Boussinesq equations.
- The Navier–Stokes hydrostatic equations.
- The Navier–Stokes non-hydrostatic equations, with various numbers of prismatic element layers.

The failure of the hydrostatic hypothesis is noticed with the Saint-Venant as well as with the Navier–Stokes equations. The Boussinesq equations and the non-hydrostatic Navier–Stokes equations give comparable results. The wave is not exactly solitary and at times leaves in its wake some undulations which must be due to the fact that our initial conditions are only an approximation of a solitary wave, and also to the fact that we have only a first-order approximation.

Plate 12 shows successive profiles of the wave at different times of 0, 10, 20, 30 and 40 seconds with four types of equations. Table 10.15 compares the calculation times with a two-dimensional mesh made of 4210 points and 7206 elements, which is reproduced along the vertical for three-dimensional equations. A remarkable result is the rather good behaviour of the non-hydrostatic equations with only two calculation planes which are the bottom and free surface. In this case the vertical velocity is linear between the bottom and free surface, which suggests an analogy with the Serre equations (Section 2.3.10). Therefore the non-hydrostatic equations appear to be less time consuming than the Boussinesq equations. Given the limitations of the domain of validity of the Boussinesq equations, and even if faster solvers can undoubtedly be constructed from this, it is perhaps wiser to concentrate research henceforth on the non-hydrostatic Navier–Stokes equations rather than to work on the Boussinesq equations with higher and higher orders which are doomed to become obsolete sooner or later due to progress in machines and numerical analysis. Li and Fleming [126], already mentioned, also reached the same conclusions in January 2001, with a finite difference model.

Table 10.15: computation time in the case of a solitary wave

Equations	Computer time
Saint-Venant	451 s
Boussinesq	6114 s
3D hydrostatic, 11 planes	3969 s
3D non-hydrostatic, 11 planes	10176 s
3D non-hydrostatic, 4 planes	5132 s
3D non-hydrostatic, 2 planes	2990 s

Guillou *et al.* [64], at the University of Caen, have also presented this case of a solitary wave as an application of a numerical 2D model with finite differences, using the sigma transform and Chorin projection method for continuity.

10.4 Flow of two fluids of different densities

A rectangular tank 30 m long, 3 m wide and 4 m deep, contains two fluids at rest with slightly different densities because of their salinity: 1 g/l in the left half and 0 g/l in the right half. The respective densities are $\rho_2 = 1000.722$ kg/m^3 and $\rho_1 = 999.972$ kg/m^3 (the water is at 4 degrees Celsius). The flow is governed by density effects. The heavy fluid sinks below the lighter one by forming a saline wedge, which, according to observations, makes an angle of about $\pi/3$ with the bottom. The light fluid spreads over the surface. This case shows the value of a non-hydrostatic simulation which gives a much more physical solution than the front of rectangular shape of the hydrostatic solution. If we consider that the whole potential energy of the initial conditions is transformed into kinetic energy, we can show that the front velocity is 0.5 α, where:

$$\alpha = \sqrt{g\frac{\rho_2 - \rho_1}{\rho_2}h}$$

The measurements show a slight asymmetry: a velocity of 0.47 α for the front at the bottom and 0.59 α on the surface. The cross-sections of Plate 23 correspond to the time t=100 s, i.e. after 100 time steps of 1 second. The mesh is composed of 12 horizontal planes each containing 4500 triangles, in such a way that the mesh size is 0.2 m on the horizontal as well as on the vertical. Here, the theoretical velocity 0.5 α is 0.086 m/s, the average velocity numerically obtained in non-hydrostatic mode being around 0.075 m/s, i.e. an error of about 12%. This underestimated result was predictable because we start from a non-established situation, with zero velocity, which makes the calculation incorrect. On the contrary, we can find the angle of $\pi/3$ between the front and the bottom. The hydrostatic option, with an average velocity of 0.05 m/s and an angle close to 90 degrees, is disqualified for this type of flow. The reader interested in mixtures of fluids of different densities and in the theory of interfacial waves will find a lot of information in references [3], [9], [184] and [193].

Appendix A

Tide-generating force

Nature, unaware of its splendour and glory,
Closed to its past, without desire or memory,
Alien to her creator,
Is nothing more than a pendulum that oscillates,
And strikes in the infinity of time and space,
The rhythm of weight.

Friedrich von Schiller

The difference between the pull exerted by a heavenly body, i.e. the moon or the sun, at the centre of the Earth and the pull at a point on the surface is responsible for the phenomenon of tides. The tide-generating force is extremely weak but is applied over enormous distances. The waves produced are, moreover, deformed and amplified by oceanic basins; thus variations in water levels of 13 m are observed while a static theory would give values not exceeding 1 m. For small domains, such as for example the English Channel, the essence of the tide is due to the waves created in the large oceans. Therefore, for numerical simulations in the English Channel, we have often neglected the tide-generating force inside the domain, to consider only the waves coming from the Atlantic Ocean or from the North Sea. The larger the domain, the less the justification for neglecting the tide-generating force in the interior of this domain. Taking into account this force in the Saint-Venant equations then becomes indispensable.

The tide-generating force derives from a potential that can be separated into a static part and a dynamic part. The dynamic part depends on the state of the tide and cannot simply be taken into account. The static part itself is split into:

- An oceanic static tide (born directly from the influence of the moon and the sun).
- A terrestrial tide, from deformation of the Earth under the influence of the tide.
- Static disturbances due to the influence of the terrestrial tide on the distribution of water bodies.

Hydrodynamics of Free Surface Flows: Modelling with the Finite Element Method J.-M. Hervouet

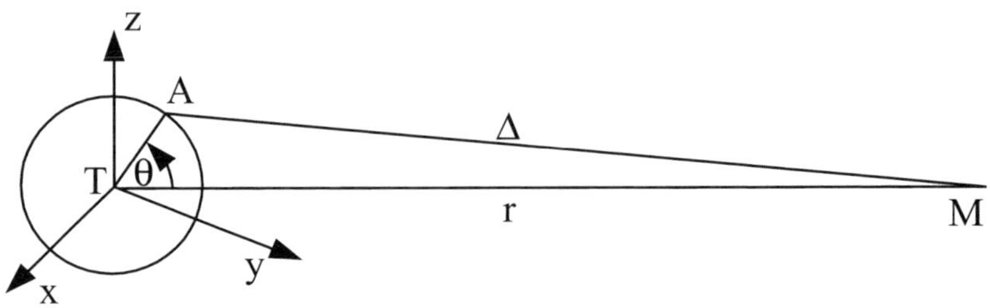

Figure A.1: zenith distance

The last two terms result from the first one in a linear manner with the help of Love numbers [130]. If F_v is the oceanic static tide, $-h_2F_v$ is the force due to the terrestrial tide ($h_2 = 0.61$). The force due to the static disturbances is k_2F_v ($k_2 = 0.30$). The total static force therefore is $0.69F_v$.

The purpose of this appendix is to show all the data, formulae, reasoning and approximations that allow us to carry out properly the calculation of the oceanic static tide. First, the general expression of forces of gravity is established for any heavenly body depending on the mass, distance from the Earth, declination and right ascension of the heavenly body. These four parameters are given in detail later on for the moon and the sun.

A.1 Oceanic static tide

A.1.1 Potential depending on the geocentric zenith distance

Exact expression

Let us suppose that we are on a plane containing the centre of the Earth T, a point on its surface A, and a heavenly body M (for Moon). The angle $(\overrightarrow{\text{TA}}, \overrightarrow{\text{TM}})$, denoted θ, is the zenith distance.

We denote Δ the distance AM and r the distance TM (see Figure A.1). a is the radius of the Earth.

A referential is selected (T, x, y, z) linked to the Earth where x and y are on the equator and z on the axis joining the centre of the Earth and the North Pole. The coordinates of A are (x_A, y_A, z_A). The coordinates of M are (x_M, y_M, z_M).

Acceleration of A is due to:

- Acceleration of T.
- Centrifugal force.
- Gravity of the Earth.
- Gravity of the heavenly body.

The tide-generating force is by definition the force acting on A and depending on M, i.e.:

$$\overrightarrow{F} = \frac{Gm_M}{\Delta^2}.\frac{\overrightarrow{\mathrm{AM}}}{\left\|\overrightarrow{\mathrm{AM}}\right\|} - \frac{Gm_M}{r^2}.\frac{\overrightarrow{\mathrm{TM}}}{\left\|\overrightarrow{\mathrm{TM}}\right\|} = \overrightarrow{F}_2 + \overrightarrow{F}_1 \tag{A.1}$$

where G is the constant of universal gravitation (equal to $6.6710^{-11}\mathrm{m}^2/\mathrm{s}^2$) and m_M the mass of the heavenly body.

G can be calculated from the terrestrial gravitation g since $g = \frac{Gm_T}{a^2}$ where m_T is the mass of the Earth.

The forces F_1 and F_2 each derive from a potential V_1 and V_2, in such a way that:

$$\overrightarrow{F} = \overrightarrow{grad}(V_1 + V_2) \tag{A.2}$$

Given (m_1, m_2, m_3) as the coordinates of the vector $\frac{\overrightarrow{\mathrm{TM}}}{\left\|\overrightarrow{\mathrm{TM}}\right\|}$. Given (p_1, p_2, p_3) as the coordinates of the vector $\frac{\overrightarrow{\mathrm{TA}}}{\left\|\overrightarrow{\mathrm{TA}}\right\|}$.

Potential V_1 at point A is:

$$V_1(x_A, y_A, z_A) = -\frac{Gm_M}{r^2}(m_1 x_A + m_2 y_A + m_3 z_A) \tag{A.3}$$

Now, $x_A = a\,p_1$, $y_A = a\,p_2$, $z_A = a\,p_3$, so:

$$V_1 = -\frac{Gm_M}{r^2}a(p_1 m_1 + p_2 m_2 + p_3 m_3) = -\frac{Gm_M}{r^2}a\cos(\theta) \tag{A.4}$$

the distance Δ is equal to:

$$\sqrt{(x_A - x_M)^2 + (y_A - y_M)^2 + (z_A - z_M)^2}$$

and one can easily conclude from this that potential V_1 is equal to $\frac{Gm_M}{\Delta}$. Therefore the potential of the generating force is:

$$V = Gm_M\left[\frac{1}{\Delta} - \frac{a}{r^2}\cos(\theta)\right] \tag{A.5}$$

with $\Delta^2 = r^2 + a^2 - 2ar\cos(\theta)$, which gives:

$$V = \frac{Gm_M}{r}\left[\frac{1}{\sqrt{1 + \frac{a^2}{r^2} - 2\frac{a}{r}\cos(\theta)}} - \frac{a}{r}\cos(\theta)\right] \tag{A.6}$$

Developed expression

Since the radius of the Earth is small in relation to the distance of the heavenly body, the development is done in relation to $\frac{a}{r}$.

If $a < r$ we can write with Legendre polynomials:

$$\frac{1}{\sqrt{1 + \frac{a^2}{r^2} - 2\frac{a}{r}\cos(\theta)}} = \sum_{n=0}^{+\infty}\left(\frac{a}{r}\right)^n P_n(\cos(\theta)) \tag{A.7}$$

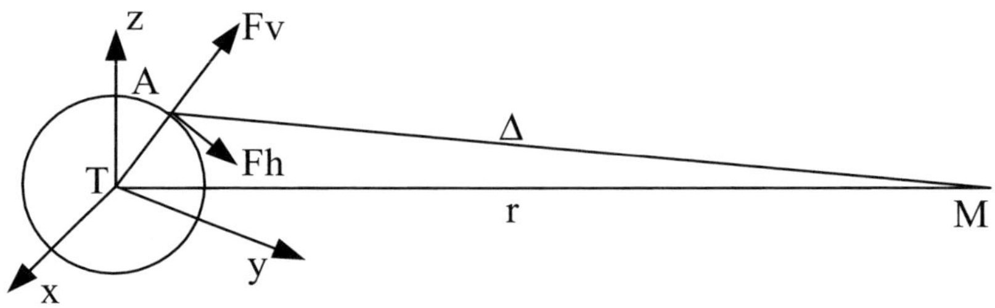

Figure A.2: horizonthal and vertical forces

with $P_0(x) = 1$, $P_1(x) = x$, $P_2(x) = \frac{3x^2-1}{2}$ and $P_3(x) = \frac{5x^3-3x}{2}$.

The recursion formula is:

$$P_{n+1}(x) = \frac{2n+1}{n+1} P_n(x) - \frac{n}{n+1} P_{n-1}(x) \tag{A.8}$$

Thus we find, stopping at the term in $\left(\frac{a}{r}\right)^3$:

$$V = \frac{Gm_M}{r} + \frac{1}{2}\frac{Gm_M}{r}\left(\frac{a}{r}\right)^2 (3\cos(\theta)^2 - 1)$$

$$+\frac{1}{2}\frac{Gm_M}{r}\left(\frac{a}{r}\right)^3 (5\cos(\theta)^3 - 3\cos(\theta)) \tag{A.9}$$

The developed expression generally chosen is:

$$V = \frac{1}{2}\frac{Gm_M}{r}\left(\frac{a}{r}\right)^2 (3\cos(\theta)^2 - 1) \tag{A.10}$$

We do not take into account the term $\frac{Gm_M}{r}$ which is independent of the position of A (the definition of V includes an arbitrary constant; we use only its derivative) and we ignore the term in $\left(\frac{a}{r}\right)^3$ responsible for a diurnal wave and third diurnal.

Vertical and horizontal components of the force, magnitude

In the TAM plane, the vertical force F_v is $\frac{\partial V}{\partial a}$ and the horizontal force F_h is $-\frac{1}{a}\frac{\partial V}{\partial \theta}$ (Figure A.2).

Thus:

Vertical force, exact expression:

$$F_v = \frac{Gm_M}{r^2}\left[\left(1 + \frac{a^2}{r^2} - 2\frac{a}{r}\cos(\theta)\right)^{-3/2}\left(\cos(\theta) - \frac{a}{r}\right) - \cos(\theta)\right] \tag{A.11}$$

Vertical force, developed expression:

$$F_v = \frac{Gm_M}{r^2}\frac{a}{r}(3\cos(\theta)^2 - 1) \tag{A.12}$$

Horizontal force, exact expression:

$$F_h = \frac{Gm_M}{r^2}\sin(\theta)\left[\left(1 + \frac{a^2}{r^2} - 2\frac{a}{r}\cos(\theta)\right)^{-3/2} - 1\right] \tag{A.13}$$

Horizontal force, developed expression:

$$F_h = \frac{3}{2}\frac{Gm_M}{r^2}\frac{a}{r}\sin(2\theta) \tag{A.14}$$

This is the horizontal component of the force that is the cause of the tide. The vertical component is in fact completely masked by gravity (it is about 10 million times weaker than gravity).

A.1.2 Potential depending on the latitude, declination and hour angle

Expression for the geocentric zenith distance

The geocentric zenith distance is the simplest parameter for establishing potential. We must now express it depending on more easily accessible angles.

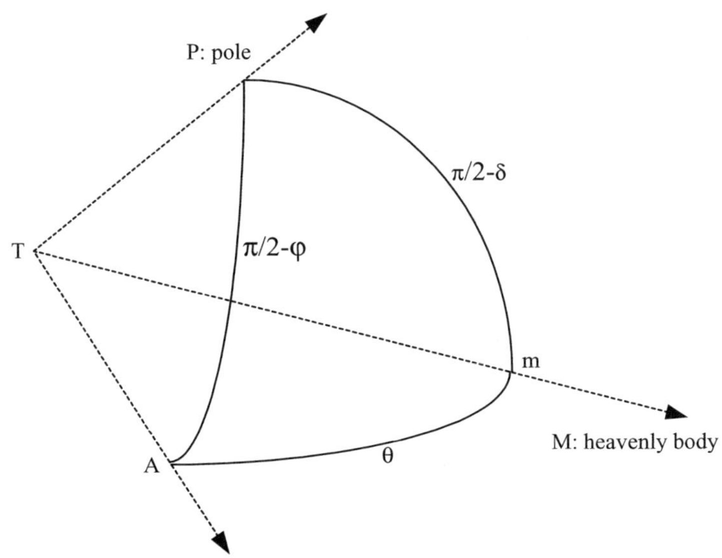

Figure A.3: declination

In Figure A.3, the points A, P and m belong to the surface of the Earth, and the spherical triangle APm has sides:

AP: angle $\frac{\pi}{2} - \varphi$ where φ is the latitude of A.

mP: angle $\frac{\pi}{2} - \delta$ where δ is the declination of the heavenly body (equivalent to a latitude).

Am: angle θ, the geocentric zenith distance.

δ varies from $-23°$ to $+23°$ approximately for the sun and from $-28°$ to $+28°$ approximately for the moon.

Hour angle AH is defined on a projection on the equator (see Figure A.4).

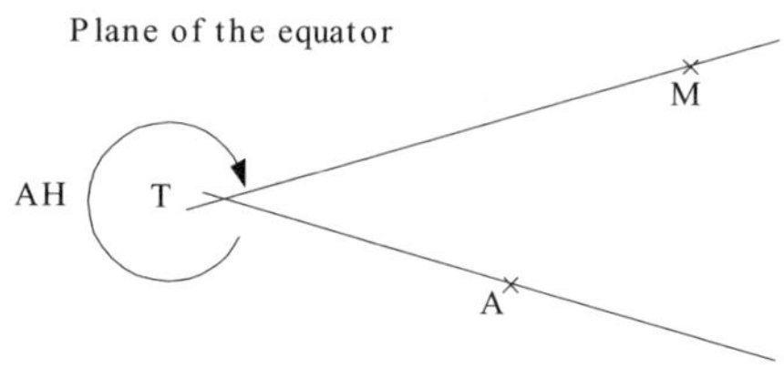

Figure A.4: hour angle

We shall see in Section A.1.3 that this hour angle AH is equal to $t_{loc} - \alpha$ where t_{loc} is the local sidereal time of place A and α the right ascension of the heavenly body (see also Figure A.6).

For the triangle APm, spherical trigonometry thus gives us:

$$\cos(\theta) = \sin(\varphi)\sin(\delta) + \cos(\varphi)\cos(\delta)\cos(AH) \tag{A.15}$$

Let us recall that for a spherical triangle with vertices A, B, C, where the sides a, b and c (see Figure A.5) are defined by angles (e.g. c is the angle AOB where O is the centre of the sphere), we have:

Sine law:

$$\frac{\sin(a)}{\sin(A)} = \frac{\sin(b)}{\sin(B)} = \frac{\sin(c)}{\sin(C)} \tag{A.16}$$

Cosine law:

$$\cos(a) = \cos(b)\cos(c) + \sin(b)\sin(c)\cos(A) \tag{A.17}$$

$$\cos(A) = -\cos(B)\cos(C) + \sin(B)\sin(C)\cos(a) \tag{A.18}$$

New formulae for the potential, wave M2

By replacing $\cos(\theta)$ by its new expression in the formula for the potential, we would obtain a new exact formula which is not written here. Complete writing of the developed

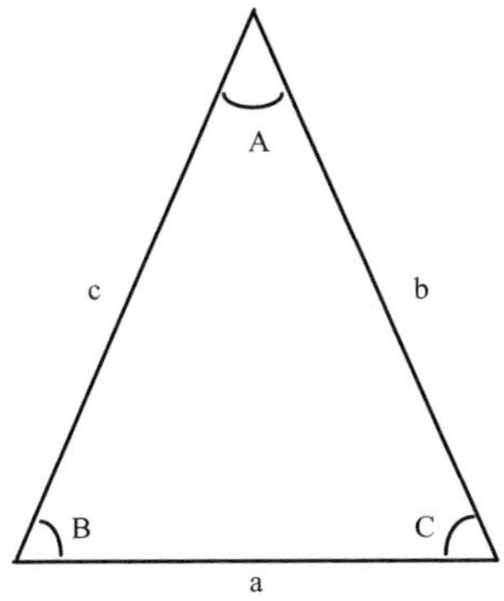

Figure A.5: spherical triangle

expression, on the other hand, allows us to show several types of tidal components. We find:

$$V = \frac{3}{4}\frac{Gm_M}{r}\left(\frac{a}{r}\right)^2\left[\frac{(1-3\sin(\varphi)^2)(1-3\sin(\delta)^2)}{3}\right]$$
$$+\frac{3}{4}\frac{Gm_M}{r}\left(\frac{a}{r}\right)^2[\sin(2\varphi)\sin(2\delta)\cos(\mathrm{AH})]$$
$$+\frac{3}{4}\frac{Gm_M}{r}\left(\frac{a}{r}\right)^2\left[\cos(\varphi)^2\cos(\delta)^2\cos(2\mathrm{AH})\right] \qquad (A.19)$$

The terms in cos(AH) and cos(2AH) are diurnal and semi-diurnal waves respectively. The semi-diurnal wave is called the M2 wave, and the value of its potential is:

$$V = \frac{3}{4}\frac{Gm_M}{r}\left(\frac{a}{r}\right)^2\left[\cos(\varphi)^2\cos(\delta)^2\cos(2\mathrm{AH})\right] \qquad (A.20)$$

Components of the generating force

The vertical component has previously been given; here is the part of this force due to the M2 wave:

$$F_v = \frac{3}{2}\frac{Gm_M}{r^2}\frac{a}{r}\cos(\varphi)^2\cos(\delta)^2\cos(2\mathrm{AH}) \qquad (A.21)$$

On the other hand, the form given for the horizontal component does not allow us to relate it to a known referential, which we are going to do now. Actually, the west–east and south–north components of the horizontal force are derivatives respectively of the potential in relation to the parallel arc and meridian.

West–east component: $F_x = -\frac{1}{a\cos(\varphi)}\frac{\partial V}{\partial AH}$ (minus sign for AH is directed towards the east–west direction).

South–north component: $F_y = \frac{1}{a}\frac{\partial V}{\partial \varphi}$.

Exact formula:

$$F_x = \frac{Gm_M}{r^2}\cos(\delta)\sin(\mathrm{AH})\, aux \tag{A.22}$$

$$F_y = \frac{Gm_M}{r^2}(\cos(\varphi)\sin(\delta) - \sin(\varphi)\cos(\delta)\cos(\mathrm{AH}))\, aux \tag{A.23}$$

where:

$$aux = \left\{\left[1 - 2\frac{a}{r}(\sin(\varphi)\sin(\delta) + \cos(\varphi)\cos(\delta)\cos(\mathrm{AH})) + \left(\frac{a}{r}\right)^2\right]^{-3/2} - 1\right\} \tag{A.24}$$

Developed form:

$$F_x = \frac{3}{2}\frac{Gm_M}{r^2}\frac{a}{r}\left[\sin(\varphi)\sin(2\delta)\sin(\mathrm{AH}) + \cos(\varphi)\cos(\delta)^2\sin(2\mathrm{AH})\right] \tag{A.25}$$

$$\begin{aligned} F_y = &-\frac{3}{4}\frac{Gm_M}{r^2}\frac{a}{r}\left[\sin(2\varphi)(1 - 3\sin(\delta)^2)\right] \\ &+\frac{3}{4}\frac{Gm_M}{r^2}\frac{a}{r}\left[2\cos(2\varphi)\sin(2\delta)\cos(\mathrm{AH})\right] \\ &-\frac{3}{4}\frac{Gm_M}{r^2}\frac{a}{r}\left[\sin(2\varphi)\cos(\delta)^2\cos(2\mathrm{AH})\right] \end{aligned} \tag{A.26}$$

M2 wave:

$$F_x = \frac{3}{2}\frac{Gm_M}{r^2}\frac{a}{r}\left[\cos(\varphi)\cos(\delta)^2\sin(2\mathrm{AH})\right] \tag{A.27}$$

$$F_y = -\frac{3}{4}\frac{Gm_M}{r^2}\frac{a}{r}\left[\sin(2\varphi)\cos(\delta)^2\cos(2\mathrm{AH})\right] \tag{A.28}$$

Example of a fixed fictitious moon in the plane of the equator

Let us calculate, in such a case, the three components of the force of the M2 wave. We have $\delta = 0$ and AH$= \omega t + \varphi_o$ where ω is the rotation of the Earth and φ_o the hour angle at $t = 0$. Let us take $\varphi_o = 0$; we get:

$$F_v = F\cos(\varphi)^2\cos(2\omega t) \tag{A.29}$$

$$F_x = F\cos(\varphi)\sin(2\omega t) \tag{A.30}$$

$$F_y = -F\sin(\varphi)\cos(\varphi)\cos(2\omega t) \tag{A.31}$$

with $F = \frac{3}{2}\frac{Gm_M}{r^2}\frac{a}{r}$, which we can also put in a form where the terrestrial gravity g appears:

$$F = \frac{3}{2}g\frac{m_M}{m_T}\left(\frac{a}{r}\right)^3 \tag{A.32}$$

with $\frac{m_M}{m_T} = \frac{1}{81.53}$ and $\frac{a}{r} = \frac{1}{60}$ from which $F \approx 8.36\,10^{-7}$.

The magnitude of the horizontal force of the M2 wave is:

$$F\cos(\varphi)\sqrt{\sin(2\omega t)^2 + \sin(\varphi)^2\cos(2\omega t)^2} \tag{A.33}$$

which gives at $t = 0$ and for $\varphi = \pi/4$ a force equal to $\frac{F}{2}$. If we are interested in total force, which is calculated with the help of exact formulae, we find:

$$F_x = \frac{Gm_M}{r^2}\sin(\omega t)\left\{\left[1 - 2\frac{a}{r}\cos(\varphi)\cos(\omega t) + \left(\frac{a}{r}\right)^2\right]^{-3/2} - 1\right\} \tag{A.34}$$

$$F_y = -\frac{Gm_M}{r^2}\sin(\varphi)\cos(\omega t)\left\{\left[1 - 2\frac{a}{r}\cos(\varphi)\cos(\omega t) + \left(\frac{a}{r}\right)^2\right]^{-3/2} - 1\right\} \tag{A.35}$$

whose magnitude is:

$$\frac{2}{3}\frac{r}{a}\left\{\left[1 - 2\frac{a}{r}\cos(\varphi)\cos(\omega t) + \left(\frac{a}{r}\right)^2\right]^{-3/2} - 1\right\} F\sqrt{\sin(\omega t)^2 + \sin(\varphi)^2\cos(\omega t)^2} \tag{A.36}$$

at t=0 and for $\varphi = \pi/4$, we get a magnitude equal to F, i.e. twice the force due to the M2 wave alone.

A.1.3 Decomposition of hour angle, sidereal time

Decomposition of hour angle

The hour angle which appears in the formulae given till now depends both on the place and on the heavenly body. Figure A.6 shows the decomposition into local sidereal time and the right ascension of the heavenly body.

The point γ is the vernal point, situated on the line at the intersection between the ecliptic plane and equator plane.

Figure A.6 shows that AH $= t_{loc} - \alpha$: local sidereal time − right ascension.

Right ascension depends only on the relative movement of the Earth and the heavenly body. It will be given in Sections A.2 and A.3 for lunar and solar movement. Now we have to connect local sidereal time to civil time.

Local sidereal time – civil time

From a date and time at a given place, the series of transformations which leads to local sidereal time is as follows:

- Passage to universal time UT. French time, for example, is UT + 1 in winter and UT + 2 in summer.
- Calculation of Julian day JD corresponding to time 0 hour of the same date (see following subsection).

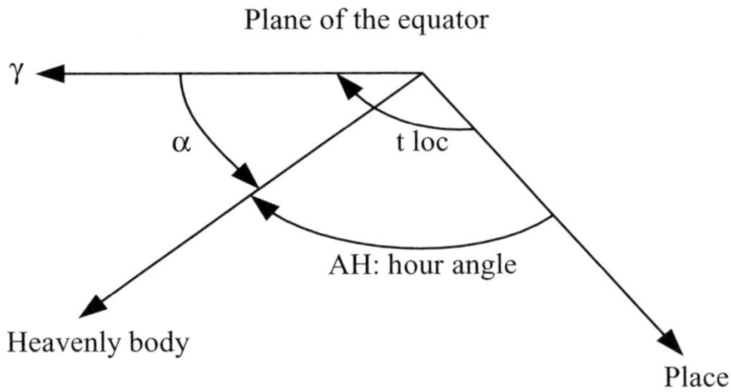

Figure A.6: right ascension and hour angle

- Calculation in Julian centuries, of T the time passed since 31 December 1899 at noon (this time serves as a reference in most astronomical calculations):

$$T = \frac{\mathrm{JD} - 2415020}{36525}$$

- Calculation of sidereal time (in hours) in Greenwich at hour 0 ([140], p. 39), θ_0:

$$\theta_0 = 6.6460656 + 2400.051262\,T + 0.00002581\,T^2 \tag{A.37}$$

  This time must be brought to a value between 0 and 24 if necessary.

- Calculation of sidereal time at Greenwich on hour UT ([140], p. 40):

$$\theta = \theta_0 + 1.002737908\,\mathrm{UT} \tag{A.38}$$

  The coefficient multiplying UT is the ratio between 24 hours and the duration of a rotation of the Earth around its own axis.

- Calculation of local sidereal time in radians:

$$t = L + \theta\frac{2\pi}{24} \tag{A.39}$$

  Here L is here the longitude of the place in radians.

  Now it remains to calculate the Julian day corresponding to hour 0 of a date from the Gregorian calendar.

Julian calendar

The Gregorian calendar, with its leap years and reforms, is not regular enough to be utilized in astronomy. Astronomers use the notion of Julian day, a period of 24 hours starting at noon in our calendar. The starting point for counting Julian days has been fixed at a date going back more than 6000 years, in such a way that 23 May 1968 at noon corresponds to Julian day 2440000. The difficulty in changing to Julian day comes from the irregularities of the Gregorian calendar. The following FORTRAN function (drawn from [158]) returns the Julian day JULDAY to noon of the day ID, of the month MM and the year IYYY of the Gregorian calendar. To obtain the Julian day JD which corresponds to hour 0, we must then deduct 0.5.

```
FUNCTION JULDAY(MM,ID,IYYY)
PARAMETER (IGREG=15+31*(10+12*1582))
IF (IYYY.EQ.0) STOP !There is no Year Zero
IF (IYYY.LT.0) IYYY=IYYY+1
IF (MM.GT.2) THEN
    JY=IYYY
    JM=MM+1
ELSE
    JY=IYYY-1
    JM=MM+13
ENDIF
JULDAY=INT(365.25*JY)+INT(30.6001*JM)+ID+1720995
IF (ID+31*(MM+12*IYYY).GE.IGREG) THEN
    JA=INT(0.01*JY)
    JULDAY=JULDAY+2-JA+INT(0.25*JA)
ENDIF
RETURN
END
```

Some rules of the Gregorian calendar (adopted on 15 October 1582) are:

- No year 0: we go from -1 to $+1$.
- Years multiples of 4 are leap years, except when they can be divided by 100 and not by 400. Hence, 1900 was not a leap year, 2000 was.

A.2 Movement of the moon

In the formulae giving the tide-generating force, a heavenly body intervenes through four parameters:

- mass,
- distance to the Earth,
- declination,
- right ascension.

Here we try not to give full explanations but simply furnish calculation formulae.

A.2.1 Mass of the moon

The ratio of the mass of the moon to the mass of the Earth is:

$$\frac{m_M}{m_T} = \frac{1}{81.53}$$

A.2.2 Distance to the Earth, longitude of the moon in its orbit

The moon describes an elliptical orbit of semi-major axis $c = 384403$ km and eccentricity $e = 0.05490$.

Wherever the term a/r comes into the formulae of the potential, we can replace it by $\frac{a}{c}\frac{c}{r}$. This term c/r (semi-major axis of the lunar orbit divided by the distance between the Earth and the moon) will be given hereafter by developing it according to the powers of e, the eccentricity. The situation is complicated because it is necessary to consider (see Figure A.7) three planes: the plane of the equator, plane of lunar orbit, plane of the ecliptic.

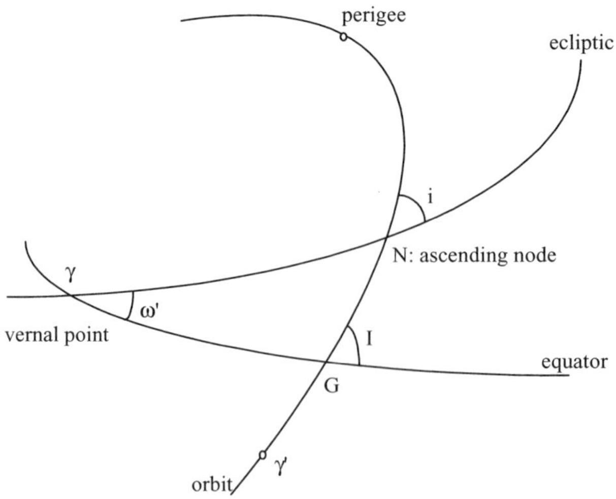

Figure A.7: movement of the Moon

Perigee is the point of lunar orbit where the moon is closest to the Earth.

The position of the moon in its orbit is measured from a fictitious point γ' in such a way that the angles γN and $\gamma' N$ are equal.

The moon forms with the point γ' an angle (measured from the centre of the Earth) called s, the mean longitude of the moon in its orbit. The mean longitude of perigee is denoted p. s and p in degrees ([140], p. 148 and [172], p. 162) are given as functions of T, the number of Julian centuries passed since 31 December 1899 at noon (see Section A.1.3):

$$s = 270.434164 + 481267.8831\,T - 0.001133\,T^2 + 0.0000019\,T^3 \tag{A.40}$$

$$p = 334.328019444 + 4069.03220556\,T - 0.01034\,T^2 - 0.0000125\,T^3 \tag{A.41}$$

In fact the mean longitude of the moon, as its name indicates, is not the real longitude denoted L. Figure A.8 shows the difference in a very exaggerated manner. The notion of the mean moon is a geometric artifice. As a first approximation, its movement is regular around O, the centre of the circle of radius c, while the moon describes an ellipse of which Earth occupies one of the foci.

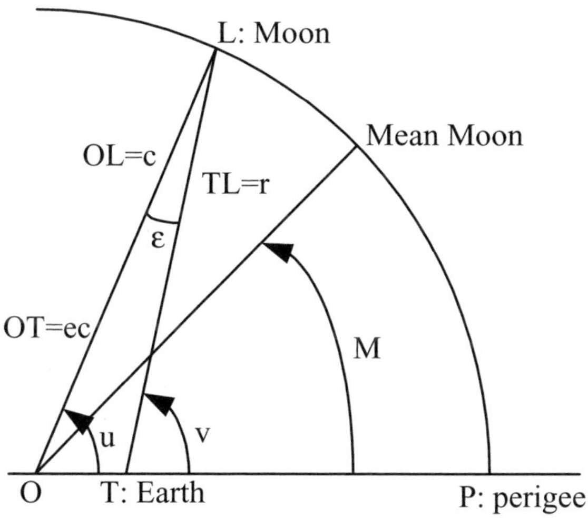

Figure A.8: mean longitude of the moon, real longitude

The longitude L is written as $p + v$ where v is the real anomaly.

M is the mean anomaly, u is the eccentric anomaly.

Kepler's equation $u - e\ \sin(u) = M$, allows us to pass from u to M. This equation is inferred from Kepler's laws. By simple geometric considerations and by ignoring the higher powers of e, we find:

$$v = M + 2e\sin(u) \tag{A.42}$$

Now $M = s - p$ and $u = M + e\sin(u)$.

Finally we find:

$$L = s + 2e\sin(s - p) + \frac{5}{4}e^2\sin(2(s - p)) \tag{A.43}$$

In the triangle OTL of Figure A.8 we have:

$$c = r\cos(e) + ec\cos(u) \tag{A.44}$$

and to second order $\cos(e) \approx 1$, hence $r = c(1 - e\cos(u))$, from which we infer:

$$\frac{c}{r} = 1 + e\cos(s-p) + e^2\cos(2(s-p)) \tag{A.45}$$

To consider the influence of the sun (evection + variation) it is appropriate to add to L the quantity:

$$\frac{15}{4}me\sin(s-2h+p) + \frac{11}{8}m^2\sin(2(s-h)) \tag{A.46}$$

and to $\frac{c}{r}$ the quantity:

$$\frac{15}{8}me\cos(s-2h+p) + m^2\cos(2(s-h)) \tag{A.47}$$

where:

m is the ratio of the average movement of the sun to the average movement of the moon, equal to 0.074804.

h is the mean longitude of the moon in its orbit ([140], p. 79):

$$h = 279.69668 + 36000.76892\,T + 0.0003025\,T^2 \tag{A.48}$$

Hence we must take:

$$L = s + 2e\sin(s-p) + \frac{5}{4}e^2\sin(2(s-p))$$

$$+\frac{15}{4}me\sin(s-2h+p) + \frac{11}{8}m^2\sin(2(s-h)) \tag{A.49}$$

$$\frac{c}{r} = 1+e\cos(s-p)+e^2\cos(2(s-p))+\frac{15}{8}me\cos(s-2h+p)+m^2\cos(2(s-h)) \tag{A.50}$$

where $\frac{c}{r}$ is called the real parallax of the moon:

$$\left(\frac{c}{r}\right)^3 = 1 + \frac{3e^2}{2} + 3e\cos(s-p) + \frac{9}{2}e^2\cos(2(s-h))$$

$$+\frac{45}{8}me\cos(s-2h+p) + 3m^2\cos(2(s-h)) \tag{A.51}$$

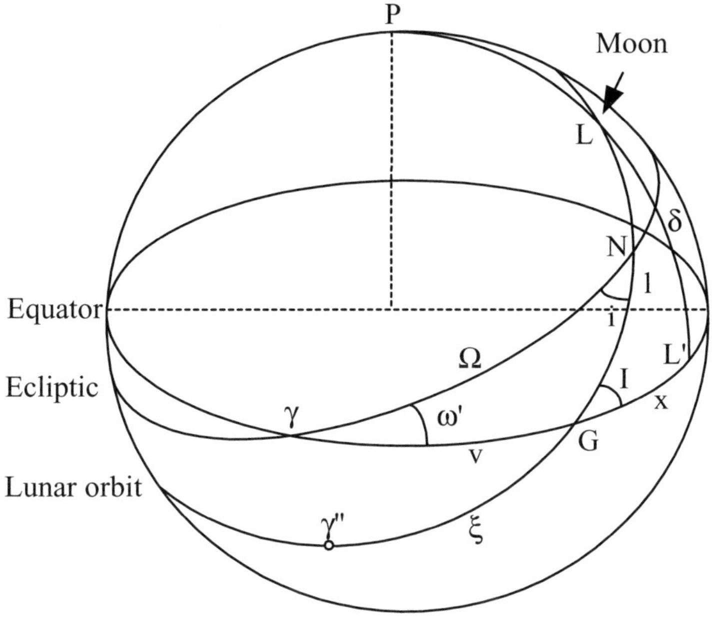

Figure A.9: declination and right ascension of the moon

A.2.3 Declination and right ascension

In Figure A.9, declination and right ascension of the moon are the angles $L'L = \delta$ and $\gamma L' = \alpha$ respectively.

Some explanations about the angles:

$\gamma N = \Omega$: longitude of ascending node.

$\gamma L' = \alpha$: right ascension of the moon.

ν: right ascension of G.

angle $\gamma' G$: ζ

angle GL: $l = L - x$

angle GL': x

We know L, equal to the angle $\gamma' L$ (see previous subsection).

Astronomy also provides us with the following angles:

ω': inclination of the equator on the ecliptic (this is the angle which defines the tropics) ([140], p. 81):

$$\omega' = 23.452294 - 0.0130125\,T - 0.00000164\,T^2 + 0.000000503\,T^3 \text{ (in degrees)} \quad (A.52)$$

i: inclination of the lunar orbit on the ecliptic, $i = 5°08'43.3546" = 5.145°$ ([172], p. 162).

Ω: longitude of the ascending node of the moon ([140], p. 162):

$$\Omega = \gamma N = 259.183275 - 1934.1420\,T + 0.002078\,T^2 + 0.0000022\,T^3$$

Calculation of the declination δ:
This is done through ζ and I.
Calculation of ζ:
In the triangle γNG, spherical trigonometry gives:

$$\sin(i)\cot(\omega') = \cot(\Omega - \zeta)\sin(\Omega) - \cos(i)\cos(\Omega) \tag{A.53}$$

as $\Omega = \gamma N = \gamma' N$, the angle GN is indeed equal to $\Omega - \zeta$.
This formula gives:

$$\cot(\Omega - \zeta) = \frac{\sin(i)\cot(\omega') + \cos(i)\cos(\Omega)}{\sin(\Omega)} \tag{A.54}$$

therefore we have ζ.
By developing $\cot(\Omega - \zeta)$ to the first order with respect to ζ, we find the following approximate formula:

$$\zeta = i\cot(\omega')\sin(\Omega) \tag{A.55}$$

ζ varies from $-12°$ to $+12°$ in 18.7 years.
Calculation of I:
Still in the triangle γNG, we have:

$$\cos(I) = \cos(\omega')\cos(i) - \sin(\omega')\sin(i)\cos(\Omega) \tag{A.56}$$

Finally we obtain ζ, because, in the triangle GLL', we have (the angle $GL'L$ is a right angle):

$$\sin(\delta) = \sin(l)\sin(I) = \sin(L - x)\sin(I) \tag{A.57}$$

Calculation of the right ascension α:
Still in the triangle GLL', we have:

$$\tan(x) = \tan(l)\cos(I) = \sin(L - x)\sin(I) \tag{A.58}$$

In the triangle γNG, spherical trigonometry again gives:

$$\cos(\Omega)\cos(\omega) = \sin(\Omega)\cot(\nu) - \sin(\omega')\cot(i) \tag{A.59}$$

hence:

$$\tan(\nu) = \frac{\sin(\Omega)}{\sin(\omega')\cot(i) + \cos(\omega)\cos(N)} \tag{A.60}$$

Knowing ν and x, we simply have $\alpha = \nu + x$.

A.3 Movement of the sun

As for the moon, we shall detail the four parameters appearing in the formulae.

A.3.1 Mass of the sun

The ratio of the sun's mass m_S to the Earth's mass is ([172], p. 162):

$$\frac{m_S}{m_T} = 331954$$

A.3.2 Distance to the Earth, longitude of the sun

Given e, the eccentricity of the terrestrial orbit ([140], p. 79):

$$e = 0.01675104 - 0.0000418\,T - 0.000000126\,T^2 \tag{A.61}$$

In the case of the sun, and denoting E as the eccentric anomaly and M the mean anomaly, we have:

$$E = M + e\sin(E) \text{ (Kepler's equation)}$$

with:

$$M = 358.47583 + 35999.04975\,T - 0.000150\,T^2 + 0.0000033\,T^3 \text{ (in degrees)} \tag{A.62}$$

E can be deduced from Kepler's equation, through successive iterations if needed, i.e. $E_{n+1} = M + e\sin(E_n)$ and $E_0 = M$.

The Earth–sun distance is therefore:

$$r = 1.0000002(1 - e\cos(E)) \tag{A.63}$$

expressed in astronomical units. One astronomical unit equals 149503899 km or 92897416 miles.

Mean value: $\frac{a}{r} = \frac{1}{23400}$.

The real anomaly of the sun v is given by ([140], p. 122):

$$\tan\left(\frac{v}{2}\right) = \sqrt{\frac{1+e}{1-e}}\tan\left(\frac{E}{2}\right)$$

The real longitude of the sun is then ([140], p. 79) $\theta = L + v - M$, with:

$$L = 279.69668 + 36000.76892\,T + 0.0003025\,T^2 \text{ (in degrees)} \tag{A.64}$$

L being the mean longitude.

A.3.3 Declination and right ascension of the sun

The latitude of the sun on the ecliptic does not exceed 1.2 seconds of arc, therefore we can consider that the sun evolves in the ecliptic. The right ascension and the declination are thus given by:

$$\tan(\alpha) = \frac{\cos(\omega')\sin(\theta)}{\cos(\theta)} \tag{A.65}$$

$$\sin(\delta) = \sin(\omega')\sin(\theta) \tag{A.66}$$

Appendix B

Diffusion matrix with tetrahedra

MAPLE V file for building the elementary diffusion matrix for a tetrahedron with linear interpolation.

> with(linalg);

Setting the values of arrays on the tetrahedra: VX, VY, VB, VT, VF, VG, VH, VU, VV, and VW

Examples: VB(ib,ip) = value of base ib at its local point ip, VX(ip) = value of X at its local point ip

Coordinates X, Y and Z (in arrays VX, VY, and VZ).

> VX:=array(1..4,[X1,X2,X3,X4]);
> VY:=array(1..4,[Y1,Y2,Y3,Y4]);
> VZ:=array(1..4,[Z1,Z2,Z3,Z4]);

Basis functions (VB), test functions (VT), functions F, G and H (arrays VF, VG and VH) for storing tensorial viscosity.

> VB:=array(1..4,1..4,[[1,0,0,0],[0,1,0,0],[0,0,1,0],[0,0,0,1]])
> VT:=array(1..4,1..4,[[1,0,0,0],[0,1,0,0],[0,0,1,0],[0,0,0,1]])
> VF:=array(1..4,[F1,F2,F3,F4]);
> VG:=array(1..4,[G1,G2,G3,G4]);
> VH:=array(1..4,[H1,H2,H3,H4]);
>

Definition of linear basis functions on the reference tetrahedron (coordinates ALPHA, BETA, GAMA)

> PSI:=array(1..4,[1-ALPHA-BETA-GAMA,ALPHA,BETA,GAMA]);
>

Resulting matrix

> MAT:=matrix(4,4);

Simplifications if origin is taken at point1.

>
> X1:=0;
> Y1:=0;
> Z1:=0;

```
> DXDALPHA:=VX[2]-VX[1];
> DXDBETA:=VX[3]-VX[1];
> DXDGAMA:=VX[4]-VX[1];
> DYDALPHA:=VY[2]-VY[1];
> DYDBETA:=VY[3]-VY[1];
> DYDGAMA:=VY[4]-VY[1];
> DZDALPHA:=VZ[2]-VZ[1];
> DZDBETA:=VZ[3]-VZ[1];
> DZDGAMA:=VZ[4]-VZ[1];
Jacobian matrix for isoparametric transformation.
> MATJAC:=matrix(3,3,[DXDALPHA,DXDBETA,DXDGAMA,DYDALPHA,
DYDBETA,DYDGAMA,DZDALPHA,DZDBETA,DZDGAMA]);
Jacobian of isoparametric transformation.
> JAC:=det(MATJAC);
Inverse of the transpose of Jacobian matrix
> TMATJAC:=matrix(3,3,[DXDALPHA,DYDALPHA,DZDALPHA,
DXDBETA,DYDBETA,DZDBETA,DXDGAMA,DYDGAMA,DZDGAMA]);
> INVJAC:=inverse(TMATJAC);
>
Computation of the 16 elementary terms of the matrix
> for i from 1 to 4
> do
> for j from 1 to 4
> do
>
> B:=VB[j,1]*PSI[1]+VB[j,2]*PSI[2]+VB[j,3]*PSI[3]+VB[j,4]*PSI[4];
> GRADB:=multiply(INVJAC,vector([diff(B,ALPHA),
                        diff(B,BETA),diff(B,GAMA)]));
> BX:=GRADB[1];
> BY:=GRADB[2];
> BZ:=GRADB[3];
>
> T:=VT[i,1]*PSI[1]+VT[i,2]*PSI[2]+VT[i,3]*PSI[3]+VT[i,4]*PSI[4];
> GRADT:=multiply(INVJAC,vector([diff(T,ALPHA),
                        diff(T,BETA),diff(T,GAMA)]));
> TX:=GRADT[1];
> TY:=GRADT[2];
> TZ:=GRADT[3];
>
> F:=VF[1]*PSI[1]+VF[2]*PSI[2]+VF[3]*PSI[3]+VF[4]*PSI[4];
> G:=VG[1]*PSI[1]+VG[2]*PSI[2]+VG[3]*PSI[3]+VG[4]*PSI[4];
> H:=VH[1]*PSI[1]+VH[2]*PSI[2]+VH[3]*PSI[3]+VH[4]*PSI[4];
>
Function to be integrated over the real element
> FONC1:=F*BX*TX+G*BY*TY+H*BZ*TZ;
> FONC1:=simplify(FONC1*JAC);
>
```

Integration over the reference tetrahedron

```
> FONC2:=int(FONC1,ALPHA);
> FONC5:=subs(ALPHA=1-BETA-GAMA,FONC2)-subs(ALPHA=0,FONC2);
> FONC6:=int(FONC5,BETA);
> FONC7:=simplify(subs(BETA=1-GAMA,FONC6)-subs(BETA=0,FONC6));
> FONC8:=int(FONC7,GAMA);
> MAT[i,j]:=simplify(subs(GAMA=1,FONC8)-subs(GAMA=0,FONC8));
> od;
> od;
```

FORTRAN printout of the result on a file named RESU

```
> writeto(`:RESU`);
> fortran(MAT);
```

Appendix C

Notations

The notations used in the course of this work are described hereafter.

C.1 Latin alphabet

a: friction coefficient at the walls.
a_{wind}: coefficient of wind influence.
b: linear friction coefficient.
c: celerity of long waves.
C: Chézy coefficient.
DT: time step.
F_x: source term along x in the momentum equation.
F_y: source term along y in the momentum equation.
$\overrightarrow{F}^a$: source term arising from atmospheric pressure.
$\overrightarrow{F}^c$: Coriolis force.
$\overrightarrow{F}^d$: force arising from spatial variations of density.
$\overrightarrow{F}^f$: friction force on the bottom.
$\overrightarrow{F}^m$: tide driving terms.
$\overrightarrow{F}^q$: source term for conservation of momentum.
$\overrightarrow{F}^t$: drag force.
$\overrightarrow{F}^v$: influence of the wind.
g: acceleration of gravity.
h: water depth (measured between bottom and free surface).
h_{prop}: height of propagation.
k: turbulent energy.
K: Strickler friction coefficient.
k_s: roughness size on the bottom.
m: Manning coefficient.
$\overrightarrow{n}_f$: normal vector at the bottom pointing outside.
$\overrightarrow{n}_s$: normal vector at the free surface pointing outside.
p: pressure.
p_h: hydrostatic pressure.

p_d: dynamic pressure.
P_{atm}: atmospheric pressure.
Q_x: flow per unit length in m^2/s, equal to hu.
Q_y: flow per unit length in m^2/s, equal to hv.
Q_{sce}: source of discharge at the bottom in m^3/s, imposed at certain points of the mesh.
R: Earth radius.
R_{ij}: Reynolds tensor.
S: salinity.
Sce: source term of the mass conservation equation, expressed in m/s, and deduced from Q_{sce}.
t: time.
T: tracer or temperature.
u: first component of velocity in the Saint-Venant equations.
U: first component of velocity in the Navier–Stokes equations.
u_{conv}: first component of advection velocity.
u_{sce}: first component of sources velocity.
U_{wind}: first component of wind velocity.
U_1,U_2,U_3: components of velocity in the Navier–Stokes equations, used in Einstein notation ($U_1 = U$, $U_2 = V$, $U_3 = W$).
v: second component of velocity in the Saint-Venant equations.
V: second component of velocity in the Navier–Stokes equations.
v_{conv}: second component of advection velocity.
v_{sce}: second component of sources velocity.
V_{wind}: second component of wind velocity.
X: Mercator abscissa.
Y: Mercator ordinate.
Z_s: free surface elevation.
Z_f: bottom elevation.

C.2 Greek alphabet

α: angle of the bottom slope (per element).
ε: turbulent dissipation.
θ_h: coefficient of semi-implicitation for the water depth.
θ_u: coefficient of semi-implicitation for the velocity.
θ_T: coefficient of semi-implicitation for the tracer.
Θ: porosity.
κ: von Karman constant equal to 0.41.
λ: latitude.
λ_0: latitude of the origin point.
ν: coefficient of molecular diffusion in m^2/s.
ν_t: coefficient of turbulent diffusion in m^2/s.
ν_e: diffusion coefficient in m^2/s, including dispersion and turbulence.
ρ: density of water.
ρ_{air}: density of air.
τ_{ij}: viscous part of stress tensor.

$\underline{\underline{\tau}}_f$: viscous stress at the bottom.
$\underline{\underline{\tau}}_s$: viscous stress at the surface.
φ: longitude.
φ_0: longitude of the origin point.
φ_i^h: test function at point i for the Saint-Venant continuity equation.
φ_i^u: test function at point i for the two Saint-Venant momentum equations.
Ψ_i^h: basis function at point i for the Saint-Venant continuity equation.
Ψ_i^u: basis function at point i for the two Saint-Venant momentum equations.
$\overrightarrow{\omega}$: vector rotation of the Earth.

C.3 Mathematical signs

$\otimes$: external product between two first-order tensors.
∇: tensorial operator "nabla".
.: between two vectors, designates the dot product.
$\mathbb{R}$: the set of real numbers.
(x, y): dot product of x and y.
$\|\overrightarrow{x}\|$: Euclidean norm of vector $\overrightarrow{x}$.

Bibliography

[1] ANDERSON M.G., BATES P.D., HERVOUET J.M., Computation of a flood event using a 2-dimensional finite element model and comparison to field data. Specialty conference on modelling of flood propagation over initially dry areas. Milan, Italy, 29–30 June 1994. 1994

[2] ATKINSON J.H., WESTERINK J.J., HERVOUET J.-M., Similarities between the quasi-bubble and the generalized wave continuity equation solutions to the shallow water equations. International Journal for Numerical Methods in Fluids, 45, 689–714. 2004

[3] BARR D., Densimetric exchange flow in rectangular channels. III. Large scale experiments. La Houille Blanche, 22, 619–632. 1967

[4] BARROS E., Estimation de paramètres dans les équations de Saint-Venant. Thèse de l'Université Pierre et Marie Curie (Paris VI). Spécialité Mathématiques Appliquées. 1996

[5] BATES P.D., HERVOUET J.-M., A new method for moving-boundary hydrodynamic problems in shallow water. Proceedings of the Royal Society, 455, 3107–3128. 1999

[6] BATES P.D., HORRITT M.S., HERVOUET J.-M., Investigating two-dimensional, finite element predictions of floodplain inundation using fractal generated topography. Hydrological Processes, 12, 1257–1277. 1998

[7] BATES P.D., STEWART M.D., SIGGERS G.B., SMITH C.N., HERVOUET J.-M., SELLIN R.H.J., Internal and external validation of a two-dimensional finite element code for river flood simulations. Proceedings of the ICE: Water and Maritime Engineering, 130, 127–141. 1998

[8] BEAM R.M., WARMING R.F., An implicit factored scheme for the compressible Navier–Stokes equations. AIAA Journal, 16, 4, 393–402. 1978

[9] BENJAMIN T., Gravity currents and related phenomena. Journal of Fluid Mechanics, 31(2), 209–248. 1968

[10] BENQUE J.-P., LABADIE G., RONAT J., Une méthode d'éléments finis pour la résolution des équations de Navier–Stokes couplées à une équation thermique. Communication au 4ième Symposium International sur les méthodes d'éléments finis pour les problèmes de mécanique des fluides, Tokyo, 26–29 July 1982. 1982

[11] BENOIST G., HERVOUET J.-M., LABADIE G., L'expérience d'EDF en matière de calcul des ondes de submersion. International meeting on dam break flood wave simulation, Montréal, 8–11 April 1991. 1991

[12] BERMUDEZ A., DERVIEUX A., DESIDERI J.A., VAZQUEZ M.E., Upwind schemes for the two-dimensional shallow water equations with variable depth using unstructured meshes. Rapport de recherche de l'INRIA n°2738. December 1995

[13] BERNARDI C., PIRONNEAU O., On the shallow water equations at low Reynolds number. Communications in Partial Differential Equations, 16(1), 59–104. 1991

[14] BIJVELDS M., Numerical modelling of estuarine flow over steep topography. Communications on Hydraulic and Geotechnical Engineering, TU Delft Faculty of Civil Engineering Report 01-2. March 2001

[15] BLUMBERG A., MELLOR G., A description of a three-dimensional coastal ocean circulation model. Three-dimensional Coastal Ocean Models, N. Heaps Ed. Volume 4 of Coastal and Estuarine Sciences. American Geophysical Union, 1-16. 1987

[16] BOULET T., ROUGE D., Telemac 2D. Version 2.0. Dossier de validation. EDF Report HE-43/92.12. 1992

[17] BOUSSINESQ J., Essai sur la théorie des eaux courantes. Mémoire présenté par divers savants à l'Académie des Sciences de Paris, 23, 380–398. 1823

[18] BROOKS A.-N., HUGUES T.J.R., Streamline Upwind Petrov Galerkin formulations for convection dominated flows with particular emphasis on the Navier–Stokes equations. Computer Methods in Applied Mechanics and Engineering, 32, 199–259. 1982

[19] BURGEAS I., CHERGUI J., GIROU D., REQUENA S., Message Passing Interface. Idris, MPI version 1.2. 1996

[20] BYATT-SMITH J.G.B., An integral equation for unsteady surface waves and a comment on the Boussinesq equation. Journal of Fluid Mechanics, 29(4), 625–633. 1971

[21] CAREY G.F., JIANG B.N., A stable least-squares finite element method for non-linear hyperbolic systems. International Journal for Numerical Methods in Fluids, 8, 933–942. 1988

[22] CASULLI V., Semi-implicit finite difference methods for the two-dimensional shallow water equations. Journal of Computational Physics, 86, 56–74. 1990

[23] CASULLI V., CATTANI E., Stability, accuracy and efficiency of a semi-implicit method for three-dimensional shallow water flow. Computers & Mathematics with Applications, 27(4), 99–112. 1994

[24] CASULLI V., CHENG R., Semi-implicit finite difference methods for three-dimensional shallow water flow. International Journal for Numerical Methods in Fluids, 15, 629–648. 1992

[25] CASULLI V., CHENG R., GARTNER J., Tidal, Residual, Intertidal Mudflat (TRIM) model and its applications to San Francisco Bay, California. Estuarine, Coastal and Shelf Science, 36, 235–280. 1993

[26] CASULLI V., STELLING G., Simulation of three-dimensional, non-hydrostatic free-surface flows for estuaries and coastal seas. Estuarine and coastal modelling. Proceedings of the 4th International Conference, San Diego, USA, M. Spaulding and R. Cheng Eds. ASCE, 1–12. 1995

[27] CHAVENT G., Analyse fonctionnelle et identification de coefficients répartis dans les équations aux dérivées partielles. Thèse de la Faculté des Sciences de Paris. Spécialité Mathématiques. 1971

[28] CHORIN A.J., Numerical solution of the Navier–Stokes equations. Mathemathics of Computation, 22, 745–762. 1968

[29] CIARLET P.G., The finite element method for elliptic problems. North-Holland. 1978

[30] COBBY D.M., MASON D., HORRITT M.S. and BATES P.D., Two-dimensional hydraulic flood modelling using a finite element mesh decomposed according to vegetation and topographic features derived from airborne scanning laser altimetry. Hydrological Processes, 17, 1979–2000. 2003

[31] COUTINHO A.L.G.A., CATABRIGA L., MARTINS M.A.D., ALVES J.L.D., An edge-based preconditioner for non-symmetric finite element equations. Proceedings of 18th Ibero-Latin American Congress on Computational Methods in Engineering, Brasilia, Brazil. Volume 3, 1249–1255. 1997

[32] DALSECCO S., GOUTAL N., HERVOUET J.-M., VUJASINOVIC Y., Recent advances in shallow water computations. Communication au 6ième colloque international sur la simulation d'écoulements par éléments finis, Antibes. 1986

[33] DAUBERT A., GRAFFE O., Quelques aspects des écoulements presque horizontaux à 2 dimensions en plan et non permanents. La Houille Blanche, 22, 847–860. 1967

[34] DAUBERT O., GEST B., LEPEINTRE F., Code 3D à surface libre MITHRIDATE, équations de base, note d'avancement. EDF Report-LNH HE.41/89.38. 1990

[35] DAVIES A., LAWRENCE J., A three-dimensional model of the M4 tide in the Irish Sea. The importance of open boundary conditions and influence of wind. Journal of Geophysical Research, 12(C8), 16197–16227. 1994

[36] DEARDOFF J., A numerical study of three-dimensional turbulent channel flow at large Reynolds numbers. Journal of Fluid Mechanics, 41(part 2), 453–480. 1970

[37] DECONINCK H., PAILLERE H., STRUIJS R., ROE P.L., Multidimensional upwind schemes based on fluctuation splitting for systems of conservation laws. Computational Mechanics, n°11, 323–340. 1993

[38] DEFINA A., D'ALPAOS L., MATTICHIO B., A new set of equations for very shallow water and partially dry areas suitable to 2D numerical domains. Proceedings of the speciality conference on modelling of flood propagation over initially dry areas, Milan, Italy, 29–30 June 1994. 1994

[39] DELFT HYDRAULICS, Delft3D-Flow manual. Delft Hydraulics, Release 3.05. 1999

[40] DHATT G., TOUZOT G., Une présentation de la méthode des éléments finis. Collection Université de Compiègne, Presses de l'Université Laval. 1981

[41] DO CARMO A., SEABRA-SANTOS J.S., On breaking waves and wave-current interaction in shallow water: a 2DH finite element model. International Journal of Numerical Methods in Fluids, 22, 429–444. 1996

[42] DONEA J., Arbitrary Lagrangian-Eulerian finite element methods. Computational Methods for Transient Analysis, 1, 101–119. 1983

[43] DONEA J., A Taylor-Galerkin method for convective transport problems. International Journal on Numerical Methods in Engineering, 20, 101–119. 1984

[44] DONNARS P., HAUGUEL A., WARLUZEL A., Utilisation d'un code d'écoulement tridimensionnel pour le calcul des évolutions du fond. EDF Report-LNH HE.41/81.25. 1981

[45] ESPOSITO P., Résolution bidimensionnelle des équations de transport par la méthode des caractéristiques. EDF Report HE-41/81.16. 1981

[46] FARHAT C., A simple and efficient automatic FEM domain decomposer. Computers and Structures, 28, 579–602. 1988

[47] FARHAT C., LANTERI S., SIMON H.D., TOP/DOMDEC, A software tool for mesh partitioning and parallel processing. Computing Systems in Engineering, 6(1), 13–26. 1994

[48] FARHAT C., LESOINNE M., Automatic partitionning of unstructured meshes for the parallel solution of problems in computational mechanics. International Journal for Numerical Methods in Engineering, 36, 745–764. 1993

[49] FISCHER H.B., Mixing in inland and coastal waters. Academic Press. 1979

[50] FLATHER R.A., Results from surge prediction model of the North-West European continental shelf for April, November and December 1973. Institute of Oceanography (UK), Report n° 24. 1976

[51] FLYNN M., Very high speed computing. Proceedings of the IEEE, 1901–1909. 1966

[52] FORTUNATO A.B., BAPTISTA A.M., Modelling near-bottom advective acceleration in surface water models. Computational Methods in Water Resources X, 2, 1045–1052. Kluwer Academic Press. 1994

[53] FREDERIKSEN C., WATTS A., Finite-element method for time-dependent incompressible free surface flow. Journal of Computational Physics, 39, 282–304. 1981

[54] GALLAND J.-C., GOUTAL N., HERVOUET J.-M., A new numerical model for solving shallow water equations. Advances in Water Resources, 14(3), 138–148. 1991

[55] GEIST A., BEGUELIN A., DONGARRA J., JIANG W., MANCHEK R., SUNDERAM V., PVM3 User's guide and Reference Manual, Oak Ridge National Laboratory. 1994

[56] GIRAULT V., RAVIART P.A., Finite element methods for the Navier–Stokes equations. Springer-Verlag. 1986

[57] GOLUB G.H. , VAN LOAN. C.F., Matrix computations (2nd edition). Johns Hopkins University Press. 1989

[58] GOUTAL N., Résolution des équations de Saint-Venant en régime transcritique par une méthode d'éléments finis. Application aux bancs découvrants. Thèse de doctorat de l'Université Paris VI. 1987

[59] GOUTAL N., Note de principe du code Ruptur. EDF Report HE-43/91-38. 1991

[60] GRAS R., Réfrigération des centrales thermiques classiques ou nucléaires par plan d'eau et rivières. EDF Report-HC091-68/n° 9. 1968

[61] GRAY W.G., KOLAR R.L., LUETTICH R.A., WESTERINK J.J., Shallow water modelling in spherical coordinates: equation formulation, numerical implementation, and application. Journal of Hydraulic Research, 32(1), 3–24. 1994

[62] GROSS E.S., KOSEFF J.R., MONISMITH S.G., Three-dimensional salinity simulations of south San Francisco bay. Journal of Hydraulic Engineering, November, 1199–1209. 1999

[63] Guide for selecting Manning's roughness coefficients for natural channels and flood plains. United States Geological Survey. Water Supply Paper 2339. 1989

[64] GUILLOU S., BARBRY N., NGUYEN K.D., Calcul numérique des ondes de surface par une méthode de projection avec un maillage eulérien adaptatif. Compte-Rendu à l'Académie des Sciences de Paris, 328, Série II b, 1–7. 2000

[65] HARLOW F., WELCH J., Numerical calculation of time-dependent viscous incompressible flow of fluid with free-surface. The Physics of Fluids, 8(12), 2182-2189. 1965

[66] HASBANI Y., LIVNE E., BERCOVIER M., Finite elements and characteristics applied to advection-diffusion equations. Computers and Fluids, 11, 2, 71–83. 1983

[67] HENDRICKSON B., LELAND R., The Chaco user's guide. Version 2.0. Sandia Technical Report. SAND95-2344. 1995

[68] HERVOUET J.-M., Application de la méthode des caractéristiques en formulation faible à la résolution des équations d'advection bidimensionnelles sur des maillages grilles. EDF Report HE-41/84.11. 1984

[69] HERVOUET J.-M., CARAC, module de convection en éléments finis, par la méthode des caractéristiques. EDF Report HE-41/86.21. 1986

[70] HERVOUET J.-M., Application of the method of characteristics in their weak formulation to solving two-dimensional advection equations on mesh grids. Computational Techniques for Fluid Flow. Volume 5 in the series "Recent advances in numerical methods in fluids". Pineridge Press. 1986

[71] HERVOUET J.-M., Comparison of experimental data and laser measurements with the computational results of Telemac code (shallow water equations). Proceedings of "Hydrocomp '89", Dubrovnik, Yugoslavia. June 1989

[72] HERVOUET J.-M., Vectorisation et simplification des algorithmes en éléments finis. EDF. Bulletin de la Direction des Etudes et Recherches. Série C, Mathématiques, Informatique n° 1, 1–37. 1991

[73] HERVOUET J.-M., Telemac, a fully vectorised finite element sofware for shallow water equations. Computer Methods and Water Resources. Rabat, Morocco, 7–11 October 1991. 1991

[74] HERVOUET J.-M., Characteristics and mass-conservation. New developments in Telemac-2D. EDF Report HE-43/92.41. 1992

[75] HERVOUET J.-M., Element by element methods for solving shallow water equations with F.E.M. IX International Conference on Computational Methods in Water Resources, Denver, Colorado, USA, 9–12 June 1992. 1992

[76] HERVOUET J.-M., Solving shallow water equations with rapid flows and tidal flats. Hydrosoft 92, Valencia, Spain, 21–23 July 1992. 1992

[77] HERVOUET J.-M., Méthodes itératives pour la résolution des systèmes matriciels. EDF Report HE-43/93.49. 1993

[78] HERVOUET J.-M., Validating the numerical simulation of dam-breaks and floods. 1st International Conference on Hydro-Science and Engineering (ICHE) Washington DC, USA, 7–11 June 1993. 1993

[79] HERVOUET J.-M., Finite Element algorithms for modelling flood propagation. Proceedings of the speciality conference on modelling of flood propagation over initially dry areas, Milan, Italy, 29–30 June 1994. 1994

[80] HERVOUET J.-M., Computation of free surface flows, a state of the art in the Telemac system. First International Conference on Hydroinformatics, Delft, The Netherlands, 19–23 September 1994. 1994

[81] HERVOUET J.-M., Recent advances in the Telemac system. "APCOM 96", Third Asian Pacific Conference. Seoul, South Korea, 16–18 September 1996. 1996

[82] HERVOUET J.-M., A numerical simulation of the Malpasset dam-break with 2D Saint-Venant equations. Proceedings of the Saint-Venant symposium, Marne la Vallée, France, 28–29 August 1997. 1997

[83] HERVOUET J.-M., Comparing numerical simulations of free surface flows using non-hydrostatic Navier–Stokes equations and Boussinesq equations. XII International Conference on Computational Methods in Water Resources. The Chersonese, Kreta, Greece, 15–19 June 1998. 1998

[84] HERVOUET J.-M., A high resolution 2-D dam-break model using parallelization. Hydrological Processes, 14(13), 2211–2230. 2000

[85] HERVOUET J.-M., On spurious oscillations in primitive shallow water equations. XIII International Conference on Computational Methods in Water Resources, Calgary, Canada, 25–29 June 2000. 2000

[86] HERVOUET J.-M., GUESMIA M., Solution of Boussinesq equations in finite elements. Coastal Engineering 1999, Lemnos, Greece, 26–28 May 1999. 1999

[87] HERVOUET J.-M., JANIN J.-M., Refined flow modelling in coastal areas. Coastal 95. Cancun, Mexico, 6–8 September 1995. 1995

[88] HERVOUET J.-M., JANIN J.-M., MARCOS F., Numerical modelling of free surface flows. Devising strategies for parallelism. Computer Modelling of Seas and Coastal Regions, La Coruna, Spain, 23–25 June 1997. 1997

[89] HERVOUET J.-M., JANIN J.-M., MOULIN C., Preconditioning and solving linear systems for the computation of free surface flows. X International Conference on Computational Methods in Water Resources, Heidelberg, Germany. 1994

[90] HERVOUET J.-M., JANKOWSKI J., Comparing numerical simulations of free surface flows using non-hydrostatic Navier–Stokes equations and Boussinesq equations. Hydroinformatics. Iowa Institute of Hydraulic Research, Iowa City, USA, 23–27 July 2000. 2000

[91] HERVOUET J.-M., LEPEINTRE F., Prise en compte des potentiels astraux dans les calculs de marée. EDF Report HE-42/92-43. 1992

[92] HERVOUET J.-M., MARCOS F., Parallelism and code coupling in the Telemac system. Proceedings of the Hydroinformatics Conference. Copenhagen, Denmark, 23–26 August 1998. 1998

[93] HERVOUET J.-M., MOULIN C., Nouveaux schémas de convection dans Telemac-2D version 2.3: apport de la méthode SUPG. EDF Report HE-43/93.27. 1993

[94] HERVOUET J.-M., PETITJEAN A., Malpasset dam-break revisited with two-dimensional computations. Journal of Hydraulic Research, 37(6), 777–788. 1999

[95] HERVOUET J.-M., SAMIE R., MOREAU B., Modelling urban areas in dam-break flood-wave numerical simulations. International Seminar and Workshop on the Rescdam Project, Seinäjoki, Finland, 1–5 October 2000. 2000

[96] HERVOUET J.M., VAN HAREN L., Code Telemac-2D - Version 3.0 - Note de Principe. EDF Report HE-43/94/051/B. 1994

[97] HERVOUET J.-M., VAN HAREN L., Recent advances in numerical methods for fluid flow. In Floodplain Processes, John Wiley & Sons Ltd. 1996

[98] HESTENES M.R., STIEFEL E., Methods of conjugate gradients for solving linear systems. Journal of Research of the NBS. V49. 1952

[99] HICKS F.E., STEFFLER P.M., A characteristic-dissipative-Galerkin scheme for open channel flow. ASCE Journal of Hydraulics Engineering, 118, 337–352. 1992

[100] HICKS F.E., STEFFLER P.M., Comparison of finite element methods for the Saint-Venant equations. International Journal for Numerical Methods in Fluids, 20, 99–113. 1994

[101] HINKELMANN R., Parallelisierung eines Lagrange-Euler-Verfahrens für Strömungs und Stofftransportprozesse in Oberflächengewässern. Institut für Strömungsmechanik und Elektronik, Rechnen im Bauwesen der Universität Hannover. Bericht n° 51. 1997

[102] HIRT C., NICHOLS B., Volume of fluid (VOF) method for the dynamics of free boundaries. Journal of Computational Physics, 39, 201–225. 1981

[103] IDEL'CIK I.E., Mémento des pertes de charge. Editions Eyrolles. 1960

[104] JANIN J.-M., Code Telemac-3D - Version 2.0 - Descriptif informatique. EDF Report LNH HE-42/94/033/A. 1994

[105] JANIN J.-M., Conservativité et positivité dans un module de transport de scalaire écrit en éléments finis - Application à Telemac-3D. EDF Report HE-42/95/054/A. 1995

[106] JANIN J.-M., DAVID E., DENOT T., Code Telemac-3D - Version 2.2 - Manuel utilisateur. EDF Report HE-42/97/048/A. 1997

[107] JANIN J.-M., DAVID E., DENOT T., Code Telemac-3D - Version 2.2 - Validation document. EDF Report LNH HE-42/97/050/A. 1997

[108] JANIN J.-M., LEPEINTRE F., PECHON P., Telemac-3D: a finite element code to solve 3D free surface flow problems. Computer Modelling of Seas and Coastal Regions, Southampton, UK. April 1992

[109] JANKOWSKI J.A., A non-hydrostatic model for free surface flows. Institut für Strömungsmechanik und Elektronik. Rechnen im Bauwesen der Universität Hannover. Bericht n°56. 1998

[110] KARYPIS G., KUMAR V., Metis mesh partitioner. Unstructured graph partitioning and sparse matrix ordering system. University of Minnesota Technical Report. 1995

[111] KAWAMURA T., MIYATA H., Simulation of nonlinear ship flows by density-function method. Journal of the Society of Naval Architects, Japan. 176, 1–10. 1994

[112] KERNIGHAN B.W., LIN S., An efficient heuristic procedure for partitioning graphs. The Bell System Technical Journal. 1970

[113] KLAASSEN G.J., van der ZWAARD J.J., Roughness coefficients of vegetated flood-plains. Journal of Hydraulic Research, n° 1. 1974

[114] KOLAR R.L., GRAY W.G., WESTERINK J.J., LUETTICH R.A., Shallow water modelling in spherical coordinates: equation formulation, numerical implementation and application. Journal of Hydraulic Research, 32(1), 3–24. 1994

[115] KOLAR R.L., GRAY W.G., WESTERINK J.J., Boundary conditions in shallow water models - an alternative implementation for finite element codes. International Journal for Numerical Methods in Fluids, 22, 603–618. 1996

[116] KORN G.A., KORN T.M., Mathematical handbook for engineers and scientists. McGraw-Hill. 1968

[117] LABADIE G., LASBLEIZ P., Quelques méthodes de résolution du problème de Stokes en éléments finis. EDF Report HE-41/83.01. 1983

[118] LABADIE G., PERON S., HERVOUET J.-M., Study of head losses in a confluence of rivers with a scale model and a numerical model solving shallow water equations. Refined flows modelling. XXIV IAHR Congress, Madrid, 9–13 September 1991. 1991

[119] LAFAURIE B., NARDONE C., SCARDOVELLI R., ZALESKI S., ZANETTI G., Modelling merging and fragmentation in multiphase flows with SURFER. Journal of Computational Physics, 113, 134–147. 1994

[120] LAITONE E., The second approximation to cnoidal and solitary waves. Journal of Fluid Mechanics, 9, 430–444. 1960

[121] LAUNDER B.E., SPALDING D.B., The numerical computation of turbulent flows. Computer Methods in Applied Mechanics and Engineering, 3. 1974

[122] LAURENCE D., VIOLLET P.-L., Cours de Simulation Numérique du LNH. Turbulence, Thermohydraulique et Stratifications. Tome 1, EDF Report-DER HE-41/91.13A. 1991

[123] LAX P.D., WENDROFF B., Systems of conservation laws. Communications in Pure and Applied Mathematics, 13, 217–237. 1960

[124] LEHFELDT R., BLOSS S., On the numerical modelling of stratified flows. Physical processes in estuaries, J. Drokers, W. Van Leussen eds, 283-284. Springer-Verlag. 1988

[125] LEPEINTRE F., Comment résoudre les équations de Saint-Venant sphériques avec un code résolvant les équations de Saint-Venant en coordonnées cartésiennes. EDF Report HE-42/90.03 and HE-43/90.03. 1990

[126] LI B., FLEMING C., Three-dimensional model of Navier–Stokes equations for water waves. Journal of Waterway, Port, Coastal and Ocean Engineering. January–February 2001

[127] LICH T.G., MUAT L.K., Boundary conditions for the two-dimensional Saint-Venant equation system. Applied Mathematical Modelling, 16, 498–502. 1992

[128] LIONS J.L., Some methods in the mathematical analysis of systems and their control. Science Press. 1981

[129] LOHNER R., MORGAN K., PERAIRE J. and VAHDATI M., Finite element flux-corrected transport (FEM–FCT) for the Euler and Navier–Stokes equations. International Journal for Numerical Methods in Fluids, 7, 1093–1109. 1987

[130] LOVE A.E.H., Some problems of geodynamics. Cambridge University Press. 1911

[131] LUCQUIN B., PIRONNEAU O., Introduction au calcul scientifique, Masson. 1996

[132] LUETTICH R.A., WESTERINK J.J., SCHEFFNER N.W., ADCIRC, An advanced three-dimensional circulation model for shelves, coasts and estuaries. Dept. of the Army, US Army Corps of Engineers. 1991

[133] LUMLEY J.L., A model for computation of stratified turbulent flows', Proceedings of International Symposium on Stratified Flows, Novossibirsk, pp. 333–343. American Society of Civil Engineers. 1972

[134] MADSEN P.A., SORENSEN O.R., A new form of the Boussinesq equations with improved linear dispersion characteristic. Part II. A slowly-varying bathymetry. Coastal Engineering, 18, 183–204. 1992

[135] MAHADEVAN A., OLIGER J., STREET R., A non-hydrostatic meso-scale ocean model. Journal of Physical Oceanography, 26(9), 1881–1900. 1996

[136] MANOHA B., Méthodes simples d'estimation de la dilution d'un rejet en mer. EDF Report-LNH HE42/80.61. 1981

[137] MARCOS F., JANIN J.-M., Code Telemac-3D - Version 2.0 - Note théorique. EDF Report-LNH HE-42/94/032. 1994

[138] MARCOS F., JANIN J.-M., Nouveaux développements dans l'étape de convection - diffusion de Telemac 3D. EDF Report HE-42/94/025. October 1994. 1994

[139] MASON D., COBBY D.M., HORRITT M.S. and BATES P.D., Floodplain friction parameterization in two-dimensional river flood models using vegetation heights derived from airborne scanning laser altimetry. Hydrological Processes, 17, 1711–1732. 2003

[140] MEEUS J., Astronomical formulae for calculators. Willmann-Bell. 1988

[141] MELLOR G.L., YAMADA T., A hierarchy of turbulence closure models for planetary boundary layers. Journal of Atmospheric Sciences, 31, 1791–1806. 1974

[142] MEWIS P., HOLZ K.-P., A quasi-bubble function approach for shallow water waves. Advances in Hydro-Science and Engineering, 1, 768–774. 1993

[143] MONAGHAN J.J., Smoothed particles hydrodynamics. Annual Review of Astrophysics, 30, 543-74. 1992

[144] MONAGHAN J.J., Simulating free surface flows with SPH. Journal of Computational Physics, 110, 399–406. 1994

[145] MUZAFERIJA S., PERIC M., Computation of free surface flows using interface-tracking and interface-capturing methods. In Nonlinear Water Wave Interaction, O. Mahrenholtz, M. Markiewicz Eds. Computational Mechanics Publications. 1998

[146] NAIM M., Modélisation numérique des effets hydrodynamiques provoqués par un glissement de terrain dans une retenue. Thèse de l'Université Joseph Fourier, Grenoble 1. 1991

[147] NAVIER C., Mémoire sur les lois du mouvement des fluides. Mémoires de l'Académie des Sciences, 6, 389–416. 1823

[148] NEZU I., NAKAGAWA H., Turbulence in open-channel flows, IAHR monograph series. Balkema. 1993

[149] NIHOUL J., JAMART B., Three-dimensional models of marine and estuarine dynamics. Volume 45 of Elsevier Oceanography Series. Elsevier. 1987

[150] NWOGU O., Alternative form of Boussinesq equations for nearshore wave propagation, Journal of Waterways, Port, Coastal and Ocean Engineering, 119, 618–638. 1993

[151] PAILLERE H., Multidimensional upwind residual distribution schemes for the Euler and Navier–Stokes equations on unstructured grids. Thèse de l'Université Libre de Bruxelles. 1995

[152] PEACEMAN D.W., RACHFORD H.H., The numerical solution of parabolic and elliptic differential equations. SIAM Journal, 3, 28–41. 1955

[153] PECHON P., Code MITHRIDATE pour les écoulements tridimensionnels à surface libre. Cahier des charges. EDF Report-LNH HE42/89.45. 1990

[154] PERTHAME B., Convergence of N-schemes for linear advection equations. Pitman Monographs Surveys in Pure and Applied Mathematics, 77. Longman. 1995

[155] PERTHAME B., SIMEONI C., A kinetic scheme for the Saint-Venant system with a source term. Rapport de Recherche DMA - 01 - 13 du Département de mathématiques et applications. Ecole Normale Supérieure CNRS UMR 8553. May 2001

[156] PIRONNEAU O., Méthode des éléments finis pour les fluides. Masson. 1988

[157] PRANDTL L., Über die ausgebidete Turbulenz. Zeitschrift für augewandte Mathematik und Mechanik. 5, 136. 1925

[158] PRESS W.H., FLANNERY B.P., TEUKOLSKY S.A., VETTERLING W.T., Numerical recipes, the art of scientific computing (FORTRAN version). Cambridge University Press. 1989

[159] QUETIN B., Modèles mathématiques de calcul des écoulements induits par le vent. 17ième congrès de l'AIRH, Baden-Baden, 15–19 Août 1977. 1977

[160] RAMETTE M., Guide d'hydraulique fluviale. EDF Report-LNH HE-40/81.04. 1981

[161] RASTOGI A.-K., RODI W., Predictions of heat and mass transfer in open channels. Journal of the Hydraulics Division, ASCE, n° HY3, 397–420. 1978

[162] RAVIART P.A., THOMAS J.M., Introduction à l'analyse numérique des équations aux dérivées partielles. Masson. 1983

[163] RAYNA G., REDUCE. Software for algebraic computation. Springer–Verlag. 1985

[164] RITTER A., Die Fortpflanzung der Wasserwellen. Zeitschrift des Vereines deutscher Ingenieure, 36(33), 947–954. 1892

[165] RODI W., Turbulence models and their applications in hydraulics. A state of the art review. Edition AIRH. 1984

[166] ROE P.L., Approximate Riemann solvers, parameter vectors, and difference schemes. Journal of Computational Physics, 43, 357–372. 1981

[167] ROE P.L., SIDILKOVER D., Optimum positive linear schemes for advection in two and three dimensions. SIAM Journal of Numerical Analysis, 29(6), 1542–1568. 1992

[168] SAAD Y., SCHULTZ M.H., GMRES, a generalized minimal residual algorithm for solving nonsymmetric linear systems. Research Report YALEU/RR 257. 1985

[169] SAINT-VENANT A.J.C., Théorie du mouvement non-permanent des eaux, avec application aux crues des rivières et à l'introduction des marées dans leur lit. Compte-Rendu à l'Académie des Sciences de Paris, 73, 147–154. 1871

[170] SELMIN V., DONEA J., QUARTAPELLE L., Finite element methods for non-linear advection. Computer Methods in Applied Mechanics and Engineering, 52, 817–845. 1985

[171] SERRE F., Contribution à l'étude des écoulements permanents et variables dans les canaux. La Houille Blanche, Juin–Juillet. 1953

[172] SCHUREMAN P., Manual of harmonic analysis and prediction of tides. US Government Printing Office. 1971

[173] SHAKIB F., HUGHES T.J.R., JOHAN Z., A multi-element group preconditioned GMRES algorithm for nonsymmetric systems arising in finite element analysis. Computer Methods in Applied Mechanics and Engineering, 75, 415–456. 1989

[174] SIMON H.D., Partitioning of unstructured problems for parallel processing. Computing Systems in Engineering, 2(2/3), 135–148. 1991

[175] SMAGORINSKI J., General circulation experiments with the primitive equations. I. The basic experiment. Monthly Weather Review, 91, 99–164. 1963

[176] SOLIVA J.-C., Modèle tridimensionnel d'écoulements méso-météorologiques. Etude de la convection-diffusion d'un polluant passif à cette échelle. Thèse ENPC. 1982

[177] SPEZIALE C.G., On non-linear k-L and k-epsilon models of turbulence. Journal of Fluid Mechanics, 178, 459–475. 1987

[178] STOKER J.J., Water waves. Wiley Interscience Publishers, 333–341. 1957

[179] STOKES G., On the theories of the internal friction of fluids motion and of the equilibrium and motion of elastic solids. Transactions of the Cambridge Philosophical Society, 8, 287–305. 1845

[180] SU C.H., MIRIE R.M., On head-on collisions between two solitary waves, Journal of Fluid Mechanics, 98, 509–525. 1980

[181] SWEERS H.E., Monograms to estimate the heat-exchange coefficient at the air-water interface as a function of wind speed and temperature; a critical survey of some literature. Journal of Hydrology. n° 30, 375–401. 1976

[182] THOMPSON K.W., Time dependent boundary conditions for hyperbolic systems. Journal of Computational Physics, 68, 1–24. 1987

[183] TSANIS I., Simulation of wind-induced water currents. Journal of Hydraulic Engineering, 115, 1113–1134. 1989

[184] TURNER J., Buoyancy effects in fluids. Cambridge University Press. 1973

[185] UNESCO, International Oceanographic Tables, Volume 4. UNESCO Technical papers in marine science, n° 40. 1987

[186] VIOLLET P.-L., Modèles de turbulence à partir de grandeurs statistiques en un point: modèle K-epsilon, EDF Bulletin de la Direction des Etudes et Recherches - Série A. Nucléaire. Hydraulique. Thermique, 1, 35–47. 1987

[187] VIOLLET P.-L., On the numerical modelling of stratified flows. Physical processes in estuaries, J. Drokers, W. Van Leussen eds, 257-277. 1988

[188] VIOLLET P.-L., CHABARD J.P., ESPOSITO P., LAURENCE D., Mécanique des fluides appliquée. Presses de l'Ecole Nationale des Ponts et Chaussées. 1998

[189] WESTERINK J.J., CONNOR J.J., STOLZENBACH K.D., A primitive pseudo wave equation formulation for solving the harmonic shallow water equations. Advances in Water Resources, Volume 10. 1987

[190] WESTERINK J.J., LUETTICH R.A. Jr, WU J.K., KOLAR R.L., The influence of normal flow boundary conditions on spurious modes in finite element solutions to the shallow water equations. International Journal for Numerical Methods in Fluids, 18, 1021–1060. 1994

[191] WU C., HUANG G., ZHENG Y., Theoretical solution of dam-break shock wave. Journal of Hydraulic Engineering, November, 1210–1215. 1999

[192] YALIN M.S., Mechanics of sediment transport. Pergamon Press. 1977

[193] YIH C.S., Stratified flows. Academic Press. 1980

Index